# NON-CONVENTIONAL ENERGY
## IN NORTH AMERICA

# NON-CONVENTIONAL ENERGY IN NORTH AMERICA

## Current and Future Perspectives for Electricity Generation

JORGE MORALES PEDRAZA

*Senior Consultant, Co-founder, and Main Investor in Morales Project Consulting; Senior Independent Consultant on International Affairs, Vienna, Austria*

ELSEVIER

Elsevier
Radarweg 29, PO Box 211, 1000 AE Amsterdam, Netherlands
The Boulevard, Langford Lane, Kidlington, Oxford OX5 1GB, United Kingdom
50 Hampshire Street, 5th Floor, Cambridge, MA 02139, United States

**Notices**
Knowledge and best practice in this field are constantly changing. As new research and experience broaden our understanding, changes in research methods, professional practices, or medical treatment may become necessary.

Practitioners and researchers must always rely on their own experience and knowledge in evaluating and using any information, methods, compounds, or experiments described herein. In using such information or methods they should be mindful of their own safety and the safety of others, including parties for whom they have a professional responsibility.

To the fullest extent of the law, neither the Publisher nor the authors, contributors, or editors, assume any liability for any injury and/or damage to persons or property as a matter of products liability, negligence or otherwise, or from any use or operation of any methods, products, instructions, or ideas contained in the material herein.

**Library of Congress Cataloging-in-Publication Data**
A catalog record for this book is available from the Library of Congress

**British Library Cataloguing-in-Publication Data**
A catalogue record for this book is available from the British Library

ISBN: 978-0-12-823440-2

For information on all Elsevier publications
visit our website at https://www.elsevier.com/books-and-journals

Publisher: Candice Janco
Acquisitions Editor: Maria Convey
Editorial Project Manager: Andrea Dulberger
Production Project Manager: Prasanna Kalyanaraman
Cover Designer: Greg Harris

Typeset by STRAIVE, India

*To my grandsons Adrián and Mikail*

# Contents

# Preface

The book has eight chapters. Chapter 1 singles out the main elements of the USA and Canada's energy policy and strategy, including the latest information on the North American regional energy integration development, and provides the reader with the latest information on the use of renewable energy sources for electricity generation at the world level and in the North American region. In addition, a list of pros and cons in using renewable energy sources for electricity generation in the North American region has been included for the reader's knowledge. One important element included in this chapter is the impact on the environment due to the use of renewable energy sources and nuclear energy for electricity generation in the North American region.

Chapter 1 also provides the reader with the latest information on net capacity additions by type of renewable energy source and the electricity generation using this type of energy source during the period 2010–2018 in the North American region and the USA and Canada. The mentioned chapter singles out the main limiting factors facing the use of this type of energy source for electricity generation in the region, the management of high-level radioactive waste in the USA and Canada, the proliferation risk associated with the use of nuclear energy for electricity generation in the countries mentioned before, the environmental impact of the use of nuclear energy for electricity generation in the USA and Canada, its economic competitiveness, and financial investment associated with the construction of nuclear power plants in the North American region. Finally,

it is important to stress that Chapter 1 provides the reader with the latest information on the level of public acceptance of the use of nuclear energy for electricity generation in the USA and Canada, and the main pros and cons associated with the use of this type of energy source for electricity generation in the North American region.

Chapter 2 provides the reader with the latest information regarding the level of hydropower electricity generation and the capacity installed in the North American region, the amount of investment associated with the construction of hydropower plants in the USA and Canada, the cost of producing electricity using hydropower plants in the region, and the number of hydropower plants constructed and under construction in both countries.

Besides, Chapter 2 includes information on the hydropower plants' operation efficiency and the types of incidents reported in hydropower plants registered in the region. The mentioned chapter includes the main advantages and disadvantages of using hydropower plants in the North American region and the future role of this type of energy source for electricity generation in the USA and Canada.

Finally, Chapter 2 includes the latest information on hydropower plant installed capacity in Canada and the USA during the period 2010–2019, the amount of hydropower electricity generation reported by the two countries during the period 2010–2018, and the current and future role of hydropower for electricity generation in the North American region.

Without a doubt, hydropower is the world's leading renewable energy source for electricity generation, with around 64.8% share (4,267,085 GWh) of the total renewable output (6,586,124 GWh) reported in 2018. In 2019 North America had a hydropower plant capacity of 183,822 MW, generating 698,754 GWh in 2018.

In the case of hydropower plant pure pumped storage, in 2019, the world capacity installed was 120,844 MW, generating 117,869 GWh in 2018. In 2019 the hydropower plant pure pumped storage capacity installed in North America reached 19,326 MW, generating 21,614 GWh in 2018.

In Canada, hydropower is the primary energy source used for electricity generation, and this situation will not change during the next decades. In 2019 Canada had a hydropower plant installed capacity of 81,053 MW, generating 381,750 GWh in 2018 (54.6% of the regional total 698,754 GWh). In 2019 the USA had a hydropower plant installed capacity of 102,769 MW, generating 317,004 GWh or 45.4% of the regional total.

In the case of hydropower plant pure pumped storage, in 2019, the capacity installed in the North American region reached 19,326 MW, generating 21,614 GWh in 2018. In 2019 the capacity installed in Canada of this type of hydropower plant reached 174 MW, generating 111 GWh in 2018 or 0.5% of the regional total. In the USA, in 2019, the capacity installed of hydropower plant pure pumped storage reached 19,152 MW, generating 21,503 GWh in 2018 or 99.5% of the regional total.

Chapter 3 includes the latest information on solar energy installed capacity and electricity generation at the world level and in the North American region, solar energy investment costs, solar energy construction and generation costs, the efficiency of the solar energy plants, and the types of incidents in solar energy plants reported in the North American region.

The mentioned chapter also provides the reader with the latest information on the use of solar energy for electricity generation in the North American region, the main advantages and disadvantages of using solar energy for electricity generation, and the main barriers to the massive use of solar energy for electricity generation in the region.

Chapter 3 also includes the latest information on the future of solar energy in the North American region, the solar energy installed capacity during the period 2010–2019 in Canada and the USA, the electricity generation using this type of renewable energy source during the period 2010–2018, and the future in using solar energy for electricity generation in both countries.

Undoubtedly, solar power climbed to number three among all renewable energy sources used for electricity generation at the world level, with a share of around 8.5% (562,033 GWh). In the North American region, the USA had, in 2019, a solar park capacity installed of 62,298 MW and generated in 2018, 85,184 GWh using this type of energy source (95.7% of the regional total, 88,986 GWh).

In the USA, solar park capacity installed in 2019 reached 264,504 MW. Solar electricity generation from solar parks provided a record of 743,177 GWh in 2018, almost 69% higher than the electricity produced in 2010 (440,677 GWh). It is important to stress that almost 90% of the increase in electricity generated in the USA during 2008–18 came from wind and solar energies. Solar power provided about 2% (85.2 TWh) of the total US electricity generated in 2019 (4401.3 TWh).

Canada has only 4.7% of the total solar PV capacity installed in the region in 2019 and has no CSP facilities in operation. Moreover, the investment trend in Canada's solar energy is to decrease during the coming years. Canada had, in 2019, a solar park capacity installed of 3310 MW and generated, in 2018, 3802 GWh, representing 4.3% of the regional total.

Chapter 4 provides the reader with the latest information on wind farm installed capacity and electricity generation at the world level and in the North American region, wind energy investment costs, wind farm construction and generation costs, the efficiency of the wind farms, and the types of incidents reported in the region. The mentioned chapter also includes information on the use of wind energy for electricity generation in the North American region, the impact of wind farms on the environment, and the public opinion about the use of wind energy for electricity generation in the USA and Canada. Chapter 4 also identifies the main advantages and disadvantages of using wind farms for electricity generation, wind farm capacity, electricity generation by wind farms in Canada and the USA, and the future in using this type of energy source for electricity generation in both countries.

Without a doubt, wind power continues to be number two within all renewable energy sources at the world level. It has a share of around 19.2% (1,262,914 GWh) of the total electricity produced in the world by all renewable energy sources. In 2019 the total wind farm capacity in the North American region reached 116,997 MW, generating 307,682 GWh in 2018.

In 2019 the USA had a wind farm capacity of 103,584 MW, generating 275,834 GWh in 2018, representing 89.6% of the total regional electricity generation using this type of renewable energy source (307,683 GWh). In 2019 Canada had a wind farm capacity of 13,413 MW, generating 31,848 GWh in 2018, representing 10.4% of the regional total.

The onshore wind farm installed capacity at the world level increased approximately 3.3-fold, rising from 177,790 MW in 2010 to 594,253 MW in 2019, generating 1,194,718 GWh in 2018. In North America, the onshore wind farm capacity increased 23.7-fold during the period 2010–19, rising

from 43,102 MW in 2010 to 116,968 MW in 2019, generating 307,581 GWh in 2018.

In the specific case of offshore wind farms, the capacity installed at the world level during the period 2010–19 increased 9.2-fold, rising from 3056 MW in 2010 to 28,155 MW in 2019, generating 68,196 GWh in 2018. Within the region, only the USA has a very small offshore wind farm capacity installed since 2016 (29 MW), generating 102 GWh in 2019. After June 2020, the US offshore wind farm capacity reached 41 MW.

Chapter 5 describes the use of geothermal energy for electricity generation at the world level and in the North American region and identifies new developments and future achievements in the geothermal energy sector. The mentioned chapter includes information on geothermal energy installed capacity and electricity generation in the region, geothermal energy investment costs, construction and generation costs associated with the construction of geothermal power plants, and the efficiency of this type of energy plant for electricity generation operating in North America.

Chapter 5 also provides the reader with the latest information on the use of geothermal energy for electricity generation in the North American region, identify the main advantages and disadvantages of using geothermal energy plants for electricity generation in the region, the geothermal energy capacity installed, and the electricity generation using this type of energy source in Canada and the USA. The mentioned chapter also describes the future of geothermal energy in the North American region, the USA, and Canada.

Undoubtedly, geothermal energy is number five among all renewable energy sources used for electricity generation at the world level, with a share of around 1.3% (88,408 GWh). However, in the North American region, only the USA has geothermal power plants in operation generating, in 2018, 18,773 GWh.

Chapter 6 provides the reader with the latest information on the use of bioenergy for electricity generation at the world level and in the North American region, the types of bioenergy sources used for electricity generation in the mentioned region, the bioenergy installed capacity in the USA and Canada, and the level of electricity generation using this type of energy sources in the region during the period 2010–2018. The mentioned chapter includes the latest information on bioenergy investment costs, bioenergy plant construction, generation costs, and the efficiency of the bioenergy plants operating in the North American region.

Chapter 6 also includes information on the use of bioenergy for electricity generation in the North American region, the main advantages and disadvantages of the use of bioenergy plants for this specific purpose in the region, on the bioenergy capacity installed in North America during the period 2010–2019, the electricity generation using this type of energy source in Canada and the USA during the period 2010–2018, and the future of the use of bioenergy for electricity generation in both countries.

It is important to single out that bioenergy falls to number four among all renewable energy sources used for electricity generation at the world level, with a share of around 7.9% (522,552 GWh). In 2019 the bioenergy capacity installed at the world level reached 124,026 MW.

In the North American region, the bioenergy capacity installed in 2019 reached 15,825 MW, representing 12.8% of the world total. The bioenergy capacity installed in the USA reached, in 2019, a total of 12,450 MW, representing 78.7% of the total at the regional level. The bioenergy capacity installed in Canada is much lower, reaching 21.3% of the total in 2019. The USA generated, in 2018, 67,885 GWh using bioenergy plants (86.4% of the regional total, 78,546 GWh). Canada generated, in the same year, 10,661 GWh, representing 13.6% of the regional total.

Chapter 7 provides the reader with the latest information on the use of nuclear energy for electricity generation at the world level and in the North American region. In addition, the mentioned chapter includes information on the next generation of nuclear power reactors that will be available in the market in the coming years.

Chapter 7 also includes information on the main steps for the introduction or expansion of a nuclear power program in any given country, the economic optimization in the use of all energy sources available in the country for electricity generation, the stability of the national energy grid, security of electricity supply, environmental issues associated with the use of nuclear energy for electricity generation, electrical grid integration policy and strategy, nuclear safety, and sharing of power plant services.

The mentioned chapter also stresses the importance of a nuclear energy program in the North American region, the number of nuclear power reactors shut down in the USA and Canada, the need for nuclear energy for electricity generation in the region, and the main limiting factors impeding the rise in the use of nuclear energy for electricity generation in both countries.

Chapter 7 provides the reader with the latest information on the nuclear power program in the USA and Canada, the nuclear policy of both countries, the evolution of the nuclear power sector in the USA, extension of the lifetime of the current nuclear power reactors in operation in the USA and Canada, reduction of the construction time of nuclear power plants, the management of high-level nuclear waste, new US nuclear regulatory commission's regulations, the nuclear industry, the uranium resources, and the mining industry in the country.

Chapter 7 also includes the latest information on the US trade within the nuclear power sector, the public opinion on the use of nuclear energy for electricity generation in the country, the nuclear power program in Canada, the uranium resources, the Canadian nuclear energy policy, the main characteristics of the CANDU technology and the Canadian nuclear industry, the evolution of the Canadian nuclear power sector, the Canadian nuclear safety commission, and decommissioning activities and trade within the nuclear sector in the country.

Finally, Chapter 7 also provides the reader with the latest information on the use of nuclear energy for electricity generation in Canada and the USA and the impact on the environment, the public opinion on the use of nuclear energy for electricity generation in Canada and the USA, and the future of nuclear energy in both countries.

Based on the information included in Chapter 7, it can be stated that Canada and the USA use different types of nuclear power reactors for electricity generation, making it very difficult to develop a regional service. In Canada's case, the only type of nuclear power reactor used for electricity generation is the so-called CANDU reactor (Pressurized Heavy Water Reactor or PHWR type). A total of 19 units with a capacity of 13,554 MW were operating in the country in April 2021, generating 95,469 GWh in 2019, or 14.9% of the total electricity generated in the country in that year. On the other hand, only two types of nuclear power reactors are used for electricity generation in the USA (Pressurized Water Reactor or PWR and Boiled Water Reactor or BWR). In April 2021, there were 94 units in operation: 62 PWR and 32 BWR with a total capacity of 96,553 MW, generating 809,409 GWh or 19.7% of the total electricity generated in the country in 2019.

Chapter 8 summarized the main topics in all previous chapters related to the different types of nonconventional energy sources used for electricity generation in the North American region, particularly in the USA and Canada.

# Acknowledgments

Without a doubt, the present book is a reality thanks to the valuable support of my lovely wife, Aurora Tamara Meoqui Puig. She had assumed other family responsibilities to give me indispensable time and a good environment to write the current book. I would also like to thank my dear friend and colleague, Eng. Alejandro Seijas, for his excellent job in revising the first draft of the book and the references used.

# CHAPTER

# 1

# General overview

## Introduction

It is an undisputed reality that there is a strong link between human progress and energy consumption. This is why energy production, particularly electricity generation and its sustained growth, constitutes indispensable components for any country's economic and social development. For this reason, the use of all types of energy sources for electricity generation available in a country should be considered during the preparation of its energy mix in order to ensure sustainable economic and social development.

However, it is important to be aware that electricity generation using fossil fuels is a significant and growing contributor to carbon dioxide emissions. This greenhouse gas contributes significantly to global warming, which produces a considerable change in the climate all over the world, affecting almost all countries in all regions in one way or another. As a result, the consumption of oil and coal for electricity generation has been reduced in several countries all over the world during the last years. It is expected that this trend will continue to be the same until 2050. Thus, natural gas is the only fossil fuel expected to increase its participation in the energy mix of many countries worldwide, particularly in the developed world.[a]

Considering all types of energy sources that can be used to satisfy the foresee increase in energy demand, there are only a few realistic options available today to further reduce $CO_2$

---

[a] Since 1970, the world has seen a rapid growth in energy demand, mainly satisfied by fossil fuels and centralized power generation (Used by permission of the World Energy Council, World Energy Issues Monitor 2019, 2019). However, the future is expected to be different: the role of fossil fuel in the energy mix of several countries during the coming decades will be lower than today and, at the same time, the role of renewables in the energy mix of these countries will be higher. In some cases, the role of renewable in the energy mix of a limited number of countries will be much higher than it is today. The future use of nuclear energy for electricity generation is today unclear. Only Asia has serious plans for the increased use of this type of energy source for electricity generation. Other regions are debating whether to continue the use of nuclear power plants for electricity generation or the closure of these power plants during the coming years.

---

emissions to the atmosphere due to electricity generation.[b] These options are, among others, the following:

**(a)** Reduce the use of oil and coal for electricity generation. The use of coal for electricity generation increased slightly in 2018, but coal consumption has been declining year by year for the last few years, and there is an increasing trend by countries, corporations, traders, and investors to shy away from coal investment (IEA, 2018b). Meanwhile, renewable energy investment continues, albeit slightly lower than in 2017, with Bloomberg New Energy Finance estimating total expenditures at US$ 332 billion in 2018 (BNEF, 2019). There are also signs that even the major oil companies are considering getting more into the electricity business. Royal Dutch Shell recently said it could develop a robust business and mentioned that it could become one of the largest electricity companies globally by 2030 (F.T., 2019);

**(b)** Increase efficiency in electricity generation and use[c];

**(c)** Expansion in the use of hydroelectricity, wind, solar, bioenergy, and geothermal, among other types of energy sources used for electricity generation;

**(d)** The massive introduction of new advanced technology like the capture carbon dioxide emissions technology at fossil-fueled (especially coal) power plants, with the purpose to permanently sequester the carbon produced by these plants;

**(e)** Closure of old and inefficiency coal-fired power plants[d];

**(f)** Increase the use of new types of nuclear power reactors inherent safe and proliferation risk-free, particularly Generation IV of nuclear power reactors and small modular reactors (SMRs);

**(g)** Increase in energy saving.

Undoubtedly, improving energy efficiency is one of the key elements to provide more electricity in countries with a high level of power consumption without increasing the use of more energy sources, particularly coal and oil. During the coming decades, the transition to a lower-carbon fuel mix is expected to continue without change, and renewables and natural gas will lead to this transition. According to BP Energy Outlook 2019 report, renewables and natural gas are expected to account for "almost 85% of the growth in primary energy, with their importance increasing in comparison with all other sources of energy." Renewable energy, with an expected increase of 7.1% per year, is likely to be the fastest-growing source of energy, contributing 50% of the growth in global power, with its share in primary power increasing by 11%, rising from 4% today to around 15% by 2040.[e]

[b] It is expected that the level of CO2 emissions from the use of some types of energy sources for electricity generation will continue to increase by 10% in 2040.

[c] It is likely that energy efficiency must be scaled up substantially during the coming years; the rate of energy intensity improvement would increase to 3.2% per year, up from recent historical averages of around 2% per year (IRENA, 2019b).

[d] Coal-fired power plants in OECD countries will be banned from 2030. A worldwide ban from 2030 on new investment in non-CCUS (Carbon capture use and storage) coal power plants has been adopted, as well as a decision to support for stronger deployment of nuclear power plants, particularly in Asia, and on hydropower plants during the coming decades (BP Energy Outlook 2019, 2019).

[e] The expected increase or decrease of fossil fuels until 2040 will be the following: natural gas (1.7% per year); oil (0.3% per year) and coal (− 0.1% per year).

According to the Global Energy Statistical Yearbook 2018 (2019), the total world electricity demand increased from 11,704 TWh in 1990 to 25,287 TWh in 2017; this means by a factor of 2.2. In 2018, the world electricity generation was 26,614.8 TWh; this represents an increase of 3.7% in 2017. In 2019, according to the BP Statistical Review of World Energy (2020) report, the world electricity generation reached 27,004.7 TWh, an increase of 1.3% concerning 2018. In the North American region, the electricity generation, in 2018, reached 5110.7 TWh or 19.2% of the world total. In 2019, the electricity generated in the North American region reached 5061.7 TWh, a decrease of 0.6% concerning 2018. According to Global Energy Statistical Yearbook 2019 (2020), the electricity demand in 2018 was 24,214.8 TWh, around 4.2% lower than in 2017.

However, according to estimates made by the World Energy Council database, the International Institute for Applied Systems (IIAS database), among other international organizations, the electricity demand probably will be tripled from now until 2050. Why this significant increase in the demand for electricity in the next 30 years? The following are, among others, the main reasons for this increase:

**(a)** Increase in the world population;
**(b)** Increase in the percentage of the world population living in big cities, rising the demand for electricity significantly;
**(c)** Improve the quality of life of the world population bringing as a consequence an increase in the demand for electricity and of other forms of energy;
**(d)** Increase in demand for electricity in the most advanced developing countries, such as Brazil, India, China, the Republic of Korea, and South Africa, among others, due to their fast economic and social development (Morales Pedraza, 2012).

The electricity generation by fuel in the North American region in 2019 is shown in Fig. 1.1.

According to Fig. 1.1, natural gas was the dominant fuel used for power generation in the North American region, in 2019, with 1770.2 TWh or 35% of the total at the regional level, followed very closely by renewables with 1192.2 TWh or 23.6% of the total at the regional level, coal with 1108.1 TWh or 21.9% of the total at the regional level, and nuclear

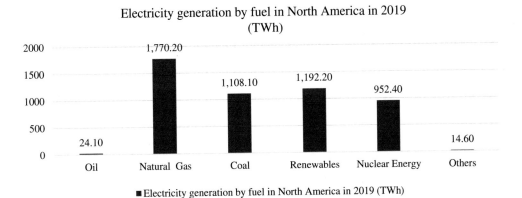

**FIG. 1.1** Total electricity produced by fuel type in 2019. *Source: 2020. BP Statistical Review of World Energy, 69th ed. and Author own calculations.*

energy with 952.4 TWh or 18.8% of the total at the regional level. In 2018, the use of fossil fuels for electricity generation was 2732.1 TWh or 53.4% of the total at the regional level. In 2019, the use of fossil fuels for electricity generation was 2902.4 TWh, increasing 6.2% with respect to 2018.

At the world level, in 2018, the dominant fuel for electricity generation was coal with 10,091.3 TWh or 37.9% of the world total, followed by renewables with 6639.4 TWh or 24.9% of the world total, and natural gas with 6082.5 TWh or 22.8% of the world total. In 2019, this situation was the same. Coal continues to be the main electricity generation energy source with 9824.1 TWh, or 36.4% of the world total, followed by renewables with 7027.7 TWh 02 26% of the world total, and natural gas with 6297.9 TWh or 23.3% of the world total. It is important to single out that hydropower provides to South and Central America more than half of its power, with a share far higher than any other region. On the other hand, in Europe, nuclear energy is the top electricity source, but not far from other energy sources. The shares of nuclear energy, coal, natural gas, and renewables within the energy mix are all in a narrow range between18% and 22%.

The world is now facing a crucial problem: how to satisfy the increasing demand for electricity using the available energy sources most efficiently and without increasing the emission of $CO_2$ to the atmosphere. How to achieve this? "One of the most effective solutions is the elaboration of national energy policy and strategy in which priorities and preferences are identified as well as the main responsibilities of the national authorities in charge of the energy sector. This energy policy and strategy should represent a compromise between expected energy shortages, environmental quality, energy security, cost, public attitudes, energy safety, available skills, and production and service capabilities. Relevant national energy authorities and the energy industry representatives must take all of these elements into account when formulating energy policy and strategy for the development of the energy sector" (Morales Pedraza, 2012). The energy mix that should be prepared due to the implementation of the energy policy and strategy adopted must consider all the country's energy options.

## Main elements of an energy policy and strategy

One of the most important policies for nations and states is that of energy security. In any country, the satisfaction of production, communication, transportation, lighting, air conditioning, water pumping, etc., demands the competition of various forms of energy. If this consumption is added to non-energy uses such as petrochemicals and other sectors, including the energy sector itself, its energy consumption will increase significantly. Therefore, sustainable energy policy should ensure a balance between energy supply and demand within the sustainable development framework, which guarantees social needs in the first place to meet the requirements of economic growth then.

According to Wikipedia Energy Policy (2019) and Flórez and Arturo (2007), energy policy is how governments have decided to address energy development issues, including energy production, distribution, commercialization, and consumption, with the purpose of supporting the sustained growth of the country. In addition, energy policy may include legislation, international treaties related to the energy sector, incentives to invest in the energy sector, guidelines for energy conservation, taxation, and other public energy policy issues.

National energy policy is also a process for the optimization of energy use. It comprises a set of measures involving the country's laws, treaties, agency directives, economic and social sectors, government levels (national, regional, and local), health, and environmental authorities. These measures are, among others, the following:

- Adoption of a statement on national energy policy regarding energy planning, energy development, energy generation, energy transmission, and energy usage;
- Approval of legislation on commercial energy activities, including energy trading, energy transport, energy storage, among others, and legislation affecting energy use, such as efficiency and emission standards;
- Instructions to be followed by state-owned energy sector assets and organizations;
- Active participation, coordination of, and incentives for mineral fuels exploration, and other energy-related research and development policy;
- Adoption of fiscal policies related to energy products and services (taxes, exemptions, subsidies, among others);
- Adoption of energy security and international policy measures such as:
  - International energy sector treaties and alliances;
  - General international trade agreements;
  - Special relations with energy-rich countries, including military presence or domination (Wikipedia Energy Policy, 2019).

The energy policy adopted must be associated with a price policy and level of income, environmental policy, and regional integration energy policy. An energy policy's main objective is to ensure security in the energy supply with quality and continuity to ensure accessibility to all forms of energy in conditions of technical and economic efficiency, with prices accessible to the population, and to support companies' financial capacity. That implies choosing a diversified energy basket and adequate storage, transport, distribution infrastructure, and appropriate marketing mechanisms (Flórez and Arturo, 2007). In other words, energy policy should promote the rational and efficient use of all types of energy sources and increase the system's reliability without reducing the level of benefits. Therefore, the implementation of energy policy should allow for systematic improvements in energy efficiency and maintain stable energy consumption levels over time, thanks to adequate energy management.

Likewise, energy policy must ensure compliance with the applicable legal requirements and other requirements related to the use and consumption of energy and energy efficiency, in addition to supporting the acquisition of energy-efficient products and services, as well as the design to improve energy performance.

Several criteria should be used during the formulation of an energy policy. These criteria are, among others, the following:

- Macroeconomics indicators and their trends;
- Energy demand;
- Availability of energy resources;
- Energy and economic development sustainability;
- Energy and its environmental impact;
- Energy development technology and new energy solutions;
- Energy infrastructure required;

- Energy market conditions;
- Energy legal framework;

The following are the main benefits of having a sound energy policy:

- Energetics, economics, and environmental benefits;
- Optimization in the use of all types of energy sources;
- Improvement of energy efficiency;
- Reduction of greenhouse gases;
- Reduction in the impact on climate change as a result of the use of clean energy for electricity generation;
- Reduction of energy costs as a result of energy-saving.

Several questions need to be answered during the preparation of a national energy policy. Some of these questions are, according to Hamilton (2013), the following:

- What is the extent of energy self-sufficiency for the country?
- Where will energy sources come from in the future?
- How will future energy be consumed (e.g., among sectors)?
- What fraction of the population will be acceptable to be classified as energy poverty, if any?
- What are the goals for future energy intensity, the ratio of energy consumed to gross domestic product (GDP)?
- What is the reliability standard for distribution reliability?
- What environmental externalities are acceptable and forecast?
- What form of "portable energy" is forecast (e.g., fuel sources for motor vehicles)?
- How will energy-efficient hardware (e.g., hybrid cars, household appliances) be encouraged?
- How can the national policy drive province, state, and municipal functions?
- What are the specific mechanisms in place to implement the complete policy?
- What future consequences will there be for national security and foreign policy? (Wikipedia Energy Policy, 2019).

## Main elements of the United States energy policy

The Trump administration's energy policy represents a shift from his predecessor's policy priorities and goals, Barack Obama. Obama's energy and environmental agenda prioritized reducing carbon emissions by closing old and inefficient coal power plants to achieve independence from fossil fuel imports from abroad. Obama's plan also:

- Guarantee high and permanent fuel reserves in the country;
- Decrease and make more efficient its industrial, automotive, and domestic energy consumption;
- Increase the areas and volumes of hydrocarbon extractions within the national territory;
- Promote new technologies and projects for their internal use and the production and use of new fuel sources;
- Increase the role of renewable energy sources within the energy mix of the country.

On the contrary, the Trump administration has decided to prioritize the use of fossil fuel, particularly coal, for electricity generation. He is also deleting or questioning the many energy and environmental regulations adopted by Obama's administration. According to President Trump's opinion, the Obama administration's energy policy impeded the USA from increasing its economic and energy output.

Regardless of what has been said above, not all Trump administration members support increasing fossil fuels for electricity generation. Instead, around 75% of President Trump supporters advocate using renewable energy sources for this specific purpose. This is because the economic reality shows that renewable energy is now among the cheapest sources of electricity. For this reason, wind and solar energy were the largest sources of energy added to US networks during the last three years, becoming a key source of employment in rural areas of the country (Robles, 2017).

It is important to stress that as a result of the implementation of President Trump's energy policy, he has pulled the USA out of the Paris climate agreement, left both the 44th G7 summit held in Canada and the 45th G7 summit held in France early, avoiding the environmental discussions (Hansler, 2018). President Trump has often said they do not believe $CO_2$ is a primary contributor to global warming. For this reason, the USA should not be part of the Paris climate agreement or any other agreement of the same characteristics. Responding to a 2018 government-funded study that warned of potentially catastrophic climate change impacts, President Trump said he had read part of the report mentioned above, but his opinion about climate change has not changed. In other words, he has said, I do not believe it (Cillizza, 2018).

In a White House speech, President Trump hailed "America's environmental leadership" under his watch, asserting his administration was "being good stewards of our public land," reducing carbon emissions and promoting the "cleanest air" and "crystal clean" water. However, it is essential to single out that many experts have indicated that the cited achievements resulted from President Obama's actions going all the way back to the Nixon administration (Rogers and Davenport, 2019).

If the Trump administration's whole period is studied in detail, it can be stress that little has changed in the US energy markets since Trump took office. States like California, New York, and Massachusetts continue to move forward with aggressive policies to reduce carbon emissions, and several companies continue to power installations with wind and solar energy (Robles, 2017).

The main elements of the Trump administration energy policy are the following:

1-Elimination of the Clean Power Plan. This plan was an Obama administration energy policy, proposed by the Environmental Protection Agency (EPA), aimed at combating climate change (global warming) (Clean Power Plan for Existing Power Plants, 2016). The main objective of the plan was to lower the carbon dioxide emitted by power generators;

2-Increase the use of coal for electricity generation. According to the National Mining Association fact sheet entitled "Coal: America's Power (2018)," coal is the US's most abundant energy resource—making up 90% of the country's fossil energy reserves. At current coal consumption rates (678 million short tons in 2016), the USA "has more than 250 years of remaining coal reserves." For this reason, coal is essential to the US economy, providing affordable electricity to households, businesses, manufacturing facilities, transportation, and communications systems and services throughout the economy. The coal industry created a total of more than 500,000 jobs and generated US$26 billion in sales and paid US$13 billion in direct wages and salaries in 2016".

According to the National Mining Association, the Clean Power Plan would cause coal production to fall by 242 million tons;

3-Making the wind industry accountable for the same environmental protections that apply to everyone else. Without a doubt, compared to the environmental impact of traditional energy sources used for electricity generation, the environmental impact of wind power is relatively minor. Wind power consumes no fuel or water and emits no air pollution. It is well-known that a wind farm may cover a large area of land. However, most of the land used to build a wind farm can also be used for other purposes because only small turbine foundations and infrastructure are unavailable for any further use. It is also true that there are reports of bird and bat mortality in wind farms, but the amount of birds killed is much lower than the number of dead birds as a result of several other human activities. It is essential to single out that prevention and mitigation of wildlife fatalities and protection of peat bogs have been adopted to minimize the rate of mortality associated with the operation of wind farms;

4-Wind and solar power disproportionate subsidies. In 2020, the current competition between whether to use conventional energy or renewable energy for electricity generation will likely evolve further to encompass not just renewable versus traditional resources but also renewables in competition with each other. In many USA regions, the Levelized Cost of energy for an onshore wind farm is less than for solar photovoltaic (PV) parks. However, costs for solar energy have been declining faster than wind energy recently. In addition, federal support for wind energy steps down in 2020, while solar energy still enjoys federal benefits. For this reason, solar energy could become increasingly cost-competitive with the use of wind energy for electricity generation. As a result, in 2020 and beyond, some wind-only customers will likely diversify and build a mixed portfolio of wind farms, solar parks, and storage facilities to fulfill commitments (Motyka, 2020).

Besides, wind and solar subsidies have dwarfed those of all other energy sources during the past decade, imposing expensive and unreliable power on US energy consumers. "The wind and solar industries claim their products are falling in price and insist they can provide power on a cost-competitive basis with conventional power. Expect the Trump administration to hold the wind and solar industries to their word, reducing subsidies and restoring a level playing field for competing for energy sources" (Taylor, 2016);

5-Ethanol gets closer scrutiny. "The ethanol industry has been under siege during Trump's administration. The biggest issue of contention has been hardship waivers that have been granted to some refiners." In addition, the Renewable Fuel Standard (RFS) "established quotas of renewable fuels that have to be blended into the fuel supply, and an enforcement mechanism to ensure those quotas were met. This enforcement mechanism essentially transfers money from oil refiners to the ethanol industry and is loathed by the nation's refiners." For this reason, it is difficult to argue that President Trump's energy policy has been beneficial to the ethanol industry. On this issue, the administration has not pleased either the ethanol industry or the oil industry (Rapier, 2019).

While President Trump's plan specifically identifies crude oil, natural gas, and coal as domestic energy sources that require deregulation and political support, it neglects to mention the important role of ethanol and RFS in securing the USA energy future. Together with the US oil and gas industry, the country "has made great strides in reducing imports and boosting domestic energy supplies. The ethanol industry alone has added nearly 3.3 billion barrels of low-cost, high-octane liquid fuel to domestic supplies since the RFS was adopted in 2005;

that's more than 1000 gal per US household." Supporting a strong RFS and eliminating regulatory barriers that restrain greater ethanol use is one of the key strategies that will help the US economy to reduce the import of fossil fuels (Oestmann, 2017);

6-Yucca Mountain finally begins accepting nuclear waste. On June 3, 2008, the US Department of Energy (DoE) applied for the US. Nuclear Regulatory Commission (NRC) for a license to construct a nuclear waste repository at Yucca Mountain. The primary purpose of this license is "to develop, build, and operate a deep-underground facility that will safely isolate spent nuclear fuel and high-level radioactive waste from people and the environment for hundreds of thousands of years. Currently, nuclear waste is stored at 121 temporary locations in 39 states across the nation" (The National Repository of Yucca Mountain, 2008);

7-Next-generation nuclear power surges forward. According to the PRIS—Power Reactor Information System (PRIS) (2021), the US had 94 nuclear power reactors operating in the country, generating 19.7% of the country's electricity in 2019.[f] The USA is the country with the highest number of nuclear power reactors in operation in 2020 in the world (21.2% of the total). However, the number of new units under construction is very small, only two units, representing 3.7% of the total units under construction. Regarding new prototypes of nuclear power reactors, the US is investigating nine different types of reactors.

The Trump administration is preparing to keep old and non-competitiveness coal and nuclear power plants online, potentially by applying rarely-used federal powers and justifying the move by arguing that these facilities are critical to national security. Based on this argument, the government should prevent the closure of more coal and nuclear power plants for economic reasons. It is important to single out that coal and nuclear power plants are retiring ahead of schedule. That is primarily due to price competition from natural gas-fired power plants and renewable solar and wind farms, which are now generating electricity at a lower price than some other energy sources (DiChristopher, 2018). According to Schroeder (2018), "White House said impending retirements of such facilities are depleting a critical part of the country's energy mix and impacting the power grid's resistance. Under the plan, grid operators will be forced to take the electricity those plants produce";

8-Hydropower reverses its long decline. According to the Department of Energy's Oak Ridge National Laboratory study (ORNL database), about 78,000–80,000 existing dams in the USA do not have hydropower connected to them. Only 3% of them are equipped to produce electricity (around 2200 dams) (Siegel, 2018). While the public may not consider hydropower in the USA as a "growth industry," President Trump is aware of hydropower potential in making the USA energy independent, if not greener. However, there will not be a hydropower rush, nor will environmental reviews be reduced. Instead, the new US administration will identify the obstacles to putting hydropower at existing dams with no power and what we can do about it (Russo, 2017). As a result of growing world concerns over climate change and falling costs associated with the use of solar and wind energies for electricity generation, the construction of new solar parks and wind farms in the North American region has increased. Simultaneously, this situation had stalled the new hydropower plants' construction, contributing zero carbon emissions until recently. Electricity generation by hydropower

---

[f] In 2019, nuclear power plants generated 19.7% of the total USA electricity production (4118.05 GWh), according to PRIS—Power Reactor Information System (PRIS) (2021).

represents, in 2019, 6.2% of all US electricity production and 6.4% of all electricity produced at the world level. In 2019, the USA occupied the fifth-place among all countries according to the level of electricity generated by hydropower plants;

9-Natural gas exports increase. According to EIA natural gas imports and exports report (2019), although most of the natural gas consumed in the USA is produced in the country, the USA imports some natural gas to help domestic supply-demand. However, it is essential to single out that total annual imports of natural gas have been declining since 2007, falling from 5 trillion cubic feet in 2007 to 2.91 trillion cubic feet in 2018,[g] mainly because the country increased its natural gas production. The USA also exports natural gas. Most natural gas imports and exports are by pipeline as gas and by ship as liquefied natural gas (LNG). In addition, small amounts of natural gas are imported and exported by trucks as LNG and compressed natural gas (CNG).

Until 2000, the USA exported relatively small natural gas volumes mostly by pipeline to Mexico and Canada. However, total annual exports have generally increased from 2000 until 2018. An increase in US natural gas production contributed to lower natural gas prices and US natural gas competitiveness. In 2018, the USA exported 3.61 trillion cubic feet of natural gas, the highest among ever, to 33 countries, converting the country into a net exporter of natural gas for the second year in a row (EIA, 2019c).

## Main elements of Canada energy policy

Without a doubt, designing an energy policy that reflects the current and future energy needs of a country is a tough job. Well-designed energy policies reduce pollution, cut consumer costs, and minimize dependence on foreign energy supplies. On the other hand, adopting the wrong energy policy could create many problems in almost all countries' economies, increase pollution, and wasted resources.

During the preparation of energy policy, all energy resources available in the country should be studied in detail to select the best combination of energy sources to conform to its energy mix. Canada has all the primary energy sources available globally, including oil and gas, coal, hydropower, biomass, solar energy, geothermal energy, wind energy, marine energy, and nuclear energy. In 2018, according to the BP Statistical Review of World Energy (2020), Canada was the world's second-largest producer of uranium and the third-largest producer of hydroelectricity. Only Russia, China, the USA, India, and Japan produce more energy than Canada.

The energy policy in Canada is based on three main principles. These principles are, according to Wikipedia Energy Policy of Canada 2019 (2019), the following:

- Competitive markets to ensure a successful and innovative energy system capable of meeting Canadian energy needs. According to Energy Policy-Natural Resources Canada (2014), "markets are the most efficient means of determining supply, demand, prices, and trade while ensuring an efficient, competitive and innovative energy system that is responsive to Canada's energy needs";

---

[g] In 2019, the USA imported 73.3 billion cubic meters of natural gas and 1.5 billion cubic meters of LNG, according to the BP Statistical Review of World Energy (2020) report.

- Respecting the jurisdictions of provinces and the federal government. According to the Canadian government structure, "provincial governments are the direct managers of most of Canada's resources and have responsibilities for resource management within their borders";
- Targeted federal interventions in the energy trading process, ensuring the specific energy-policy objectives are achieved (Laverty, 2015). These policy objectives include issues of health and safety (e.g., pipeline regulation) and environmental sustainability.

The following are the main federal decisions adopted in the framework of the implementation of Canada's energy policy:

- The creation of the National Energy Board to promote in the public interest safety and security, environmental protection, and efficient energy infrastructure and markets in the regulation of pipelines, transmission lines, energy development, and trade;
- The creation of the Canadian Nuclear Safety Commission to regulate all aspects of the nuclear power industry in Canada;
- The establishment of the Atomic Energy of Canada Limited to foster the advancement of nuclear energy and nuclear technology;
- The Program's funding on Energy Research and Development supports the expansion of energy technologies (Energy Policy-Natural Resources Canada, 2014).

In 2019, renewable energy sources provided about 17.3% of Canada's total primary energy supply (Energy Fact Book 2019–2020 Canada, 2019; Renewable Energy Facts, 2019) and about 67% of its electricity production (Electricity Facts, 2019).[h] In 2019, according to the BP Statistical Review of World Energy (2020) report, renewable energy sources provided 27.7% of Canada's primary energy consumption.

The majority of renewable energy produced in Canada comes from hydroelectricity (382 TWh). That makes Canada, in 2019, the third-largest producer of hydroelectric power globally, after China, with a production of 1269.7 TWh or 30.1% of the total electricity produced by hydropower plants in the world in that year, and Brazil with 399.3 TWh or 9.5% of the world total (BP Statistical Review of World Energy, 2020).

Wind power is a fast-growing sector of the world energy market, generating, in 2019, 1429.6 TWh, representing an increase of 12.5% with respect to 2018. Canada was the eighth largest producer of wind power in 2019 (34.2 TWh). Canada has also built many solar PV power plants, mainly in Ontario, with one in Sarnia, the largest solar PV park globally at the time of its construction. The solar PV park production reached 4.3 TWh in 2019, representing 0.06% of the world total (BP Statistical Review of World Energy, 2020).

The Canadian government has expressed interest in increasing the percentage of Canada's electricity generated by renewable energy sources and has announced that it will stop using coal to produce electricity in 2030. It is important to stress that coal power plants produce 8.3% of the electricity generated in Canada in 2019. But these plants are responsible for more than 70% of greenhouse gas emissions in the electricity generation sector and 8% of the country's total emissions. The government expects that by 2030, around 90% of the electricity generation will be produced without greenhouse gas emissions.

---

[h] According to the BP Statistical Review of World Energy report, this percentage is 65.3%.

## North America energy integration

At the highest levels of governments in the USA and Canada, there has been a sustained interest in the benefits of enhancing North America's energy integration. Several recent developments make this discussion more relevant than ever before, including:

- New administrations took office in Canada and the USA with a stated interest in bringing greater focus and more global engagement to discuss clean energy issues;
- The adoption of the Paris Climate Convention (COP 21) and measures needed to implement its provisions globally;
- The development of new renewable energy technologies, which raises new questions about grid management and the benefits of integration;
- The USA's shale gas boom will increase natural gas trade opportunities, raising new questions about land use, trade opportunities, and air emissions.

The significant electricity integration that already exists between the USA and Canada suggests that international policymakers cannot overlook these developments. "Identifying the most efficient opportunities in the changing North American paradigm will require extensive effort and thought leadership from governments, which must balance national priorities and regional benefits while utilizing existing institutional and regulatory frameworks with asymmetrical federal, state, and local jurisdictions. To navigate this process effectively, it is critical that policymakers have a clear understanding of the current status of integration" (Vidangos et al., 2016).

However, it is vital to single out that "though enhancing regional energy integration does not automatically result in benefits for all parties, it is a central concept in economics that international integration has the potential to provide new opportunities for growth, largely through trade and increased productive efficiency" (Pineau, 2013).

Despite the broad consensus among governments and energy industry representatives that enhanced integration can bring significant benefits to society, there are myriad challenges as well as pitfalls for inefficient integrative energy policy. The barriers can be grouped into three categories:

- **System Complexity**: While significant integration already exists between the USA and Canada, such integration evolved historically from an evolving "bottom-up" approach among provincial and state authorities without a "top-down" strategic vision or standardized process. As a result, US-Canadian integration reflects both countries' market structures, which are primarily fragmented and involve a range of federal, provincial, and state government jurisdictions as well as a range of industry, nongovernment, and society stakeholders (Resources for the Future, 2016), resulting in a system of complex systems. While the USA did try to simplify its market through a standard market design initiative led by Federal Energy Regulatory Commission in 2005, this effort did not proceed as designed (Federal Energy Regulatory Commission, 2005), and no equivalent action has been attempted in Canada to adopt a national common market framework;
- **Regulatory Policy**: As with any sector that requires infrastructure development, the policy and regulatory framework are critical to the pace of development. While regulatory policy naturally varies regionally to reflect the specific priorities of different

jurisdictions (i.e., economic development, environmental protection, sustainable development, social engagement), it can also lead to significant inefficiencies in permitting, siting, licensing, as well as tax, tariff, and incentives policies that reduce the possibilities of integration. In all countries, the private sector has complained that permitting/siting/licensing requirements are slow, costly, and onerous. However, some industries speak more positively about their experience complying with Canadian cross-border regulations (Resources for the Future, 2016).

Tariff policies can also provide disincentives for cross-border energy trade. Even nationally-focused energy incentives, while not always a direct deterrent to cross-border investment, may also influence cross border energy trade opportunities (Vidangos et al., 2016);

- **Social and Political Considerations**: Finally, it is important to mention that regional energy integration can inspire passionate discussions about national sovereignty, interdependency, environmental safety and sustainability, local industries, and jobs, among other relevant issues. While there is no easy solution to many of these concerns, fair and transparent regulatory processes, extensive community stakeholder engagement, and precise quantification of the costs and benefits for a community of the regional energy integration can enhance the chance of a successful process. It is important also to stress that despite that no federal authority in the USA or Canada can unilaterally mandate movement toward greater energy integration, federal governments still play a vital role in the sector's thought leadership, strategic planning, analysis, tool development, and convening of critical stakeholders to explore new energy opportunities. In addition, federal governments are responsible for maintaining a prosperous economy, delivering on international commitments (including emissions reductions relating to climate change), and ensuring energy security for their citizens (Vidangos et al., 2016).

## The use of renewable energy sources for electricity generation at the world level and in the North American region

One of the most relevant characteristics of a country's electrification using different renewable energy sources is that it can start to reduce energy-related $CO_2$ emissions immediately and substantially. Without a doubt, energy in the form of electricity would increasingly become the predominant energy carrier, growing from a 20% share of final consumption today to an almost 50% share by 2050. It is expected that renewable energy "would be able to provide the bulk of global power demand (86%) economically in 2050. As a result, gross electricity consumption is projected to reach more than double. The pairing is also getting cheaper than fossil fuel-based alternatives, lowers local air pollution, increases health benefits, results in positive socio-economic benefits, and will be a key enabler to building a connected and digitalized economy and society. Electrification, when paired with renewables, goes hand-in-hand with energy efficiency, resulting in lower overall energy demand" (IRENA 2019, Global Energy Transformation: A Roadmap to 2050 (2019 edition) and International Renewable Energy Agency, 2019).

The use of renewable energy for electricity generation in the North American region is an excellent business opportunity that cannot be ignored today. It is an absolute truth that there has never been a better opportunity to replace carbon-based fossil fuels with alternatives energy sources such as wind, hydroelectric, solar power, bioenergy, and geothermal energy. For this reason, the governments of Canada and the USA are reevaluating their long-term energy policy. This reevaluation is particularly true in the USA's case, which during the Trump administration promoted and supported the use of all available fossil fuels in the country for electricity generation, particularly coal and nuclear energy, instead of renewables. Despite the favorable environment for changing an energy policy that makes emphasis, in the case of the USA, on the use of fossil fuels for electricity generation, however, businesses are only tentative consumers of renewable energy, and many energy producers are still waiting to see if there are significant opportunities to generate new revenue by using renewables or nuclear power instead of fossil fuels for electricity generation during the coming years.

The cost of renewable energy has been reduced significantly in recent years. Still, it remains expensive for both consumers and suppliers, particularly in Canada, impeding the shift to use new renewable energy sources for electricity generation.

The evolution of the renewable energy capacity in the North American region during the period 2010–19 is shown in Fig. 1.2.

According to Fig. 1.2, the following can be stated: renewable energy capacity increased 67.2% in the North American region during the period under consideration, rising from 218,540 MW in 2010 to 365,501 MW in 2019. Moreover, it is expected that this trend will continue without change during the coming years, particularly under the Biden administration's new energy policy.

The evolution of electricity generation using all available renewable energy sources in the North American region during the period 2010–18 is shown in Fig. 1.3.

Based on the data included in Fig. 1.3, the following can be stated: the electricity generation in the North American region using all available renewable energy sources increased by

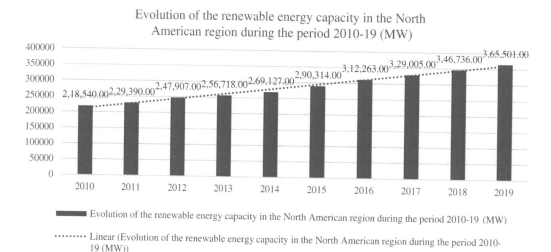

**FIG. 1.2**  Evolution of the renewable energy capacity in the North American region during the period 2010–19. *Source: IRENA, 2020. Renewable capacity statistics 2020. International Renewable Energy Agency (IRENA), Abu Dhabi and Author own calculations.*

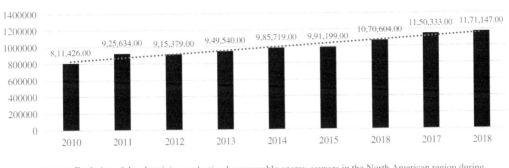

Evolution of the electricity production by renewable energy
sources in the North American region during 2010-2018 (GWh)

FIG. 1.3    Evolution of electricity generation using all available renewable energy sources in the North American region during the period 2010–18. *Source: IRENA, 2020. Renewable Energy Statistics 2020. The International Renewable Energy Agency, Abu Dhabi and Author own calculations.*

44.3% during the period 2010–18, rising from 811,426 GWh in 2010 to 1,171,147 GWh in 2018. Furthermore, it is expected that the electricity generation in the region using all available renewable energy sources will continue to increase during the coming years, with higher participation of wind and solar energies in the US and the Canadian energy mix.[i]

Some of the problems that need to be overcome to increase the use of renewable energy sources for electricity generation in the North American region are, among others, the following:

- Massive investments in infrastructure. Investment needs to be focused on using low-carbon technology for electricity generation, sustainability, and long-term solutions that embrace electrification and decentralization. The leading investment is required for the increasing use of "smart energy systems, power grids, recharging infrastructure, storage, hydrogen, and district heating and cooling in cities" (Global Energy Transformation: A Roadmap to 2050, 2019);
- A reduction in the cost of renewables. The cost of renewable energy has continued to decline rapidly during the last few years. Overall, the fall in electricity cost from utility-scale solar PV projects since 2010 has been remarkable, with the global average price dropping 73% (IRENA, 2018). Cost declines have been registered in diverse countries, including Saudi Arabia, the United Arab Emirates, Brazil, and the USA, where wind and solar PV costs are now approaching US$2–3 cents per kWh (CleanTechnica, 2018;

[i] In the latest long-term projections (January 2020), the US Energy Information Administration (EIA) projects electricity generation from renewable energy sources such as wind and solar to surpass nuclear and coal by 2021 and to surpass natural gas in 2045. In the Annual Energy Outlook 2020 (AEO, 2020) Reference case, the share of renewables in the US electricity generation increases from 19% in 2019 to 38% in 2050.

GTM, 2019; IRENA, 2018). In the USA, the Energy Information Administration expects non-hydroelectric renewable energy resources, such as solar and wind, to be the fastest-growing source of electricity generation nationwide in 2019 and 2020 (EIA, 2019a) and possibly beyond;

- Lack of interest and resources, particularly in the case of the USA, to replace the use of fossil fuel for renewable energy sources for electricity generation;
- Insufficient support of the public opinion in the USA and Canada to increase the use of renewable energy sources for electricity generation;
- The USA and Canadian governments are given high subsidies to the energy sector to promote the use of some types of renewable energy sources.[j]

It is a fact that the world continues to electrify, with power consumption growing strongly, especially in emerging economies led by China and India. The steady growth of power demand in developing economies, which is expected to account for more than 80% of the expansion in world output, is a critical factor for the increase in the penetration of different renewable energy sources in many countries' energy mixes.[k]

According to the World Energy Outlook 2018 report, a doubling of electricity demand in developing countries puts the use of renewable energy sources for electricity generation at the center of strategies for economic development and $CO_2$ emissions reduction. "In the absence of a greater policy focus on energy efficiency,[l] almost one-in-every-three dollars invested in global energy supply, across all areas, goes to electricity generation and networks in developing economies. Today there is around 350 GW of excess capacity in regions including China, India, Southeast Asia, and the Middle East, representing additional costs that the system, and consumers, can ill afford" (World Energy Outlook 2018, 2018).

Without a doubt, renewable energy sources are, among all available energy sources at the world level, the fastest-growing source of energy, 7.6% per year, accounts for around 50% of the increase in energy generated at this level, and it is responsible for 66% of the rise in global power generation. For this reason, renewable energy becomes "the single largest source of global power generation by 2040" (BP Energy Outlook 2019, 2019).[m] According to the mentioned report, around 75% of the entire growth in primary energy is expected

---

[j] According to the Global energy transformation: A roadmap to 2050 (2019 edition) report, energy subsidies totaled at least US$605 billion in 2015 and are projected to increase to over US$850 billion annually by 2050. However, it is also possible a decline in subsidies to U$470 billion in 2050. "The types of subsidies would change drastically, moving away from fossil fuels and renewable power technologies to technologies needed to decarbonize the transport and industry sectors. In total, the savings from avoided subsidies and reduced environmental and health damages are about three to seven times larger than the additional energy system costs. In monetary terms, total savings resulting could amount to between US$65 trillion and US$160 trillion over the period to 2050."

[k] Global energy grows in the coming decades is expected to be 1.2% per year, which is a little bit down from over 2% per year registered in the last 20 years (BP Energy Outlook 2019, 2019).

[l] Energy efficiency must be scaled up substantially. The rate of energy intensity improvement is projected to increase up to 3.2% per year, or 1.3% more from recent historical averages of around 2% per year.

[m] "Both wind and solar power grow rapidly – increasing by a factor of 5 and 10, respectively – accounting for broadly similar increments to global power. This rapid growth is aided by continuing pronounced falls in the costs of wind and solar power as they move down their learning curves (BP Energy Outlook 2019, 2019).

to be used for power generation, with roughly 50% of all primary energy is expected to be absorbed by the power sector by 2040. The mix of fuels in global power generation is moving away from the use of conventional energy to renewable energy sources. Renewables are gaining share at the expense of coal and nuclear power and account for almost 75% of the increase in power generation, with their share in the global power sector increasing to approximately 30%. It is important to stress that the percentage of coal within the world energy mix is expected to "declines significantly, such that by 2040 it is surpassed by renewables as the primary source of energy in the global power sector" (BP Energy Outlook 2019, 2019).

On the other hand, the slower growth of energy demand in OECD countries will slow down the speed at which renewables can penetrate many countries' energy sectors. This outcome is because, without government support, it is hard for a new renewable power plant to compete commercially against an existing facility using fossil fuels for electricity generation in many countries. On the contrary, the strong growth of power demand in developing economies will facilitate the increased penetration of renewables in these countries' energy mix (BP Energy Outlook 2019, 2019).[n]

It is important to be aware that today's power market designs are not always up to the task of coping with rapid changes in the generation mix.[o] Revenue from wholesale markets is often insufficient to trigger new investment in firm generation capacity. That could compromise supply reliability if not adequately addressed. On the demand side, efficiency gains from more stringent energy performance standards have played a pivotal role in holding back energy demand. Growth prospects depend on how fast electricity can gain ground in providing heat for homes, offices, factories, and power for transportation.

However, regardless of what has been said above, the following can be stated: "renewables are set to penetrate the global energy system more quickly than any fuel previously in history. Historically, it has taken many decades for new fuels to penetrate the energy system. For example, it took almost 45 years for the share of oil to increase from 1% of world energy to 10% in the late 1800s/early 1900s. For natural gas, it took over 50 years from the beginning of the 20th century. This slow pace of change stems in large part from the capital intensity of the energy system" (BP Energy Outlook 2019, 2019).

[n] The growth in renewable energy is dominated by the developing world, with China, India and other Asian countries accounting for almost 50% of the growth in global renewable power generation (BP Energy Outlook 2019, 2019).

[o] It is important to be aware of the fact that a rapid, least-cost energy transitions require an acceleration of investment in cleaner, smarter, and more efficient energy technologies. The changes underway in the electricity sector require constant vigilance to ensure that market designs are robust even as power systems decarbonize. More than 70% of the US$2 trillion required in the world's energy supply investment each year, across all domains, should either comes from state-directed entities or responds to a full or partial revenue guarantee established by regulation. Frameworks put in place by the public authorities also shape the pace of energy efficiency improvement and of technology innovation. Government policies and preferences will play a crucial role in shaping where to go from here. But policy makers also need to ensure that all key elements of energy supply, including electricity networks, remain reliable and robust (World Energy Outlook 2018, 2018).

The share of renewables in world energy is expected to increase from 1% to 10% in around 25 years; this means 20–25 years less than in oil and natural gas, respectively. That is far quicker than any fuel that has ever penetrated the energy system in history.

To increase the role of the different renewable energy sources in the energy mix of the USA and Canada during the coming decades, the following multiples acceptance factors should be considered:

- Sociopolitical acceptance: this means acceptance of policymakers and key stakeholders;
- Market acceptance: this means acceptance of investors and consumers;
- Community acceptance: pertaining to procedural and distributional justice and trust (Rand and Hoen, 2017; Wüstenhagen et al., 2007).

However, as Sovacool (2009) points out, these social, technical, economic, and political dimensions of acceptance influence each other in an integrated manner. "For example, community acceptance of wind energy can affect market acceptance and vice versa. Indeed, this has been the case when local opposition has delayed or derailed proposed wind projects" (Rand and Hoen, 2017; Corscadden et al., 2012).

It is essential to single out that during several years, intensive debates around the acceptance of wind energy for electricity generation in North America focused on the following issues:

- Sociopolitical and market acceptance;
- Technological innovation;
- Economic incentives;
- Impacts on the operations;
- The resistance of the electric grid.

Simultaneously, less attention was given to the social effects of wind energy use for electricity generation (Lantz and Flowers, 2011; Phadke, 2010). However, the rapid growth in the use of wind energy for electricity generation in the North American region has significantly increased the footprint of wind developments, increasing local conflicts and bringing the issue of community acceptance to the forefront (Lantz and Flowers, 2011).

Considering all available renewable energy sources, without a doubt, wind and solar power dominated overall renewable energy additions in the power sector again in 2018, with an estimated 51 GW of wind power (GWEC, 2019) and 109 GW of solar PV power (BNEF, 2019) installed. For the seventh successive year, renewable sources' net additional power generation capacity exceeded non-renewable sources. Growth rates in renewable power have averaged 8–9% per year since 2010 (IEA, 2018a).

On the other hand, in 2018, 28% of the world's electricity was generated from renewable energy sources (in 2019, the percentage was 26%), most of which (96%) were produced by the three main types of renewables: hydropower, wind, and solar energies. In the document entitled "International Energy Outlook 2019 (IEO 2019)" (EIA, 2019b), the US Energy Information Administration (EIA) projects that renewables will collectively increase to 49% of global electricity generation by 2050 (see Fig. 1.4). EIA expects the solar energy share of electricity generation of the top three renewable energy sources to grow the fastest and hydroelectric share to grow the slowest among all renewable energy sources.

The evolution of the global renewable share in electricity generation during the period 2000–50 is shown in Fig. 1.4.

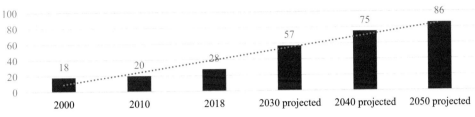

FIG. 1.4 Evolution of the global renewable share in electricity during the period 2000–50 (percentage). *Source: IRENA, 2019. Global Energy Transformation: A Roadmap to 2050 (2019 edition). International Renewable Energy Agency and EIA, 2019. International Energy Outlook 2019 (IEO 2019). US Energy Information Administration (EIA), USA.*

According to Fig. 1.4, the following can be stated: the global renewable share in electricity generation increased by 10% during the period 2000–18. It is projected to reach 86% in 2050; this represents an increase of 4.8-fold during the whole period under consideration. Besides, it is expected that the global renewable share in electricity generation will continue to grow even beyond 2050.

From Fig. 1.5, the following can also be stated: in 2000, the only type of renewable energy source used for electricity generation at the world level was hydropower. After that year, hydropower continued to be one of the main renewable energy sources used by several countries for electricity generation, but wind and solar energies began to be a significant component of the energy mix of many countries. It is expected that, in 2050, the use of wind and solar energies for electricity generation will be higher than the use of hydropower. It is also projected that, in 2050, the use of fossil fuels for electricity generation will continue to be the primary type of energy source at the world level. However, it is most likely that, by 2050, some

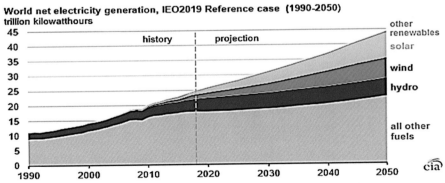

FIG. 1.5 World Net electricity generation during the period 1990–2050 (trillion kWh). *Source: US Energy Information Administration. International Energy Outlook 2019.*

countries will depend entirely, or almost entirely, on the use of renewable energy sources for electricity generation and for satisfying their foreseen increase in the future energy demand.

According to Fig. 1.6, the region with the highest participation of renewable energy sources in its energy mix is the Latin American region with 52.4%, followed by the European region with 34.2%, and the North American region with 27.7%. Within the North American region, electricity generation by energy source in 2019, is shown in Fig. 1.7. Based on the data included in Fig. 1.7, the following can be stated:

- Natural gas is the main energy source used for electricity generation in the region: 1770.2 TWh (35% of the total (5061.7 TWh));
- Coal is the second energy source used for electricity generation in the region: 1108.1 TWh (21.8% of the total);
- Renewables is the third energy source used for electricity generation in the region: 1192.2 TWh (23.6% of the total);
- Nuclear energy is the fourth energy source used for electricity generation in the region: 952.5 TWh (18.8% of the total);

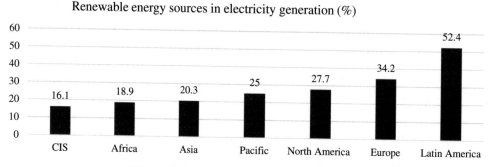

FIG. 1.6   Renewable energy source in electricity generation. *Source: 12 March 2018. Analysis and Mitigation of Power Quality Issues in Renewable Energy Based Distributed Generation Systems Using Custom Power Devices. Institute of Electrical and Electronics Engineers (IEEE). https://ieeexplore.ieee.org/document/8314672/authors#authors; Author own work.*

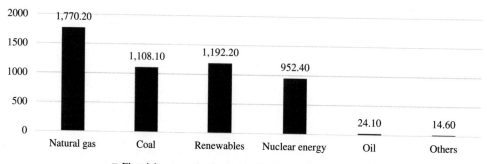

FIG. 1.7   Electricity generation by type of fuel in the North American Region. *Source: 2020. BP Statistical Review of World Energy, 69th ed. and Author own calculations.*

- Oil is the sixth energy source used for electricity generation in the region: 24.1 TWh (0.5% of the total);
- Others: 14.7 TWh (0.3% of the total) (BP Statistical Review of World Energy, 2020).

From the above data, the following can be stated:

- The participation of natural gas in electricity generation in 2019 in the North America region increase by 3%;
- The participation of coal for the same purpose and the same year decreased by 3.7%;
- The participation of renewables in electricity generation in the region in 2019 increase by 0.5%;
- The participation of nuclear energy in the region energy mix in 2019 increase by 0.1%;
- The participation of oil in the region energy mix in 2019 was the same as in 2018;
- The participation of other energy sources in the region energy mix in 2019 increases by 0.1%.

Finally, it is crucial to be aware of the following: the world is in a transit period from the use of fossil fuels for electricity generation to the use of renewable energy sources in the highest possible proportion. During this transitional period, two crucial factors that need to be considered are the so-called "systemic innovation and energy efficiency." The North American region's energy regulations have focused strongly on implementing different actions during the last 30 years to introduce and promote energy efficiency to ensure its energy supply. An example of these actions is the Energy Efficiency Policy of 2005 in the USA, which offers federal consumers and companies tax credits (tax benefits) for the purchase of hybrid electric vehicles that use fuel efficiently, for the construction and remodeling of buildings, and the purchase of energy-efficient appliances and products. On the other hand, "countries need to devote more attention to enabling smarter energy systems through digitalization, through the coupling of sectors via greater electrification, and by embracing decentralization trends. This innovation also needs to be expanded beyond technology and into markets and regulations as well as new operational practices in the power sector and business models" (IRENA, 2019b).

It is important to stress that for every US$1 spent during this energy transition period, "there would be a payoff of between US$3 and US$7- or, put in cumulative terms over the period 2050, a payoff of between US$65 trillion and US$160 trillion. The energy transition requires fewer overall subsidies, as total energy sector subsidies can be reduced by US$10 trillion over the period" until 2050. The level of additional investments needed for the energy transition period to move the world on a more climate-friendly path is US$15 trillion by 2050. This significant amount of resources can be reduced due largely to rapidly falling renewable energy costs, as well as opportunities to electrify transport and other end-uses. "Overall, total investment in the energy system would need to reach US$110 trillion by 2050, or around 2% of the average annual gross domestic product (GDP) over the period" (IRENA, 2019b).

## The renewable energy market in the North American region

According to the Global energy transformation: A Roadmap to 2050 (2019 edition) report, the North American region's energy sector has been suffering a significant change in the last few years. Renewable energy sources "are dominating the global market for new generation capacity, the electrification of transport is showing early signs of disruptive acceleration, and key enabling technologies such as batteries are experiencing rapid reductions in costs. Despite these positive developments, however, deployment of renewable solutions in energy-consuming sectors, particularly buildings and industry, is still well below the levels needed, and progress in energy efficiency is lagging."

Based on the content of the report mentioned above, renewable energy investment continued in 2018, but at a lower level than in 2017, with Bloomberg New Energy Finance estimating total investments in new renewable energy facilities at US$332 billion in 2018 (BNEF, 2019). There are also signs that even the oil majors producers are considering increasing the use of more renewable energy sources for electricity generation. Investment in energy infrastructure needs to be focused on using low-carbon technologies for electricity generation and other purposes and sustainable and long-term solutions that embrace electrification and decentralization. "Investment is needed in smart energy systems, power grids, recharging infrastructure, storage, hydrogen, and district heating and cooling in cities" (IRENA, 2019b).

Considering current and planned energy policies, it is foreseeable that the global energy sector will see cumulative investments of US$110 trillion over the period until 2050. The transition toward a decarbonized global energy system is expected to require scaling up investments in the energy sector by a further 16%; this means an additional US$15 trillion by 2050. It is foreseen that until 2050 "a total of US$110 trillion would be invested in the energy system, representing, on average, 2% of global gross domestic product (GDP) per year." The types of investments in the energy sector will shift away from the fossil fuel sector "towards energy efficiency, renewables, and enabling infrastructure." Significantly, the additional energy investments[P] that the energy sector requires are 40% lower than was estimated earlier. That level is due mainly to a rapid decreased in "renewable power costs and the potential for further cost reductions, as well as the emergence of electrification solutions that are getting cheaper and more efficient" (IRENA, 2019b).

In 2017, the North American renewable energy market had total revenues of US$118.2 billion, representing an annual growth rate of 7.8% between 2013 and 2017. Market electricity generation level increased by 4.2% between 2013 and 2017, to reach a total of 1161.2 TWh in 2017. In 2018, the electricity generation by renewables reached 1174.1 TWh, an increase of 1.1% with respect to 2017, and 1192.2 TWh in 2019, an increase of 1.5% concerning 2018. Canada (382 TWh) and the USA (271.2 TWh) are among the top six producers of electricity using hydropower plants in the world, along with China (1269.7 TWh), Brazil (399.3 TWh), Russia (194.4 TWh), and India (161.8 TWh) (Market Line 2018, 2018; BP Statistical Review of World Energy, 2020).

The costs of renewable energy technologies became cheaper in the last years and made this electricity source most competitive in many parts of the world. According to the International Agency for Renewable Energy (IRENA) report entitled "Renewable Power Generation Costs in 2018", many countries in all regions of the world today are using renewable energy sources for electricity generation because it is the lowest-cost energy source available for this purpose. "As costs for solar and wind technologies continue falling, this will become the case in even

---

[P] The additional energy investments needs are front loaded. "While additional investments are required in the first period of the transition (to 2030), as the year 2050 approaches, technology progress, better understanding of the power system. and increasing electrification of end-use applications result in more optimistic, lower investment estimates" (IRENA, 2019b).

more countries. The cost of electricity from bioenergy, hydropower, geothermal, and wind farms was within the range of fossil fuel-fired power generation costs between 2010 and 2018. Since 2014, the global-weighted average cost of solar PV electricity has also fallen into the fossil-fuel cost range" (IRENA, 2019a).

From Fig. 1.8, the following can be stated: the average cost of solar energy dropped significantly during the period 2009–18, falling from a little bit more than US$350 per MWh in 2009 to US$50 per MWh in 2018; this means a decrease of more than seven-fold. The second renewable energy source that decreases in the period considered is wind energy. The average cost of wind energy fell from around US$135 per MWh in 2009 to US$45 per MWh in 2018; this means a decrease of three-fold. Finally, according to Fig. 1.8, the only energy source that increases its average cost during the period considered is nuclear energy. The average price of nuclear power rises from around US$125 per MWh in 2009 to US$148 per MWh in 2018; this means an increase of 18.4%.

According to the IRENA Renewable Power Generation Costs 2018 report, the average global cost of electricity from the concentration of solar energy (CSP) decreased, in 2018, by 26%, bioenergy by 14%, solar PV and onshore wind energy by 13%, hydroelectric power cut costs by 11% and geothermal and offshore wind by 1%, respectively (see Table 1.1) (Anuta et al., 2018; IRENA, 2019a). Furthermore, with the forecast that the prices of renewables for electricity generation are expected to continue to decline during the coming years, this type of energy source will likely increase their advantage in terms of cost concerning other types of energy sources, especially in the case of solar and wind technologies.[q]

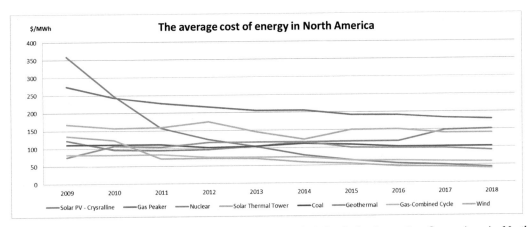

**FIG. 1.8** Levelized Cost of Energy Comparison—Historical Utility-Scale Generation Comparison in North America. *Source: Lazard's Levelized Cost of Energy Analysis, Version 12.0. https://www.lazard.com/media/450784/lazards-levelized-cost-of-energy-version-120-vfinal.pdf.*

[q] Onshore wind and solar PV are set by 2020 to consistently offer a less expensive source of new electricity than the least-cost fossil fuel alternative, without financial assistance. Among projects due to be commissioned in 2020, a total of 77% of the onshore wind and 83% of the utility-scale solar PV project capacity have electricity prices that are lower than the cheapest fossil fuel-fired power generation option (Anuta et al., 2018).

**TABLE 1.1** Global electricity costs of different renewables in 2018.

| | Global weighted-average cost of electricity (US$/kWh) | Cost of electricity (US$/kWh) | Change in the cost of electricity 2017–18 (%) |
|---|---|---|---|
| Bioenergy | 0.062 | 0.048 | − 14 |
| Geothermal | 0.072 | 0.060 | − 1 |
| Hydropower | 0.047 | 0.030 | − 11 |
| Solar PV | 0.085 | 0.058 | − 13 |
| Solar CSP | 0.185 | 0.109 | − 26 |
| Offshore wind | 0.127 | 0.102 | − 1 |
| Onshore wind | 0.056 | 0.044 | − 13 |

*Source: IRENA, 2019. Renewable Power Generation Costs in 2018. International Renewable Energy Agency, Abu Dhabi.*

Costs of renewable energy used for electricity generation have continued to decline rapidly during the past years. Overall, the fall in electricity costs from utility-scale solar PV projects since 2010 has been remarkable, with the global average cost dropping 73% (IRENA, 2019a). Moreover, cost declines have been registered in the USA, where wind and solar PV costs are now approaching US$0.02–0.03 cents per kWh (CleanTechnica, 2018; GTM, 2019; IRENA, 2019a).

It is also expected that more than 75% of the onshore wind projects and 80% of the solar PV capacity scheduled to be put into service next year will produce energy at lower prices than the new options of cheaper coal, oil, or natural gas. It is now possible to obtain onshore wind and solar PV costs of between US$0.03–0.04 cents per kWh in areas with good resources and favorable regulatory and institutional frameworks. By 2020, the potential value for onshore wind is likely to be US$0.045 per kWh, while solar PV is likely to be US$0.048 per kWh (Anuta et al., 2018).

Considering the data included in Table 1.1, the following can be stated: The type of renewable energy source with the lower cost of electricity in 2018 was hydropower with US$0.030 per kWh, followed by onshore wind with US$0.044 per kWh and bioenergy with US$0.048 per kWh. Furthermore, the type of renewable energy source with a significant reduction in electricity cost in 2018 was solar CSP with − 26%, followed by bioenergy with − 14%, and solar PV and onshore wind with − 13% each.

Without a doubt, renewable energy is the backbone of any development that aims to be sustainable and meet the Paris Agreement's climate change objectives.

## Pros and cons in the use of renewable energy sources for electricity generation in the North American region

Renewable energy sources are abundant and environmentally friendly types of energy sources currently used for electricity generation. Unlike fossil fuels, they are not going to expire easily and soon as they are constantly replenished.

Like fossil fuels, renewable energy sources have their own shortcomings. One of its major deficiencies is that they are highly dependent on the weather. Any significant change in weather can reduce the production of energy from these sources. The world is now in a transit period from the use of fossil fuels for electricity generation to increase renewable use for the same purpose. Many countries will not be in a position to switch over to renewable energy sources anytime soon completely. However, getting a significant portion of their daily energy needs from this type of energy source can undoubtedly positively impact their economy and environment.

Today, there is a depth of energy debate among government and energy industry representatives arguing the pros and cons of a significant increase in the use of renewable energy sources for electricity generation. The following are the main pros and cons of the use of renewable energy sources for electricity generation in the North American region:

A-Pros in the use of renewable energy sources for electricity generation

According to Joshi (2018), Rinkesh (2019), and Admin (2019), the following are the main pros in the use of renewable energy sources for electricity generation:

- **Stable Energy Prices:** An increase or decrease in the supply of fossil fuels directly results in inflation. The cost of producing energy from renewable energy sources depends on the amount of financial resources spent on the infrastructure and not on the inflated cost of natural resources. For this reason, it can be expected much more stable energy prices when the bulk of the energy is coming from renewable energy sources;
- **Constant Source of Energy:** Renewable energy sources are an inexhaustible type of energy source and can be quickly replenished. They are plentiful and are available all over the world. Also, being non-consumable, one does not have to worry about their reserves getting declined or exhausted in the future;
- **Clean Energy Source:** Most renewable energy sources do not involve the combustion or burning of fossil fuels or other substances, which would result in the release of toxic chemicals or other harmful atmospheric byproducts. Therefore, they are clean sources of energy and offer numerous environmental benefits. In addition, almost all renewable energy plans emphasize that they have a much smaller carbon footprint than any fossil fuel options available.

  Thus, the use of renewable energy sources for electricity generation makes the environment healthier as they do not pollute it with $CO_2$ and other toxic gases produced by the burns of fossil fuels for electricity generation. In other words, if more and more renewable energy sources are used for electricity generation, it will lead to a healthy lifestyle. No more spending on health risks associated with the use of fossil fuels for electricity generation like cancer and heart diseases;
- **Reliability:** Renewable energy sources are reliable sources of energy, which will never run out. It is derived from natural resources that are readily available and can be sustainable for a lifetime. If the Sun always rises and the wind still blows, renewable energy types' reliability can exceed fossil fuels. When a fossil fuel source runs dry, the whole process has to be moved. Once in place, many of the renewable energy plants have a constant and permanent source of fuel. Unlike fossil fuels, where supply can be affected by wars, strikes, trade disputes, and political instabilities, renewable energy sources do not have such cons. Sun shines, and the wind blows everywhere, and each country can take advantage of that type of energies to produce clean energy on a large scale.
- **Large Scale Job Creation:** It is estimated that the adoption of renewable energy technologies for electricity generation will create a large number of jobs worldwide. The development of renewable energy sources leads to economic growth through the creation of job opportunities in manufacturing and installation of energy sources;

- **Low-Cost Operation:** Once in place, most of the renewable energy plants have a much lower overall cost of operations than the use of fossil fuel technologies require. That could balance out their higher cost of development and implementation;
- **Micro-plants Options Possible:** From solar panels on homes to small wind farms, many different renewable energy sources can be used in remote areas or even urban ones with low-cost micro plant options. That radically reduces the waste incurred in transporting energy from major plants too;
- **Diversify Energy Supply:** Different renewable energy sources can be used for electricity generation, act as an alternative to non-renewable energy supply, and reduce dependency on imported fossil fuels.

B-Cons in the use of renewable energy sources for electricity generation
The following are the main cons in the use of renewable energy sources for electricity generation:

- **High Development and Investment Cost:** It costs a lot to develop a renewable energy plant in both types of research and manufacturing the components needed to succeed. The known ways of using fossil fuels are less costly because all manufacturing and construction processes are already in place. The initial investment or setup cost is significantly high and acts as a deterrent in government and energy industry for switching over to the use of this type of energy source for electricity generation;
- **Weather Dependent:** Renewable energy sources are highly dependent on the weather. A significant change in weather conditions has an impact on energy produced by this type of energy source. Almost all of the suggested renewable energy sources are very vulnerable to weather and other climate conditions. Renewable energy sources depend heavily on the Sun and wind to produce energy.
For this reason, abundant rain or slow wind can reduce the production of power. Atmospheric conditions and geographical locations make a considerable impact on the efficacy in the use of these energy sources for electricity generation;
- **Unable to Produce Energy in Large Quantities:** Unlike fossil fuels-power plants that produce an abundant supply of power, renewable sources cannot provide that much energy in a short period. The technology used in producing energy is new, and the main other factors like weather play spoilsport that hamper the production of electricity on a large scale. This means that either energy consumption is reduced, or new power plants must be built that could produce energy at a faster rate;
- **Not Available in all Areas:** One of the major cons of many renewable energy plants is that the raw material – solar intensity, wind, or water, is not available in all locations. That means that governments and the energy industry still have to create an infrastructure for transporting the energy that may not be better than what is already in place;
- **Large Areas Required:** Another disadvantage of using renewable energy sources for electricity generation is that to produce a large amount of energy, a large number of solar panels and wind turbines have to be set up. For this, large areas of land are needed to produce such a massive amount of energy on a large scale that could have a negative impact on food production;

- **Low-Efficiency Levels:** The technologies used for generating electricity by renewable energy are still new in the market. They lack the efficiency needed to produce the power required for a quick return on investment. For this reason, investors may fail to invest in these technologies due to fear of low yields;
- **Air Pollution**: Harnessing energy from biomass may result in air pollution, and also it requires a lot of energy to produce it;
- **Environmental Concerns**: Production of electricity by the use of hydroelectric power plants from flowing water or dam water can create ecological concerns where interference with water flow can affect the growth of marine species and force the movement of the population in the area of the dam;
- **Storage Capabilities**: Due to the intermittency of renewable energy supply, there is a need for energy storage, which may be expensive for large-scale energy production from power plants. Besides, there are no adequate storage facilities associated with the use of some types of renewable energy sources such as solar or wind energy;

## The impact on the environment due to the use of renewable energy sources for electricity generation in the North American region

All types of energy sources have some impact on the environment and population. Some types of energy sources have a high negative impact on the environment and population, while others have less or almost none. For example, the use of fossil fuels for electricity generation, particularly oil and coal, do substantially more harm than the use of renewable energy sources for the same purpose, including air and water pollution, damage to public health, wildlife and habitat loss, water use, land use, and global greenhouse gas emissions (Union of Concerned Scientists, 2013).

According to the report entitled "IRENA 2019 Global energy transformation: A Roadmap to 2050 (2019 edition)", annual energy-related $CO_2$ emissions are projected to decline 70% below today's level. An estimated 75% of this reduction can be achieved through the increased use of renewable energy sources for electricity generation and new electrification technologies; if energy efficiency is included, this share rises to over 90%. "However, the world is on a much different path: energy-related emissions have increased by over 1% annually,[r] on average, over the last five years. Current plans and policies result in a similar level of annual emissions in 2050 compared to today, which risks putting the world on a pathway of 2.6 degrees Celsius of temperature rise or higher already after 2050." The report mentioned above indicates "that emissions would need to be reduced by around 3.5% per year from now until 2050. Energy-related emissions are expected to peak in 2020 and begin to decline thereafter."

Why is this happening? The answer to this question is the following: in many countries, energy policies are not sufficiently aligned with climate goals, and policies often lag market developments. The energy transition also needs to be viewed within the broader framework of economic growth and sustainability. It is also clear that a politically viable energy transition

---

[r] According to the same report, the precise figure is 1.3% annually, on average, over the last five years.

must be fair to succeed on a global level. It is crucial to be aware of the following concept: the global energy transition is more than a simple transformation of the energy sector; it transforms societies and economies. The energy transition is multi-faceted and evolving in terms of technologies, socio-economics, institutional drivers, and finance firms. For this reason, governments and the energy industry need to be prepared for the adoption of an energy system with much higher shares of renewable energy sources for electricity generation, broader innovation, strategic planning to increase investments, and to avoid social stresses and economic problems (IRENA, 2019b).

The intensity of the environmental impact caused by the use of renewable energy sources for electricity generation depends on the specific technology used and the geographic location selected for constructing the power plant, among other factors. By knowing the current and potential environmental impact caused by the use of each renewable energy source for electricity generation, specific steps can be implemented to effectively avoid or minimize these impacts on the negative environment and population.

Without a doubt, the use of renewable energy sources for electricity generation is the type of energy source with a less negative impact on the environment and population, besides nuclear energy, in comparison with fossil fuels. However, the use of renewable energy sources for electricity generation entails some potential environmental impacts that affect the environment and the population as well.

The use of renewable energy sources for electricity generation "can make large tracts of land unusable for competing for agriculture uses, disrupt marine life, birdlife and flora/fauna, and produce visual and noise pollution. Generally, though, these potential negative environmental impacts are site-specific, and there are several ways to minimize the effects, which are usually small and reversible". But there are also environmental benefits from the use of renewable energy sources for electricity generation other than reducing $CO_2$ and other air gas emissions. For example, hydropower plants[s] "can improve water supplies and facilitate the reclamation of degraded land and habitat" (Edvard, 2011). Hydropower involves the conversion of gravitational energy in falling or fast-running water to electricity. There are many benefits to the use of hydropower plants for electricity generation in the North American region. The reason is simple: water is a plentiful resource in many areas of Canada and the USA. In fact, in Canada, the use of hydropower plants for electricity generation produced 57.8% of the electricity generated in the country in 2019.

It is important to single out that large-scale hydropower plants' environmental and social effects are site-specific and subject to much controversy in several countries worldwide. Large-scale hydropower projects "may disturb local ecosystems, reduce biological diversity, or modify water quality. They may also cause socio-economic damage by displacing local populations" (Edvard, 2011).

---

[s] "Hydroelectric power includes both massive hydroelectric dams and small run-of-the-river plants. Large-scale hydroelectric dams continue to be built in many parts of the world (including China and Brazil), but it is unlikely that new facilities will be added to the existing US fleet in the future. Instead, the future of hydroelectric power in the USA will likely involve increased capacity at current dams and new run-of-the-river projects" (Union of Concerned Scientists, 2013).

One alternative solution to the construction of large-scale hydropower plants for electricity generation is building mini and micro-hydropower plants with relatively modest and localized effects on the environment and population. However, the kWh cost of this type of power plant is generally higher in comparison with large-scale hydropower plants. Besides, it is vital to be aware that although hydropower plants emit some greenhouse gases on a life-cycle basis, mainly methane generated by decaying bioenergy in reservoirs, in most cases, these greenhouse emissions are far less than the $CO_2$ and other gas emissions caused by the burning of fossil fuels (Edvard, 2011).

The use of bioenergy for electricity generation and heating can have many environmental benefits if this type of renewable energy source is produced and used most effectively. If the land used for bioenergy production is replanted, then the carbon released will be recycled into the next generation of growing plants. Bioenergy plants have lower emissions of $CO_2$ than coal and oil-fired power plants, but they may produce more particulate matter. However, biomass power plants share some similarities with fossil fuel power plants because both types of plants "involve the combustion of a feedstock to generate electricity" (Union of Concerned Scientists, 2013).

For this reason, it raises similar concerns about air emissions and water use as fossil fuel power plants. Sources of biomass for electricity generation and heating are diverse, ranging from "energy crops, like switchgrass, to agricultural waste, manure, forest products, waste, and urban waste. Both the type of feedstock and how it is developed and harvested significantly affect land use and life-cycle global warming emissions impacts of producing power from biomass." (Union of Concerned Scientists, 2013).

The transition to increased use of electrified forms of transport and heat, combined with the rise in the use of renewable energy sources for electricity generation, can reduce around 60% of the energy-related $CO_2$ emissions needed to meet the main objectives of the Paris Agreement on Climate Change. When these measures are combined with the direct use of renewable energy sources for electricity generation, the share's emissions reductions could reach 75% of the total required. However, this reduction is not enough, and further $CO_2$ emissions reduction will still be needed. Without a doubt, bioenergy could play a leading role "in sectors that are hard to electrify, such as shipping, aviation, and certain industrial processes. Biofuel consumption must be scaled up sustainably to meet this demand. Efforts are also needed to reduce non-$CO_2$ greenhouse gas emissions and non-energy use emissions (such as waste-to-energy, bioenergy, and hydrogen feedstocks) to reduce industrial process emissions and fugitive emissions by the coal, oil, and gas industries. Efforts are needed outside of the energy sector to reduce greenhouse gas emissions in agriculture and forestry" (IRENA, 2019b).

The geothermal energy source is an abundant renewable energy source in the North American region. In the USA, geothermal energy sources are located in most eastern and southern California, southeastern Idaho, western Oregon, northwestern Utah, northern Louisiana, south of Arkansas, and northeastern Texas. The use of geothermal energy sources for electricity generation has several benefits, including a small impact on the environment and very few greenhouse emissions.

Geothermal power plants most widely developed, also known as "hydrothermal plants," are "located near geologic "hot spots," where hot molten rock is close to the Earth's crust and produces hot water." In other regions, a different type of geothermal power plant, known as "enhanced geothermal systems or hot dry rock geothermal," can be found in other regions. This type of geothermal power plant involves drilling into the Earth's surface to reach more in-depth geothermal resources, allowing broader access to a geothermal energy source (Hennick, 2018). Geothermal power plants also differ in terms of the technology they use to convert the resource to electricity (direct steam, flash, or binary) and the type of cooling technology they use (water-cooled and air-cooled) (Union of Concerned Scientists, 2013).

Environmental impacts in geothermal energy use for electricity generation differ depending on the conversion and cooling technology used. It is crucial to know that although geothermal energy is a renewable energy source, geothermal power plants may release gaseous emissions into the atmosphere during their electricity generation. These gases are mainly carbon dioxide and hydrogen sulfide with traces of ammonia, hydrogen, nitrogen, methane, radon, and the volatile boron, arsenic, and mercury species. The environmental impact of the use of geothermal power plants for electricity generation could have a negative effect on the future development of this type of energy resource and its use for electricity generation in some countries.

The power generation from wind farms is one of the cleanest and most sustainable ways to generate electricity. The use of wind energy for this specific purpose produces no toxic pollution or global warming emissions. Wind energy is also abundant, inexhaustible, and affordable, making it a viable and large-scale alternative to many countries' fossil fuels. The most significant effects on the environment of the use of wind energy for electricity generation are, according to Edvard (2011) and the Union of Concerned Scientists (2013), the following:

- **Land Use**: For the construction of a wind farm, a large amount of land is needed, and this could be a challenge to wildlife and habitat;
- **Visual Effects**: In a wind farm, several turbines must be built in a limited area and are, therefore, highly visible for all surrounded regions in all directions;
- **Noise**: Wind turbines produce aerodynamic noise from air passing over the propellers and mechanical noise from the turbine's moving parts. New wind turbine designs have reduced noise, and research continues to reduce further the noise associated with the operation of a wind farm;
- **Electromagnetic Interference**: Wind turbines may spread out electromagnetic signals causing interference to communication systems. For this reason, it is vital to select the site where a wind farm will be constructed, avoiding military zones or airports. The correct site selection can minimize electromagnetic interference.

The USA has, in 2019, the largest installed wind capacity within the North American region (second at the world level, after China with 210,478 MW) with 103,584 MW or 83.8% of the total at the regional level. Yet only 6.3% of the electricity generated in the USA came from wind power in 2018. In Canada, wind energy produced 4.8% of the total electricity generated by the country in 2018 (31.8 TWh). The highest wind potential in the North American region is found over the US's central plains, along the northwestern and northeastern coastlines of the USA and Canada, and along the coast of the Gulf of Mexico and the Yucatan peninsula.

The environmental impacts associated with the use of solar power for electricity generation include "land use and habitat loss, water use, and the use of hazardous materials in manufacturing. Although the types of impacts vary greatly depending on the scale of the system and the technology used, photovoltaic (PV) solar cells or concentrating solar thermal plants (CSP)" (Union of Concerned Scientists, 2013). "If the sunlight only falling on a 30-mile square was harnessed, it could generate 3125 GW of power. That is slightly more than the USA's average primary energy rate of 3000 GW consumed by electricity and transportation. There are virtually no environmental effects associated with solar energy other than the use of water for cooling, circulation, and mirror cleaning. Not to mention, there are no geographical restrictions with solar energy, because wherever there is sunlight, there is potential for solar power" (Renewables in North America, 2019). In Canada, solar energy generated only 0.6% of the country's total electricity in 2018.

Finally, it is important to stress the following: cross-border electricity integration is not guaranteed to be a tool that will enhance climate commitments, but with specific, well-designed policies and project developments, it could provide significant benefits. It is generally understood that cross-border electricity integration could be a valuable tool for governments to achieve climate goals if electricity integration leads to:

- More substantial commitments for climate emissions reductions;
- An increase in the deployment of clean energy;
- A reduction in the implementation of high-emitting energy;
- A reduction in energy demand (such as through energy efficiency).

However, if electricity integration enhances the deployment or slows fossil fuel-fired generation retirements or leads to significant demand increases, such development could increase emissions. The Canadian and USA administrations have made strong statements on cooperating to enhance their climate commitments further.

## Renewables and net capacity additions by type in the North American region

Since the beginning of using different conventional energy sources for electricity generation, humankind has used fossil fuels to meet its energy requirements. "Coal, oil, and natural gas have lit homes and powered machinery for centuries, driving civilization forward. But as human development accelerated, the unsustainability of such energy became apparent. Global fuel supplies deteriorated, and the atmosphere became more polluted" (Tanti, 2018). The search for clean energy sources for electricity generation and heating began to ensure a sustainable future with a less negative impact on the environment and population. The following is a summary of the existence of renewable energy sources in the North American region.

A-Solar energy

According to the IEA report entitled "Solar Energy Perspectives. Executive Summary 2011", solar energy is radiant light and heat from the Sun (see Fig. 1.9) that is harnessed using a range of ever-evolving technologies such as solar heating, solar PV, solar thermal energy, solar architecture, molten salt power plants, and artificial photosynthesis. The two main solar power

**FIG. 1.9**   Concentrated solar park. Part of the 354 MW SEGS solar complex in northern San Bernardino County, California. *Source: Courtesy fabersam Pixabay.*

types are solar PV[t] and solar thermal or concentrated solar power or CSP[u]. Solar PV transforms the light from the Sun into an electric current using what is known as the photoelectric effect. A solar panel system uses photovoltaic cells made of semiconductor cells to capture the Sun's energy to produce electricity. It is essential to single out that the solar PV cells do not require constant direct sunlight to produce electricity; even on a cloudy day, they can still generate sufficient electricity for a standard residential house. Solar PV systems can be "categorized by various aspects, such as grid-connected vs. stand-alone systems, building-integrated vs. rack-mounted systems, residential vs. utility systems, distributed vs. centralized systems, rooftop vs. ground-mounted systems, tracking vs. fixed-tilt systems, and new constructed vs. retrofitted systems. Other distinctions may include systems with microinverters vs. central inverter, systems using crystalline silicon vs. thin-film technology, and systems with modules from Chinese vs. European and US-manufacturers" (Wikipedia Photovoltaic System, 2020).

Solar thermal technology captures the Sun's heat, and this heat is used directly or converted into mechanical energy and, in turn, electricity, known as CSP.

The evolution of the solar energy capacity in the North American region during the period 2010–19 is shown in Fig. 1.10.

[t] "The photovoltaic effect (or photoelectric effect) converts light into electricity. It was discovered by French physicist Edmond Becquerel in 1839 and was first used in industrial applications in 1954. The principle is: an electric current occurs when electrons are displaced. For this to happen, photons (light particles) excite the outermost electrons of the atoms of certain semiconductor elements. In practice, light hitting a photovoltaic cell is converted into electricity by a semiconductor, generally silicon. A photovoltaic panel is made up of several cells producing direct current, which is then converted into alternating current by an inverter" (Planete Energies, 2014).

[u] The thermal solar power technology "concentrates the warmth of the Sun's rays using collectors to heat a transfer fluid (gas, oil or molten salt, for example) to a high temperature. The fluid heats a network of water, which produces steam and drives a turbine (mechanical energy), thereby generating electricity. Heat from the Sun's rays is collected in large power plants where flat or curved mirrors are installed over vast areas. The technology is best suited to countries where the sunlight is intense, for example in desert regions" (Planete Energies, 2014).

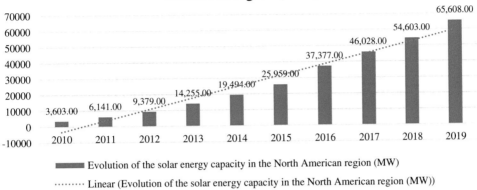

## Evolution of the solar energy capacity in the North American region (MW)

**FIG. 1.10** Evolution of the solar energy capacity in the North American region during the period 2010–19. *Source: IRENA, 2020. Renewable Capacity Statistics 2020. International Renewable Energy Agency (IRENA), Abu Dhabi and Author on calculations.*

According to Fig. 1.10, the region's solar capacity increased 18.2-fold during the period under consideration, rising from 3603 MW in 2010 to 65,608 MW in 2019. It is expected that this trend will continue during the coming years. Without a doubt, the increase in solar energy capacity will be mainly in solar PV, but much lower than the increase expected in wind energy capacity in the same period.

There are two types of solar power installations, depending on the type of solar system used: individual systems for homes or small communities. In this case, solar PV panels can power electrical devices, while solar thermal collectors can heat homes or hot water. Solar PV or CSP plants that cover hundreds of acres produce electricity on a large scale, can be fed into power grids.

Considering the data included in Fig. 1.11, the following can be stated[v]: from a total of 179 PW[w] of incoming solar energy, only 89 PW is absorbed by land and oceans; this represents 49.7% of the total incoming solar energy. A total of 52 PW is reflected by the atmosphere, clouds, and Earth's surface, and a total of 121 PW is radiated to space from the atmosphere and from Earth to space.

Without a doubt, solar energy is an essential source of energy, "and its technologies are broadly characterized as either passive solar or active solar, depending on how they capture and distribute solar energy or convert it into solar power. Active solar techniques include the use of solar PV systems, concentrated solar power (CSP), and solar water heating to harness the energy. Passive solar techniques include orienting a building to the Sun, selecting

---

[v] The Earth receives 179 PW of incoming solar radiation (insolation) at the upper atmosphere (see Fig. 1.11) Approximately 30% is reflected back to space while the rest is absorbed by clouds, oceans, and land masses. The spectrum of solar light at the Earth's surface is mostly spread across the visible and near-infrared ranges with a small part in the near-ultraviolet (Intergovernmental Panel on Climate Change, 2007). Solar radiation is absorbed by the Earth's land surface, oceans – which cover about 71% of the globe – and atmosphere. Sunlight absorbed by the oceans and land masses keeps the surface at an average temperature of 14 °C (Somerville, 2007).

[w] Petawatts (PW). One PW represents $10^{15}$ watts.

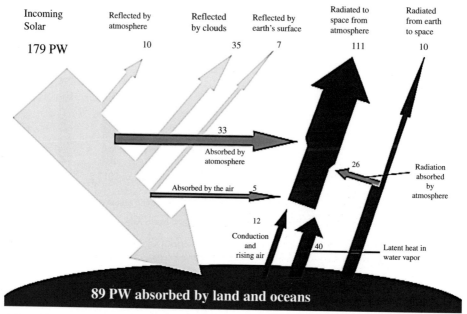

FIG. 1.11    Breakdown of the incoming solar energy. *Source: User A1 Wikipedia CCBY-SA 3.0.*

materials with favorable thermal mass or light-dispersing properties, and designing spaces that naturally circulate air" (Wikipedia Solar Energy 2019, 2019).[x]

Table 1.2 includes data on solar energy potential by region. According to that table, North America ranks fourth, respecting the solar potential by region with 181.1 EJ (minimum) and 7410 EJ (maximum) after the Middle East and North Africa, Sub-Saharan Africa, and the former Soviet Union.

In the last years, the increasing competitiveness of solar PV pushes, according to World Energy Outlook 2018 report, its installed capacity beyond the installed capacity of wind energy before 2025; it is likely to pass hydropower installed capacity around 2030 and the installed capacity of coal before 2040. Although investment in households and businesses, most of these new solar PV capacities play a strong supporting role. However, and despite the latest solar energy capacities installed in the North American region in 2019, the established solar power capacity in the area is the lowest compared to the capacities installed of other types of energy sources. The evolution of annual solar PV additions during the period 2010–50 is shown in Fig. 1.12.

According to Fig. 1.12, the annual solar PV additions increased 6.4-fold during the period 2010–18. It is expected that yearly solar additions will increase 3.3-fold during the period 2018–50, much lower than the increase registered during the period 2010–18.

Based on the information included in the report mentioned above, renewables and coal will switch places in the future in the energy mix at the world level. In the specific case of

---

[x] Although solar energy refers primarily to the use of solar radiation for practical ends, all renewable energy sources, other than geothermal and tidal power, derive their energy either directly or indirectly from the Sun. Active solar techniques use PV, CSP, solar thermal collectors, pumps, and fans to convert sunlight into useful outputs. "Passive solar techniques include selecting materials with favorable thermal properties, designing spaces that naturally circulate air, and referencing the position of a building to the Sun. Active solar technologies increase the supply of energy and are considered supply side technologies, while passive solar technologies reduce the need for alternate resources and are generally considered demand side technologies" (Philibert, 2005).

**TABLE 1.2** Annual solar energy potential by region.

| Region | North America | Latin America and the Caribbean | Western Europe | Central and Eastern Europe | The former Soviet Union | The Middle East and North Africa | Sub-Saharan Africa | Pacific Asia | South Asia | Centrally planned Asia | Pacific OECD |
|---|---|---|---|---|---|---|---|---|---|---|---|
| Minimum | 181.1 | 112.6 | 25.1 | 4.5 | 199.3 | 412.4 | 371.9 | 41.0 | 38.8 | 115.5 | 72.6 |
| Maximum | 7410 | 3385 | 914 | 154 | 8655 | 11,060 | 9528 | 994 | 1339 | 4135 | 2263 |

Note:
- Total global annual solar energy potential amounts from 1575 EJ (minimum) to 49,837 EJ (maximum)
- Data reflects assumptions of annual clear sky irradiance, annual average sky clearance, and available land area. All figures are given in exajoules (EJ)

Quantitative relation of global solar potential vs. the world's primary energy consumption:
- The ratio of potential vs. current consumption (402 EJ) as of year: from 3.9 (minimum) to 124 (maximum)
- The ratio of potential vs. projected consumption by 2050 (590–1050 EJ): from 1.5–2.7 (minimum) to 47–84 (maximum)
- The ratio of potential vs. projected consumption by 2100 (880–1900 EJ): from 0.8–1.8 (minimum) to 26–57 (maximum)

*Source: 2020. World Energy Assessment. United Nations Development Programme.*

solar energy, it is important to single out that solar systems are expected to generate more than 20 times the power needed to meet the world's energy demand every year.

The distribution of energy generation capacity in the North American region by energy source in 2019 is shown in Fig. 1.13.

According to Fig. 1.13, in 2019, natural gas was the energy source with the highest capacity installed at the regional level, with 43.91% of the total, followed by coal with 22.15%, nuclear energy with 8.79%, hydropower with 8.33%, wind energy with 7.68%, oil with 3.31%, solar energy with 2.96% and others with 2.69%. Solar energy is the renewable energy source with the lowest capacity installed in the region compared to hydropower and wind energy.

From Fig. 1.13, the following can be stated: in 2019, fossil fuels (oil, natural gas, and coal) installed capacity at the regional level represents 69.37% of the total, renewable energy sources 19.15%, and nuclear energy 8.97%. It is projected that the structure of the distribution of energy generation capacity reported in 2019 will not change significantly during the coming years. For this reason, fossil fuel will remain the energy source with the highest capacity installed at the regional level but with a lower percentage. Without a doubt, renewable energy installed capacity will continue to be the second-largest energy source installed, but with an estimated rate higher than fossil fuels.

The evolution of the electricity generation by solar energy in the North American region during the period 2010–18 is shown in Fig. 1.14.

Based on the data included in Fig. 1.14, the following can be stated: electricity generation increased 20.2-fold in the North American region using solar energy during the period 2010–18, rising from 4197 GWh in 2010 to 88,986 GWh in 2018. It is expected that this trend will continue without change during the coming years. The major increase is expected to be registered in onshore solar energy.

Without a doubt, the use of solar energy for electricity generation has certain benefits and some limitations. According to the EIA source, these are the main advantages:

- Solar energy systems do not produce air and water pollutants or carbon dioxide[y];
- Solar PV systems can supply electricity in locations where electricity distribution systems (power lines) do not exist or are difficult to build. They can also supply electricity to an electric power grid;
- Solar PV arrays can be installed quickly and can be any size.
- Solar energy systems on buildings have minimal effects on the environment (EIA Energy From the Sun, 2019; EIA Photovoltaics and Electricity, 2019).

Besides, the use of solar energy for electricity generation can also have a positive, indirect effect on the environment and the population because it replaces or reduces the use of other energy sources that have a more significant negative impact on the environment and the population.

The main limitations in the use of solar energy for electricity generation are, according to the EIA Energy from the Sun 2019 report, the following:

- The amount of sunlight that arrives at the Earth's surface is not constant. The amount of sunlight varies depending on location, time of day, the season of the year, and weather conditions;

---

[y] However, some toxic materials and chemicals are used to make the PV cells that is used in solar parks with the purpose of converting sunlight into electricity. Some solar thermal systems use potentially hazardous fluids to transfer heat, and the leaks of these materials could be harmful to the environment and population.

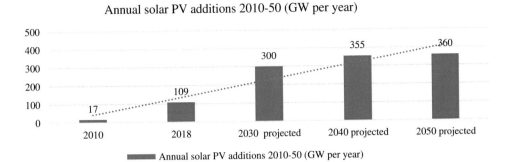

**FIG. 1.12** Annual solar PV additions reported and projected during the period 2010–50 (GW per year). *Source: IRENA, 2019. Global Energy Transformation: A Roadmap to 2050 (2019 edition). International Renewable Energy Agency.*

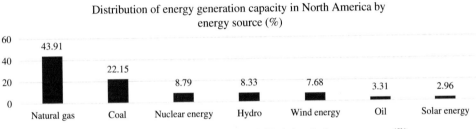

**FIG. 1.13** Distribution of energy generation capacity in North America in 2019 by energy source. *Source: 2019. Statista Database.*

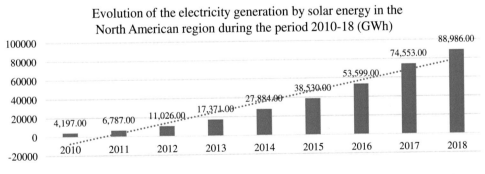

**FIG. 1.14** Evolution of the electricity generation by solar energy in the North American region during the period 2010–18. *Source: 2020. Renewable Energy Statistics 2020. IRENA and Author own calculations.*

- The amount of sunlight reaching a square foot of the Earth's surface is relatively small, so a large surface area is necessary to absorb or collect a reasonable amount of energy (EIA Energy From the Sun, 2019).

Large solar-power plants' construction is not free of other negative impacts on the environment near their locations. Clearing land for the construction and placement of solar parks may have long-term effects on native plants and animals' habitats. "Some solar power plants may require water for cleaning solar collectors and concentrators or for cooling turbine generators. Using large volumes of groundwater or surface water in some arid locations may affect the ecosystems that depend on these water resources. Besides, the beam of concentrated sunlight a solar power tower creates can kill birds and insects that fly into the beam" (EIA Solar Energy and the Environment, 2019).

One of the main problems that need to be solved for the future expansion of the use of solar power for electricity generation is the low efficiency at which PV cells convert sunlight to electricity. The efficiency level could be higher or lower, depending on the type of semiconductor material and PV cell technology used (see Fig. 1.15). In the middle of the 1980s, "the efficiency of commercially available PV modules averaged less than 10%, increased to around 15% by 2015, and is now approaching 20% for state-of-the-art modules. Experimental PV cells and PV cells for niche markets, such as space satellites, have achieved nearly 50% efficiency" (EIA Photovoltaics and Electricity, 2019). However, some types of FV cells have reached 30% efficiency. During 2019, the capacity factor of solar energy facilities in the USA reached 24.5%, being the lowest capacity factor among all other energy sources available in the country.

Solar energy for electricity generation has an important characteristic that distinguishes it from other energy sources. In essence, there are virtually no environmental effects associated with the use of solar energy for electricity generation. Besides, there are no geographical restrictions for constructing a solar park; there is a possibility to use it for electricity generation wherever there is sunlight.

At the end of 2019, global solar PV installations reached 578.6 GW, an annual increase of 97.7 GW from 2018. In 2019, the leading markets in terms of total capacity installed were China (205.1 GW), the USA (60.5 GW), and Japan (61.8 GW). At the end of 2019, cumulative global CSP installations reached 6.3 GW, up 5.7 GW from 2018 (BP Statistical Review of World Energy, 2020). Solar installations represented 22% of all new US electric generation capacity in 2018, second to natural gas (58%).

In 2018, solar energy represented 4.6% of net summer capacity and 2.3% of the annual net generation. However, ten states generated more than 4% of their net yearly electricity from solar energy, with California leading the way at 19% (Feldman and Margolis, 2019). The USA installed 9.1 GW of solar PV in 2019 compared to 2018. Analysts also expect US solar PV capacity to double by 2022. In 2018, the USA produced approximately 1 GW of c-Si modules and 0.4 GW of a thin film. The USA expanded its solar PV manufacturing capacity to 6 GW in 2019 (up from 2.5 GW in 2017), and it is expected to add another 3 GW in the near future.

B-Wind energy

Over the last three decades, wind energy in North America has evolved from a fringe, isolated, experimental concept into a mainstream and viable source of electricity, and representing the largest source of new power capacity additions during the last years (CanWEA, 2016; Wiser and Bolinger, 2016). Wind energy is widely seen as an inexhaustible energy source with the potential to provide a wide range of environmental and social benefits (Intergovernmental Panel on Climate Change, 2011). State and provincial-level mandates, federal incentives,

# Inside a photovoltaic cell

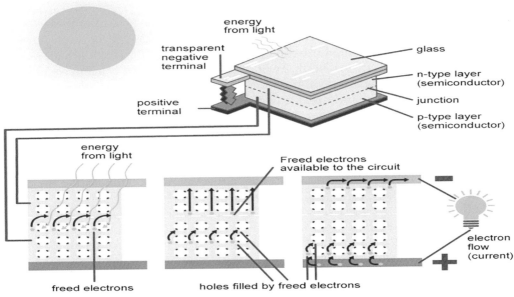

Source: U.S. Energy Information Administration

**FIG. 1.15** Inside a solar PV cell. *Source: US Energy Information Administration (US EIA). https://www.eia.gov/energy-explained/solar/photovoltaics-and-electricity.php.*

declining wind energy costs, and relatively favorable economics have spurred the aggressive North American wind deployment of the past 10–15 years (Wiser and Bolinger, 2016).

Despite what has been said above, it is expected that growth in wind energy deployment will likely continue during the coming years. In the USA, for example, recent market analysis suggests that annual wind power capacity additions are expected to increase rapidly during the coming five years (Wiser and Bolinger, 2016), driven by expected lower prices (Wiser et al., 2016). Meanwhile, the US Department of Energy's Wind Vision Report (US DoE, 2015) outlines wind energy pathways to provide up to 35% of the nation's electrical demand by 2050. According to this report, wind farms have primarily been developed, and, for this reason, future wind development likely will happen increasingly near communities. The report underlines the need to understand better the drivers of wind facility acceptance among affected communities (US DoE, 2015).

One important issue that should not be ignored while planning the expanding use of wind energy for electricity generation in the North American region during the coming years is the impact of operations and maintenance expenses for wind power assets. According to 2018 IHS Markit Wind O&M Benchmarking in North America: Aging Turbines, Rising Costs—A Study of O&M Costs for North American Wind Power Plants (2018), the operations and maintenance cost of wind farms operating in the region will cost the industry about US$7.5 billion annually by 2021. The 2018 IHS Markit estimates the current North American operations and maintenance market "is valued at approximately US$5 billion to US$6 billion annually but is expected to exceed US$8.3 billion by 2027, an increase of nearly 40%". It is projected that for

2021, the industry will reach a critical turning point, as operating expenditure for the North American region wind industry will eclipse capital expenditure for the first time.

Undoubtedly, North American wind installations' operations and maintenance market is a steadily growing industry, driven both by more turbines and—as those turbines age—more spending per turbine. According to the 2018 IHS Markit Wind O&M Benchmarking in North America: Aging Turbines, Rising Costs—A Study of O&M Costs for North American Wind Power Plants (2018) report, "the North American wind turbine fleet is aging overall – the average age of installed capacity will rise from seven years in 2018 to 14 years in 2030. As projects age, they cost more, making the operations and maintenance business even more intriguing than it is today". More than 50,000 utility-scale wind turbines comprising approximately 100,000 MW of generating capacity are installed in 42 US states and 12 Canadian provinces and territories. By the year 2028, the 2018 IHS Markit report expects those numbers to increase to more than 75,000 wind turbines with a capacity of more than 150,000 MW.

The evolution of wind farm capacity in the North American region during the period 2010–19 is shown in Fig. 1.16.

According to Fig. 1.16, the following can be stated: wind farm capacity in the North American region increased 2.7-fold during the period under consideration, rising from 43,102 MW in 2010 to 116,997 MW in 2019. It is expected that this trend will continue without change during the coming years.

One of the key findings from the 2018 IHS Markit Wind O&M Benchmarking in North America: Aging Turbines, Rising Costs is that larger, newer wind projects have operations and maintenance costs averaging 25% fewer per MWh than ones using smaller turbines installed before 2010. Based on the report mentioned above, the operations and maintenance costs of larger projects "are not only lower compared to smaller projects, but also demonstrate more cost stability as they age. Most of these larger projects benefit from economies of scale, potential procurement efficiencies, and lower exposure to individual turbine failures." Another key finding is that first-year operations and maintenance costs decline as the industry continues to learn. "First-year costs fell from an average of US$46,000 per installed MW during the period 2008–13 to US$38,000 per MW during the period 2014–17", a decrease of 17.4%.

Evolution of wind farm capacity in the North American region during
the period 2010-19 (MW)

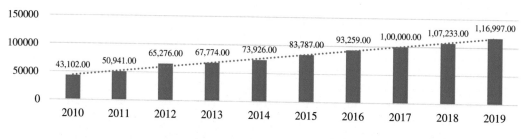

Evolution of wind farm capacity in the North American region during the period 2010-19 (MW)

**FIG. 1.16**   Evolution of the capacity of wind farms in the North America region during the period 2010–19. *Source: 2020. Renewable Energy Statistics 2020. IRENA and Author own calculations.*

Within the North American region, the USA has, in 2019, the largest installed wind capacity in the region with 103.6 GW. Canada has, in 2019, around 13.4 GW of wind capacity installed. The highest wind potential for North America is found over the USA's central plains, along the northwestern and northeastern coastlines of the country, and along the coast of the Gulf of Mexico and Yucatan peninsula. The USA has, in 2019, a total of 29 MW of offshore wind capacity, generating 102 GWh in that year. Finally, it is important to know the evolution of the wind farms' electricity generation in the region during the period 2010–18. Fig. 1.17 shows this evolution.

According to Fig. 1.17, the following can be stated: electricity generation by wind farms in the North American region increased almost 3-fold during the period 2010–18, rising from 103,872 GWh in 2010 to 307,682 GWh in 2018. As a result of the expected increase in wind energy capacity in the region until 2030 and the closure of old and ineffective coal power plants, it is foreseeable that electricity generation from this type of installation will increase further during the coming years.

In 2019, wind farms' capacity factor in the USA reached 34.8%, the second-lowest capacity factor among all other energy source capacity factors in the country.

C-Hydropower

Hydropower is the region's largest renewable energy source for electricity generation. It is expected to continue to play an essential role in the USA and Canada's energy mix during the coming decades. Without a doubt, the heart of Canadian electricity exports to the USA is hydropower. The USA and Canada have similar installed hydropower capacity (102.8 GW and 81.1 GW, respectively), but Canada uses hydropower as a higher percentage of total electricity generated (59.2% versus 6.5% in the USA in 2018). There is potential to increase Canada's hydropower capacity significantly. The Canadian Hydropower Association estimates that

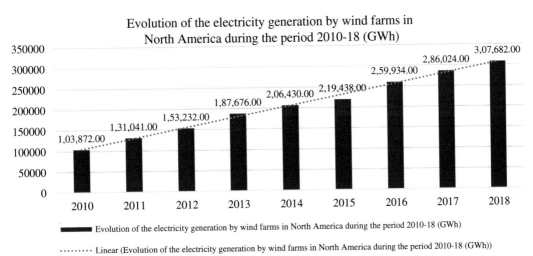

FIG. 1.17 Evolution of the electricity generation by the wind farms operating in the region during the period 2010–2018. *Source: 2020. Renewable Energy Statistics 2020. IRENA and Author own calculations.*

Canada could increase hydropower generation capacity two-fold, to 160 GW (Canadian Hydropower Facts, 2016; BP Statistical Review of World Energy, 2019; Renewable Energy Statistics 2019, 2019).

The controversies relating to hydropower development generally focus on new large-scale hydropower developments and concerns over the environmental damage and displacement of communities due to flooding to establish reservoirs and the influence on freshwater, marine birds mammalian life, both in the repository and downstream of the facility. While the Canadian Hydropower Association estimates that most dams have lifetimes of 100 or more years, this lifetime can be severely reduced by silicification in some world regions. In an era of climate change, variations in the geographical distribution of water and the greater incidences of drought worldwide may also change certain hydropower developments' viability. However, while social and environmental concerns have affected several hydropower developments globally, Canadian hydropower developments have continued to grow robustly in recent years. Besides, there are hydropower plants in all fifty states in the USA. Some states in the Pacific northwest generate the majority of their electricity using this type of power plant.

On the other hand, and according to the Canadian Hydropower Association, Canada is one of the few countries in the world to generate the majority of its electricity from hydropower plants. Besides, approximately 1% of all US electricity is supplied by Canadian hydropower and is used in New York, New England, the Midwest, and the Pacific Northwest. Together, Canadian and US hydropower resources represent, in 2018, approximately 80% of total renewable electricity generation in the North American region with an installed capacity of 183,851 MW.

According to the Canadian Hydropower Association, around 250,000 MW of potential capacity is still available in Canada and the USA. The use in Canada and the USA of hydropower plants for electricity generation avoid the emission of 350,000,000 tons per year of greenhouse gas emissions.

Although growth in hydropower in North America remains modest compared to other regions of the world, there is an increased focus on pumped storage hydropower projects added in the last years. It is important to stress that hydropower's role in the USA and Canada's clean energy mix will evolve during the coming years. More intermittent renewable energy sources such as wind and solar are deployed, particularly in the USA. "In Canada, major storage projects under construction include Keeyask generating power plant in Manitoba, Site C in British Columbia, Muskrat Falls in Newfoundland and Labrador, and Romaine-4 in Quebec." In the USA, "140 MW of installed capacity was added through retrofits to existing facilities. Of particular interest for business strategies in North America are the opportunities for power export within the region" (World Hydropower Congress, 2019).

According to the Renewable Energy and Jobs Annual Review 2019 (2019) report from IRENA, hydropower has the largest installed capacity of all renewables, accounting for almost 50% of renewable energy in the world, but is now expanding slowly. The sector employs 2.05 million people directly, nearly 75% of whom are in operations and maintenance. The number of employees in the construction and installation sectors represents around 23% of the total.

There are many benefits due to the use of hydropower for electricity generation within the North American region because water is a plentiful resource in many areas of Canada and the USA, but the hydropower plants are relatively easy to construct are easily maintained. According to the BP Statistical Review of World Energy (2020), renewable energy accounted for 23.6% of total energy production in 2019 in the region and 17.3% of the domestically produced electricity in the USA in that year. Hydroelectric power is currently the largest renewable electricity producer in the USA, generating around 6.2% of the nation's total electricity in 2019 and 35.6% of the entire renewable electricity generation. The USA is the fourth-largest producer of hydroelectricity in the world after China (1260.7 TWh), Brazil (399.3 TWh), and Canada (382 TWh). In Canada's case, renewable energy accounted for 65.3% of total energy generation in 2019 in the country. Hydropower is also the largest renewable electricity producer in the country, generating 57.8% of the nation's total electricity in 2019. Canada is the third-largest producer of hydroelectricity in the world.

However, it is important to know that the construction of large hydropower plants in the North American region could negatively impact the environment and population. For this reason, there is strong opposition to the construction of this type of power plant in the region.

The evolution of the hydropower capacity in the North American region during the period 2010–19 and the electricity generation in the region during the period 2010–18 is shown in Figs.1.18 and 1.19.

Based on the data included in Fig. 1.18, the following can be stated: the region's hydropower capacity increased 4.4% during the period under consideration, rising from 176,101 MW in 2010 to 183,822 MW in 2019. It is expected that this trend will continue without change during the coming years, despite the small decrease reported in 2019.

The evolution of the electricity generation using hydropower plants in the region during the period 2010–18 is shown in Fig. 1.19.

Based on the data shown in Fig. 1.19, the following can be stated: the electricity generation using hydropower plants in the North American region increased 9.6% during the period 2010–18, rising from 637,794 GWh in 2010 to 698,754 GWh in 2010. However, looking at Fig. 1.19, it is easy to verify that the electricity produced by the hydroelectric plants operating in the region varied considerably with periods of growth and decreased depending on the prevailing climate. For this reason, it is difficult to predict how electricity generation will behave through existing hydroelectric plants in the region over the next few years.

In 2019, the capacity factor of hydropower plants in the USA reached 39.1%, the third-lowest capacity factor among all energy source capacity factors in the country.

### D-Bioenergy

Bioenergy has long been accepted as a useful renewable energy source, especially in mitigating greenhouse gases, nitrogen oxides, and sulfur oxide emissions. Bioenergy is carbon neutral and is usually low in both nitrogen and sulfur.

Bioenergy use falls into two main categories: "traditional" and "modern." Traditional bioenergy use refers to biomass combustion[z] in such forms as wood, animal waste, and traditional charcoal. Modern bioenergy technologies include liquid biofuels produced from bagasse and other plants, bio-refineries, biogas produced through anaerobic digestion of residues, wood pellet heating systems, and other technologies. About 75% of the world's renewable energy use involves bioenergy, with more than half of traditional biomass use (Bioenergy, 2019).

---

[z] "Biomass has significant potential and can be directly burned for heating or power generation, or it can be converted into oil or gas substitutes. Liquid biofuels, a convenient renewable substitute for gasoline, are mostly used in the transport sector." (Bioenergy, 2019).

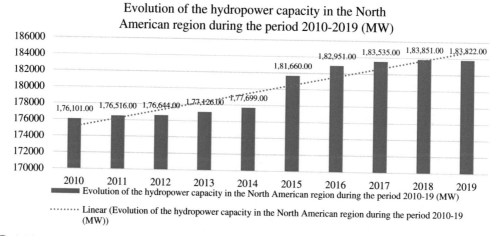

**FIG. 1.18** Evolution of the hydropower capacity in the region during the period 2010–19. *Source: 2020. Renewable Energy Statistics 2020. IRENA and Author own calculations.*

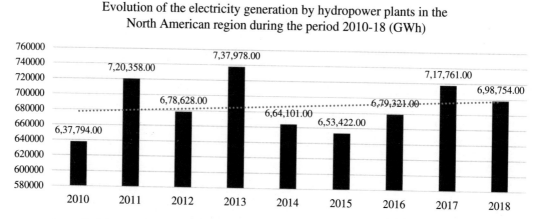

**FIG. 1.19** Evolution of the electricity generation using hydropower plants in the region during the period 2010–18. *Source: 2020. Renewable Energy Statistics 2020. IRENA and Author own calculations.*

For the past decade, bioenergy has been used as a supplemental fuel to substitute for up to 10% of the base fuel in most coal and natural gas-fired power plants. There are several successful co-firing projects in many parts of the world, particularly in Europe and North America. However, despite remarkable commercial success in Europe, most of the bioenergy co-firing in North America is limited to demonstration levels, using primarily wood for solid fuel and cereal crops, which are converted to biofuels or biogas. According to the level of bioenergy production reached in 2018, North America occupied third place with 81,174 GWh (15.5% of the total), after the EU with 188,053 GWh (36%), and Asia with 164,901 GWh (31.6% of the total).

The USA is the world's largest bioenergy producer with 67,885 GWh in 2018 (13% of the world total) and the leading producer within the North American region (86.4% of the regional total). The USA primarily produces ethanol from corn but also produces biodiesel from grain stock. It is also a leader in the development of advanced biofuels, particularly algae-based fuels.

In 2010, Canada implemented a federal mandate that 5% of the gasoline pool would be replaced by bioethanol by 2011 and that 2% of diesel fuel would be replaced by biodiesel or other renewable diesel fuel. Canola oil is the major biofuel feedstock in Canada. The country has long supported biofuel production for its environmental benefits.[aa]

The evolution of the bioenergy capacity within the North American region during the period 2010–19 and the evolution of electricity generation during the period 2010–18 are shown in Figs.1.20 and 1.21.

According to Fig. 1.20, the following can be stated: the North American region's bioenergy capacity increased by 31.9% during the period under consideration, rising from 11,997 MW in 2010 to 15,826 MW in 2019. It is expected that this trend will continue without change during the coming years. For this reason, it is foreseeing that the bioenergy capacity of the region in 2021 will be 18.1 GW[ab]; this means an increase of 14.6% with respect to the capacity installed in 2019.

Based on the data included in Fig. 1.21, the following can be stated: bioenergy electricity generation in the North American region increased by 8.9% during the period 2010–18, rising from 72,136 GWh in 2010 to 78,546 GW in 2018. However, it is important to single out that electricity generation through the use of bioenergy within the North American region during the period 2010–18 shows two opposite trends. During the period 2010–14, bioenergy electricity generation grew 15.6%, reaching the production peak in 2014. However, from that year and until 2018, bioenergy electricity generation decreased by 5.8%, a trend that appears to continue at least for the next few years.

Without a doubt, bioenergy can play a transformative role in the transition to a decarbonized economy within the North American region, with potential applications in electricity, heat, chemicals, and transportation fuels. Deploying bioenergy with carbon capture and sequestration can be one of the few cost-effective carbon-negative opportunities available if climate change becomes worse than anticipated or emissions reductions in other sectors prove particularly difficult to achieve. Advanced bioenergy power generation employs a similar system design to advanced coal technology, enabling a transition strategy to low-carbon energy.

E-Geothermal

Geothermal power is abundant in North America. The region holds diverse geologic conditions, economic and political contexts, and geothermal experiences using this renewable energy source for electricity generation and heating.

---

[aa] A former coal-fired thermal power plant in northwestern Ontario is now operating on biomass, making it the largest power plant in North America fueled completely by biological material.

[ab] It is expected that the USA will have 15.3 GW of bioenergy capacity and Canada 2.8 GW in 2021 (Cumulative Capacity of Bioenergy in North America in 2015 and 2021, by Country 2019, 2019).

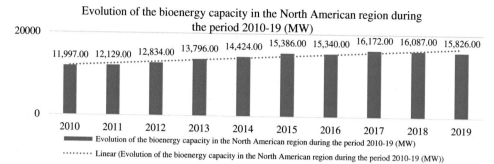

**FIG. 1.20** Evolution of the bioenergy capacity in the North American region during the period 2010–19. *Source: 2020. Renewable Energy Statistics 2020. IRENA and Author own calculations.*

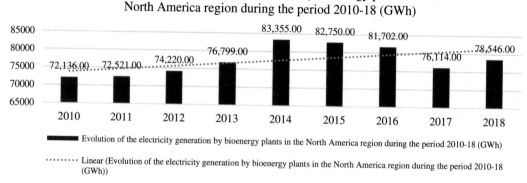

**FIG. 1.21** Evolution of the bioenergy electricity generation in the North American region during the period 2010–18. *Source: 2020. Renewable Energy Statistics 2020. IRENA and Author own calculations.*

In the USA, prime locations are clustered in most eastern and southern California, southeastern Idaho, western Oregon, northwestern Utah, northern Louisiana, southern Arkansas, and northeastern Texas. California is the national and world leader in using geothermal energy sources for electricity generation and heating. "The state market has taken hits due to subsidies aimed at intermittent renewables (geothermal is considered baseload) and competition with natural gas but is considering two pieces of legislation that could bring about a geothermal revival in the state" (Blodgett, 2014). Nevada is also a strong market with plans to retire old and ineffective coal-fired power plants and replace them with renewables, including geothermal power plants. "The industry is robust and innovative but not without its challenges, including policy barriers, inadequate transmission infrastructure, and a stiff energy market" (Blodgett, 2014).

In 2019, Canada had no geothermal power plant capacity installed. However, according to several experts' opinions, the country "could have up to 5000 MW lying in wait and accessible with currently available technology". Geothermal Energy Association April 2014 international projects list "showed nine current geothermal power generation projects, all in early or prospective stages: six in British Columbia, two in Saskatchewan, and one in Alberta"

(Blodgett, 2014). The benefits of geothermal energy are numerous, including a small environmental footprint, low water usage, and very few emissions, while being a completely renewable, continuous, and dependable resource.

The USA has 8331 records of geothermal energy zones on public land managed by the Bureau of Land Management. A total of 539 zones for geothermal energy exploitation have been authorized, and 223 zones for geothermal exploitation are still waiting for correspondent government authorization. A total of 7495 zones has already been closed for geothermal energy exploitation, and 16 zones have been rejected.

Unlike other renewable energy sources such as wind and solar, geothermal energy is dispatchable, meaning that it is available whenever needed and can quickly adjust output to match demand. According to the Annual Energy Outlook 2013 (2013), geothermal generators have the highest capacity factor of all types of new electrical generation plants, measuring how much power a facility generates as a percent of its maximum capacity. The US EIA rates new geothermal plants as having a 92% capacity factor, comparable to those of nuclear (between 70% and 90% average), and higher than natural gas (87%) or coal (between 50% and 65%), and much higher than those of intermittent sources such as onshore wind (between 20% and 40%) or solar PV (between 10% and 30%).

The evolution of geothermal power plants' capacity in North America during the period 2010–19 and the evolution of electricity generation in the same region during the period 2010–18 are shown in Figs.1.22 and 1.23.

According to Fig. 1.22, the following can be stated: geothermal installed capacity in the North American region increased by 6.2% during the period 2010–19, rising from 2405 MW in 2010 to 2555 MW in 2019. As shown in Fig. 1.22, the region's geothermal capacity suffered up and down during the period considered. From 2010 to 2013, geothermal installed capacity grew by 8.4%, reaching its highest capacity in 2013. Until 2017, geothermal installed capacity decreased by 4.8%, increasing 2.9% again during the period 2017–19. Due to these changes in the geothermal installed capacity recorded in recent years, it is difficult to predict which trend will prevail in the future, even more so if this type of energy source's peculiar characteristics are considered. However, geothermal installed capacity is likely to grow in the long term but at a much slower pace than other renewable energy sources' growth.

Regarding the evolution of the geothermal electricity generation shown in Fig. 1.23, the following can be stated: geothermal electricity generation in the North American region increased by 6.8% during the period under consideration, rising from 17,577 GWh in 2010 to 18,733 GWh in 2018. It is expected that this trend will continue in the future but at a low pace.

## The use of renewables energies for electricity generation in the North American region

Without a doubt, the current climate change poses significant risks for the environment and population. Overall, scientists generally concur that the rapid changes experienced in the global climate during the last several decades are mostly caused by human activities, which have led to increased emissions of greenhouse gases (Intergovernmental Panel on Climate Change

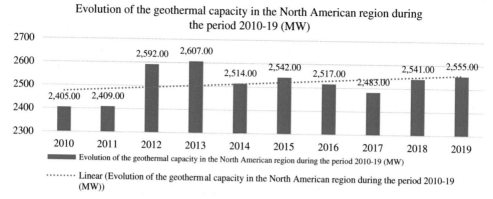

**FIG. 1.22** Evolution of the geothermal capacity in the North American region during the period 2010–19. *Source: 2020. Renewable Energy Statistics 2020. IRENA and Author own calculations.*

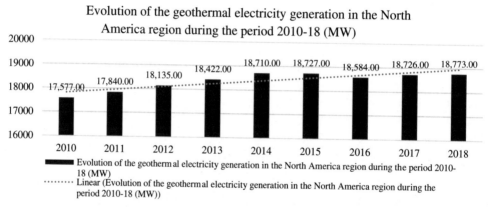

**FIG. 1.23** Evolution of the geothermal electricity generation in the North American region during the period 2010–18. *Source: 2020. Renewable Energy Statistics 2020. IRENA and Author own calculations.*

(IPCC), 2013). The need to stop climate change from affecting the environment and population is recognized as a significant challenge for scientists, decision-makers, and the general public.

It is projected that global temperatures will continue to increase, affecting more and more countries all over the world, if practical actions are not implemented to reduce the total annual greenhouse gas emissions by all countries in all regions. In contrast to climate change adaptation, "scientists and decision-makers recognize that mitigation efforts are less dependent on local responses and more on national and international cooperation" (Hagen and Pijawka, 2015). For this reason, "an increasing number of local, regional, national, and international organizations have decided to establish and promulgate carbon emission reductions" (Hagen and Pijawka, 2015). Besides, many countries have signed various international treaties, setting greenhouse gas reduction goals and strengthening international collaboration in combating climate change. However, several states will not be able to fulfill their international commitments and greenhouse gas reduction targets without adopting and implementing policies "that provide opportunities for involving local jurisdictions" and support of the population (Hagen and Pijawka, 2015).[ac]

Despite the increasing use of different conventional energy sources for electricity generation in North America, such as domestic shale gas and fracking for oil, there is significant interest among citizens and policymakers to increase the use of renewable energy sources for this specific purpose in order to reduce greenhouse gases emissions and their increasing climate change impacts. Without a doubt, the use of renewable energy sources for electricity generation offers local energy independence, job creation and economic development, improved environmental quality, ecological integrity, enhanced human health, and other benefits.[ad] At national and regional levels, it offers energy and climate security (North American Center for Transborder Studies (NACTS), 2011).

In 2019, about 5061.7 TWh of electricity was generated at utility-scale electricity generation facilities in the North American region. In Canada, the electricity generation reached 660.4 TWh (13% of the total), and in the USA, 4401.3 TWh (87% of the total). At the regional level, about 58.1% of this electricity generation was from fossil fuels (coal, natural gas, petroleum, and other gases); approximately 18.6% was from nuclear energy, and about 23.3% was from renewable energy sources. The BP Statistical Review of World Energy (2019) reported that at the regional level, the energy capacity and electricity generation by sources are included in Table 1.3.

Some areas of North America have no indigenous conventional energy resources, and, for this reason, the use of renewable energy sources for electricity generation is an attractive option. It can be a more productive and efficient way to produce the necessary energy to meet local needs. "While adding renewable energy to the transmission grid can be initially challenging, it offers the opportunity to increase overall electrical reliability and local energy." (Hagen and Pijawka, 2015).

The electricity demand by selected region in 2016 and its growth to 2040 is shown in Fig. 1.24.

According to Fig. 1.24, in 2016, the significant growth in electricity demand was registered in China, followed by India and the EU. In 2040, China is expected to report the major increase in electricity demand, followed by India, Southeast Asia, and Africa.

## The use of nuclear energy for electricity generation in the North American region

One of the available energy sources that have been probed in the past that can be effectively used for electricity generation is nuclear energy. The nuclear industry has proclaimed that nuclear energy for electricity generation played an important role in developing the

---

[ac] Well-designed renewable energy policies are less successful if public perception factors are not recognized. The level of public support for this type of policies is often impacted by the way the public processes information and how it perceives threats and other perceptional factors such as trust in government and the private sector (Bord et al., 1998).

[ad] "Given its increasing affordability, the applications and use cases of renewable energy have broadened. Alongside electricity production, it is providing new solutions for mobility and energy security worldwide." (Tanti, 2018).

**TABLE 1.3** North America electricity generation by source, amount, and share of total (2018 and 2019).

| Energy source | USA TWh | Share of the total (%) | Canada TWh | Percentage of the total (%) | Regional TWh |
|---|---|---|---|---|---|
| Total all-sources (2019) | 4401.3 | 87 | 660.4 | 13 | 5061.7 |
| Fossil fuels (Total) (2019) | 2774.4 | 95.7 | 128 | 4.3 | 3002.1 |
| Natural gas (2019) | 1700.9 | 96 | 69.3 | 4 | 1770.2 |
| Coal (2019) | 1053.5 | 95 | 54.6 | 5 | 1108.1 |
| Petroleum (2019) | 20 | 83 | 4.1 | 17 | 24.1 |
| Nuclear energy (2019) | 852 | 89.4 | 100.5 | 10.6 | 952.5 |
| Renewables (Total) (2019) | 761 | 63.8 | 431.3 | 36.2 | 1192.3 |
| Hydropower | 271.2 | 41.5 | 382 | 58.5 | 653.2 |
| Pumped storage hydropower | 19.2 | 99 | 0.2 | 1 | 19.4 |
| Wind energy | 275.8 | 89.7 | 31.8 | 10.3 | 307.6 |
| Onshore wind energy | 275.7 | 89.7 | 31.8 | 10.3 | 307.5 |
| Offshore wind energy | 0.1 | 100 | – | – | 0.1 |
| Bioenergy (Total) | 67.9 | 86.4 | 10.7 | 13.6 | 78.6 |
| Solid biofuels and renewable waste | 54.3 | 84.8 | 9.7 | 15.2 | 64 |
| Liquid biofuels | 0.4 | 100 | – | – | 0.4 |
| Biogas | 13.3 | 93 | 1 | 7 | 14.3 |
| Renewable Municipal solid waste | 8.4 | 97.7 | 0.2 | 2.3 | 8.6 |
| Other solid biofuels | 45.9 | 82.9 | 9.5 | 17.1 | 55.4 |
| Solar energy (Total) | 85.2 | 95.7 | 3.8 | 4.3 | 89 |
| Solar photovoltaic | 81.2 | 95.5 | 3.8 | 4.5 | 85 |
| Solar thermal | 3.9 | 100 | – | – | 3.9 |
| Geothermal | 18.8 | 100 | – | – | 18.8 |

*Sources: 2020. Renewable Energy Statistics 2020. IRENA and 2020. BP Statistical Review of World Energy, 69th ed.*

energy sector in more than 30 countries in the 1970s and 1980s in North America, Latin America, Europe, Asia, and at least one country in Africa. They believe that nuclear energy will continue to play an important role in electricity production in the future, not only in the same countries but also in other countries as well. However, it is essential to single out that in 2000, more than 40% of global energy demand was in Europe and North America and some 20% in Asia's developing economies. By 2040, this situation is completely reversed (see Figs.1.25–1.27). In the coming years, nuclear energy participation

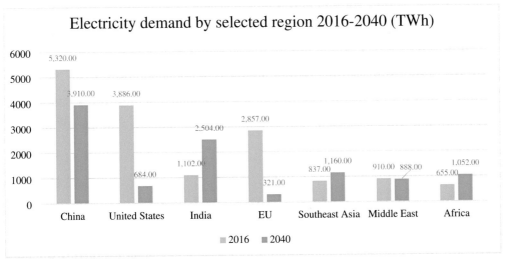

**FIG. 1.24** Electricity demand by selected region. *Source: 2017. World Energy Outlook 2017. International Energy Agency and Author own work.*

in several countries' energy mix is expected to be lower than today, except for the Asian and the Pacific region.

According to Fig. 1.25, the USA (around 2250 Mtoe) and the EU (approximately 1700 Mtoe) were the areas of significant energy demand in 2000, followed by China (about 1200 Mtoe), Africa (around 500 Mtoe), and India (around 450 Mtoe). In 2017, the situation changed, and China became the country with the highest energy demand (a little more than 3000 Mtoe), followed by the USA (around 2150 Mtoe) and the EU (approximately 1600 Mtoe) (see Fig. 1.26). It is expected that by 2040 China will continue to be the country with the highest energy demand (around 3800 Mtoe), followed by the USA (around the same level as in 2017), India

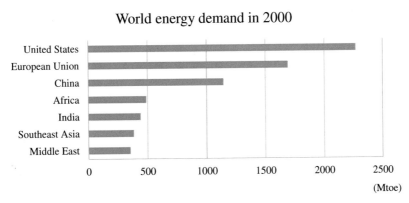

**FIG. 1.25** World energy demand in 2000. *Source: 2018. World Energy Outlook 2018. Executive Summary. International Energy Agency and Author own work.*

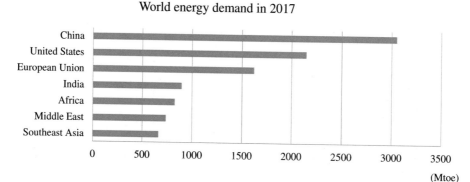

**FIG. 1.26** World energy demand 2017. *Source: 2018. World Energy Outlook 2018. Executive Summary. International Energy Agency and Author own work.*

(about 1900 Mtoe), Africa (approximately 1300 Mtoe) and the EU (around 1500 Mtoe) (see Fig. 1.27).

The different energy sources' participation in satisfying the world energy demand during the period 2017–40 is shown in Figs.1.26 and 1.27.

Over the last years, several international assessments of the possible future of conventional and non-conventional energy sources, particularly the role of nuclear power in world energy generation, have been adjusted to more optimistic prospects for the horizon of 2040. According to Fig. 1.28, advanced economies will increase natural gas use for energy generation and decrease oil and coal use for this specific purpose. The use of nuclear energy for electricity generation will also decrease in this group of countries. However, in the case of developing economies, the situation will be somehow different. This group of countries will significantly increase the use of natural gas for power generation in the industry. For other purposes, this group will continue to use coal for electricity generation and reduce the role of oil in their energy mix. The role of nuclear energy for electricity generation will be higher than today, particularly in Asia and the

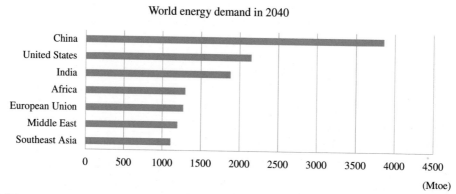

**FIG. 1.27** World energy demand 2040. *Source: 2018. World Energy Outlook 2018. Executive Summary. International Energy Agency and Author own work.*

Pacific, the region with the major program related to nuclear power plants' construction. Out of the 54 nuclear power plants under construction at the world level in April 2021 (see Fig. 1.29), 34 units are under construction in the Asian and the Pacific region (63% of the total). According to the IAEA information database in China (16 units) and India (6 units) alone are 22 units under construction (PRIS—Power Reactor Information System (PRIS), 2021).

According to Table 1.4, there are only two nuclear power reactors under construction in the North American region (3.8% of the world total), both in the USA. There are no concrete plans for the building of more units in the region. Based on the World Nuclear Industry Status Report 2018, it is important to stress that nine nuclear power reactors started in 2018, seven in China and two in Russia. In 2019, four units started electricity production, two of them in China. It is crucial to single out that the number of units under construction fall for the sixth year in a row, declined from 68 units at the end of 2013 to 54 units in April 2021. Besides, 27 units under construction are behind schedule, mostly by several years (51.9% of the total), and 11 of them increased delays.

World Energy Outlook 2018 presents a reference scenario (see Fig. 1.30) where the EU will implement an effective closure program of nuclear power plants during the coming years (from a little more than 120 GW of installed capacity in 2017 to around 20 GW of installed capacity in 2040, a reduction of 83.3%), followed by the USA (with a little more than 100 GW installed capacity in 2017 to a little more than 40 GW of installed capacity installed in 2040, a reduction of 60%), and Japan (with 40 GW of capacity installed in 2017 to around 10 GW of capacity installed in 2040, a reduction of 75%).

The most effective program at the world level for the use of nuclear energy for electricity generation will be implemented in China (with a little less than 40 GW of installed capacity in 2017 to around 150 GW of capacity to be installed in 2040, an increase of 3.8-fold), followed by India (with about 10 GW of capacity installed in 2017 to 40 GW of capacity to be installed in 2040, an increase of four times), and Russia (with around 30 GW of capacity installed in 2017 to approximately 40 GW of capacity to be installed in 2040, an increase of 33.3%). There are no concrete plans for the increased use of nuclear energy for electricity generation and heating in the North American region during the coming years.

However, it is important to stress that nuclear power will only continue to play an important role in the energy mix in those countries where:

- Governments support the use of nuclear power for electricity generation;
- Promote and encourage private investment, especially in liberalized markets;
- And if concerns about plant safety, nuclear waste disposal, and proliferation risk can be solved to the public's satisfaction (Schneider and Froggatt, 2007).

According to the PRIS-IAEA database, in April 2021, 444 nuclear power reactors were in operation in 33 countries, including Taiwan, with a net capacity of 394,098 MW (see Fig. 1.31) and 53 nuclear power reactors under construction in 19 countries with a net capacity of 56,941 MW. In the North American region, there are 113 nuclear power reactors in operation (25.5% of the world total) in April 2021, being the third region with the highest number of nuclear power reactors under operation, after Europe (179 units, including Russia, Ukraine, and Belarus), and Asia and the Pacific (139 units). In April 2021, 192 nuclear power reactors were permanently shut down in 20 countries, totalizing 87,248 MW. The countries with the

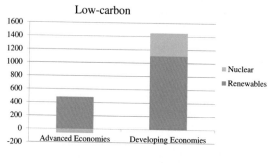

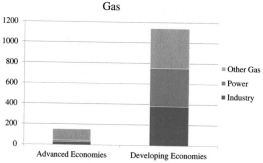

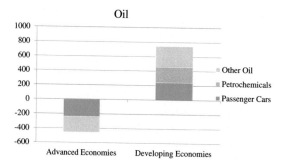

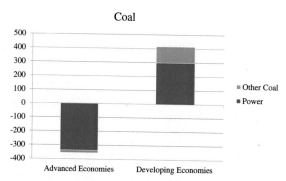

**FIG. 1.28** Change in global energy demand 2017–40. *Source: 2018. World Energy Outlook 2018. Executive Summary. International Energy Agency and Author own work.*

## Number of Reactors Under Construction

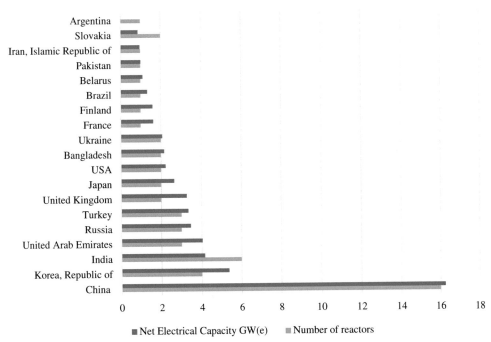

**FIG. 1.29**  Number of nuclear power reactors under construction in April 2021. *Source: April 2021. PRIS—Power Reactor Information System. www.iaea.org/pris, © IAEA.*

highest number of nuclear power reactors in operations are the USA with 94 units, France with 57 units, China with 50 units, Russia with 38 units, Japan with 33 units, the Republic of Korea with 24 units, and India with 23 units (see Fig. 1.32 and Table 1.5). In 2019, the capacity factor of nuclear power plants in the USA reached 93.5%, being the highest capacity factor among all power plants operating in the country.

The number of nuclear power reactors in operation and the country's capacity are shown in Table 1.5.

The countries with the highest number of nuclear power reactors permanently shut down until April 2021 are the following:

- The USA, with 39 units with a net capacity of 18,141 MW;
- The UK, with 30 units with a net capacity of 34,715 MW;
- Germany, with 30 units with a net capacity of 18,262 MW;
- Japan, with 27 units with a net capacity of 17,119 MW;
- France, with 14 units with a net capacity of 5549 MW (see Fig. 1.32)

From the 33 countries that in 2020 operate nuclear power plants, seven of them, the USA, France, China, Japan, India, Russia, and South Korea, run 71.9% of the total units in operation

**TABLE 1.4** Number of nuclear power reactors under construction by country (April 2021).

| Country | Number of reactors | Total net electrical capacity MW |
|---|---|---|
| Argentina | 1 | 25 |
| Bangladesh | 2 | 2160 |
| Belarus | 1 | 1110 |
| Brazil | 1 | 1340 |
| China | 16 | 16,243 |
| Finland | 1 | 1600 |
| France | 1 | 1630 |
| India | 6 | 4194 |
| Iran, Islamic Republic of | 1 | 974 |
| Japan | 2 | 2653 |
| Korea, Republic of | 4 | 5360 |
| Pakistan | 1 | 1014 |
| Russia | 3 | 3459 |
| Slovakia | 2 | 880 |
| Turkey | 3 | 3342 |
| Ukraine | 2 | 2070 |
| United Arab Emirates | 3 | 4035 |
| United Kingdom | 2 | 3260 |
| United States of America | 2 | 2234 |
| **Total** | **54** | **57,583** |

Source: April 2021. PRIS—Power Reactor Information System. www.iaea.org/pris, © IAEA.

and have 73.9% of the world nuclear-installed capacity. However, the role of nuclear power in the overall energy sector remains minimal, even in these seven countries. France is the country with the highest participation of nuclear energy in its energy balance globally and generated, in 2019, a total of 70.6% of its electricity with nuclear power plants (379,500 GWh).

According to the regional classification of its member states included in the IAEA Statute, approved in 1957, the number of nuclear power reactors in operation by region is shown in Fig. 1.33 and Table 1.6.

Based on the information included in Table 1.6, the following can be stated: North America is the region with the highest number of nuclear power reactors in operation with 113 units, followed by Asia and the Far East with 111 units and Western Europe with 106 units. However, if the traditional classification of the regions within the United Nations system is used, then

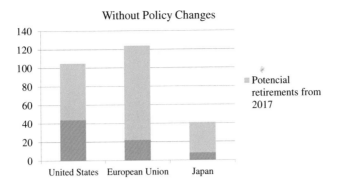

FIG. 1.30  Change in the use of nuclear energy for electricity generation during the period 2017–40. *Source: 2018. World Energy Outlook 2018. Executive Summary. International Energy Agency and Author own work.*

Europe is the region with the highest number of nuclear power reactors in operation with 180 units, followed by Asia and the Pacific with 142 units and North America with 113 units. In the future, it is expected that the region with the highest number of nuclear power reactors in operation will be Asia and the Pacific.

By type of nuclear power reactor, the number of units in operation is shown in Fig. 1.34.

According to Fig. 1.34, a total of 302 nuclear power reactors in operation are of the PWR type (68% of the total), followed by the 63 BWR type (14.2%), 49 PHWR type (11%), 14 GCR type (3.2%), 12 LWGR type (2.7%), and 3 FBR type (0.7%). In the North American region, the number of nuclear power reactors by type is shown in Table 1.7.

Based on the data included in Table 1.7, the following can be stated: the nuclear power reactors constructed by Canada are all of the same type (PHWR), while in the case of the USA, 62 units are of the PWR type (66% of the total), while 32 units are of the BWR type (34%).

In 2019, six units with a total capacity of 5174 started their construction in Bangladesh (one unit), the Republic of Korea (one unit), Russia (one unit), Turkey (one unit), and the UK (one unit). Besides, nine units with a total capacity of 10,358 MW were connected to the electrical grid in China (seven units) and Russia (two units).

According to the World Nuclear Industry Status Report 2018 (Schneider and Froggatt, 2018), in 2017, construction began on five nuclear power reactors and in the first half of 2018

## Number of Nuclear Power Reactors in Operation in the World

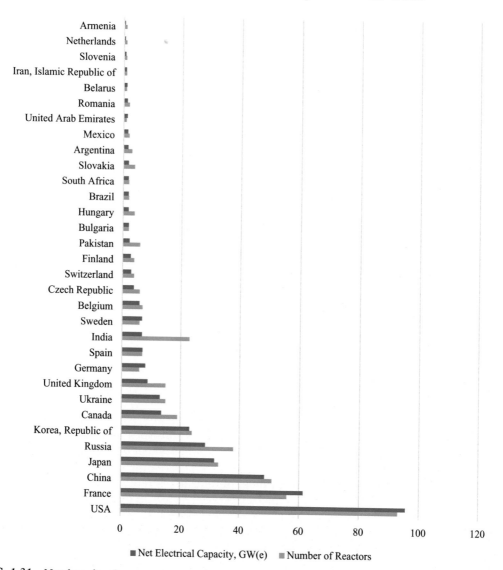

FIG. 1.31   Number of nuclear power reactors in operation in the world in April 2021. *Source: April 2021. PRIS— Power Reactor Information System. www.iaea.org/pris, © IAEA.*

on two. This figure is two and three times lower than 15 new units construction that started in 2010 and the ten new units construction that began in 2013. Historical analysis shows that the construction of new nuclear power reactors began in the world peaked in 1976 at 44 units.

Based on the World Nuclear Industry Status Report 2019, in 2018, nuclear power generation in the world increased by 2.4%, of which 1.8% was due to a 19% increase in China.

## Permanent Shutdown Reactors

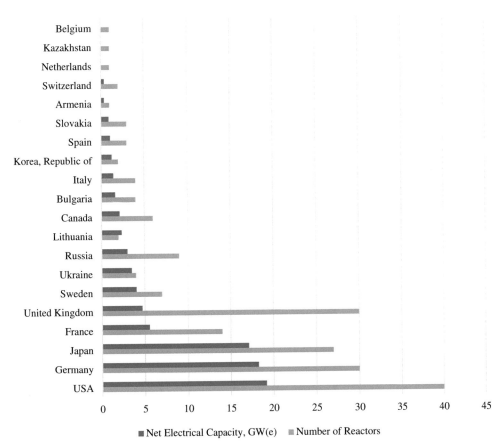

**FIG. 1.32** Total number of nuclear power reactors permanently shut down in April 2021. *Source: April 2021. PRIS—Power Reactor Information System. www.iaea.org/pris, © IAEA.*

Global nuclear power generation, excluding China, increased by 0.6% for the first time after decreasing three years in a row without making up for the decline since 2014. Nine nuclear power reactors started up in 2018, of which seven were in China and two in Russia. Four units started up in the first half of 2019, of which two were in China. The number of units under construction globally declined for the sixth year in a row, from 68 reactors at the end of 2013 to 46 by mid-2019, of which ten units are under construction in China.

It is important to note that no new commercial nuclear power reactor has started in China since December 2016. For this reason, it will by far be missing its Five-Year-Plan 2020 nuclear targets of 58 GW installed and 30 GW under construction. The nuclear share of global electricity generation has continued its slow decline from a historical peak of about 17.5% in 1996 to 10.15% in 2018. Japan had restarted nine units by mid-2018 but none since. As of mid-2019, 28 nuclear power reactors, including 24 units in Japan, are in Long-Term Outage. At least 27 of the 46 units under construction

**TABLE 1.5**  Number of nuclear power reactors by country (April 2021).

| Country | Number of reactors | Total net electrical capacity MW |
|---|---|---|
| Argentina | 3 | 1641 |
| Armenia | 1 | 375 |
| Belarus | 1 | 1110 |
| Belgium | 7 | 5930 |
| Brazil | 2 | 1884 |
| Bulgaria | 2 | 2006 |
| Canada | 19 | 13,554 |
| China | 50 | 47,518 |
| Czech Republic | 6 | 3932 |
| Finland | 4 | 2794 |
| France | 56 | 61,370 |
| Germany | 6 | 8113 |
| Hungary | 4 | 1902 |
| India | 23 | 6885 |
| Iran, Islamic Republic of | 1 | 915 |
| Japan | 33 | 31,679 |
| Korea, Republic of | 24 | 23,172 |
| Mexico | 2 | 1552 |
| Netherlands | 1 | 482 |
| Pakistan | 6 | 2332 |
| Romania | 2 | 1300 |
| Russia | 38 | 28,578 |
| Slovakia | 4 | 1814 |
| Slovenia | 1 | 688 |
| South Africa | 2 | 1860 |
| Spain | 7 | 7121 |
| Sweden | 6 | 6859 |
| Switzerland | 4 | 2960 |
| Ukraine | 15 | 13,107 |
| United Arab Emirates | 1 | 1345 |
| United Kingdom | 15 | 8923 |
| United States of America | 94 | 96,553 |
| Total | **444** | **394,098** |

*Source: April 2021. PRIS—Power Reactor Information System. www.iaea.org/pris, © IAEA.*

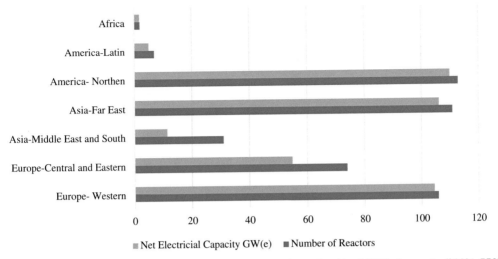

**FIG. 1.33** Total number of nuclear power reactors in operation by region (April 2021). *Source: April 2021. PRIS—Power Reactor Information System. www.iaea.org/pris, © IAEA.*

**TABLE 1.6** Total number of nuclear power reactors in operation by region (April 2021).

| Region | Number of reactors | Total net electrical capacity (GW) |
| --- | --- | --- |
| Europe – Western | 106 | 104.55 |
| Europe – Central and Eastern | 74 | 54.81 |
| Asia – Middle East and South | 31 | 11.48 |
| Asia – Far East | 111 | 106.21 |
| America – Northern | 113 | 110.11 |
| America – Latin | 7 | 5.08 |
| Africa | 2 | 1.86 |
| **Total** | **444** | **394.04** |

*Source: April 2021. PRIS—Power Reactor Information System. www.iaea.org/pris, © IAEA.*

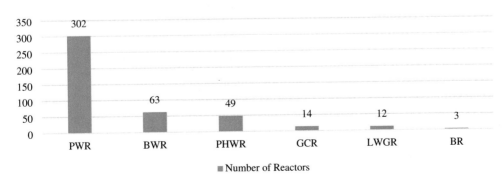

**FIG. 1.34** Number of units by type in operation in April 2021. *Source: April 2021. PRIS—Power Reactor Information System. www.iaea.org/pris, © IAEA.*

**TABLE 1.7** Number of nuclear power reactors in operation in North America per type in April 2021.

| Region | PWR | BWR | PHWR | GCR | LWGR | FBR |
|--------|-----|-----|------|-----|------|-----|
| United States | 62 | 32 | – | – | – | – |
| Canada | – | – | 19 | – | – | – |
| Total | 62 | 32 | 19 | | | |

*Source: April 2021. PRIS—Power Reactor Information System. www.iaea.org/pris, © IAEA.*

(58.7% of the total) are behind schedule, mostly by several years; 11 units (23.9%) have reported increased delays, and three units (7%) have had documented delays for the first time over the past year. Only nine units (52.9%) of the 17 nuclear power reactors scheduled for startup in 2018 were connected to the grid (Schneider and Froggatt, 2019). The evolution of the beginning of the construction of nuclear power reactors during the period 2007–17 is shown in Fig. 1.35.

On the other hand, between 1970 and mid-2018, a total of 94 (12%) of all nuclear power plant construction projects were abandoned or suspended in 20 countries at various stages of development. The latest was the two AP1000 units at V.C. Summers in the USA abandoned in 2017 after spending some US$5 billion on the project. At the same time, four new countries are building four nuclear power reactors. These countries are Bangladesh, Belarus, Turkey, and United Arab Emirates (UAE). The first nuclear power reactor startup in UAE, which was recently reported on time and budget as of late 2016 for a startup in 2017, is at least three years behind schedule. Belarus's first unit was about one year delayed after the reactor pressure vessel was dropped and replaced. The projects in Bangladesh and Turkey only started a few months ago. During 2019 is expected that four new units with a total capacity of 1844 MW be connected to the grid in the Republic of Korea (one unit), Russia (two units), and Slovakia (one unit).

According to Fig. 1.35, the period considered can be divided into three sections. In the first section, covering the period 2007–10, the number of units that began their construction increased by 100%, rising from eight units in 2017 to 16 units in 2010. In the second section, covering the years 2011–15, the number of units that began their construction increased only 100% from four units in

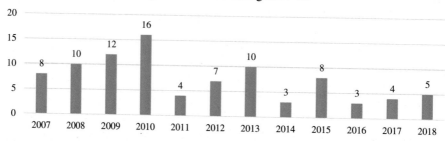

**FIG. 1.35** Evolution of the beginning of construction of new nuclear power reactors during the period 2007–18. *Source: 2019. Nuclear Power Reactors in the World. Reference Data Series No. 2. 2019 edition. ISBN 978-92-0-102719-1. IAEA, Vienna.*

2011 to eight units in 2015. In the third section, covering the period 2016–18, the number of units that began their construction increased only by 67%, rising from three units in 2016 to five units in 2018. If the whole period is considered, then the number of units that began their construction decreased by 37.5%, decreasing from eight units in 2017 to five units in 2018. The peak in the building of new nuclear power reactors during the period 2007–18 was reached in 2010 with 16 units.

Concerning the construction of nuclear power reactors at the world level, it is relevant to single out the following. During the period of peak construction of nuclear power reactors in the 1970s and 1980s, there were several major nuclear supply companies in Canada, France, Germany, Japan, the former Soviet Union (now the Russian Federation), Sweden, Switzerland, UK, and the USA. In 2010, the leading nuclear system supplier companies were located in Canada, China, France, India, Japan, the Republic of Korea, the Russian Federation, and the USA. In addition, other potential nuclear system suppliers have designs in development, such as Argentina and South Africa. Still, the designers of currently available nuclear steam supply systems have reduced to a small group who increasingly are working very closely together. This group includes Areva and Mitsubishi, GE and Hitachi, and Toshiba and Westinghouse.

One of the significant problems that several countries have to deal with in the coming years, particularly in North America and Europe, is the increasing aging power generation capacity, even in the field of electricity generation using nuclear energy (see Fig. 1.36).

According to Fig. 1.36, most nuclear power reactors currently in operation worldwide (296 units) are 30 or more years old. This number represents 66.7% of the total nuclear power reactors now in service in the world. Therefore, to cope with the aging of nuclear power reactors currently in operation at the world level, a significant investment in the energy sector

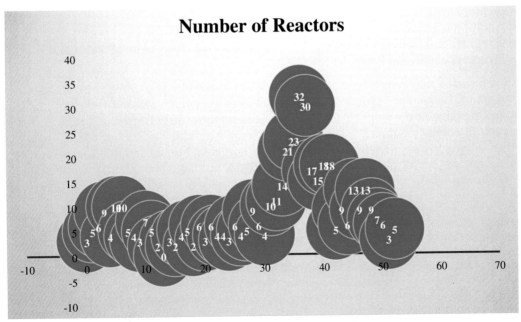

**FIG. 1.36** Age of the current nuclear power reactors in operation (April 2021). *Source: April 2021. PRIS—Power Reactor Information System. www.iaea.org/pris, © IAEA.*

is needed as soon as possible to meet the expected increase in the electricity demand and to replace aging infrastructures in the energy sector.

The number of nuclear power reactors in operation in the world in April 2021 with less than 20 years connected to the grid is 85, representing 19.1% of the total. For this reason, the age of nuclear power reactors is one of the urgent problems that need to be faced by those countries in which these reactors are operating. Two possibilities must be considered. First, extending their lifetime. Second, by initiating the shut down and decommissioning of these units. Considering the current energy situation in the world, it is expected that the majority of the countries will decide to extend the lifetime of the majority of their units now in operation to satisfy the foresee increase in the demand for electricity in the coming years in the most cost-effective manner.

The extension of the lifetime of the majority of the current nuclear power reactors in operation does not represent a risk that could affect the safe operation of these reactors. The reactor technology available for use today is fundamentally based upon previous designs but with significant improvement and takes into account the following design characteristics:

**(a)** A 60-year life;
**(b)** Simplified maintenance – online or during an outage;
**(c)** Easy and short construction;
**(d)** Inclusion of safety and reliability considerations at the earliest stages of design;
**(e)** Modern technologies in digital control and man-machine interface;
**(f)** Safety system design guided by risk assessment;
**(g)** Simplicity by reducing the number of rotating components;
**(h)** Increased reliance on passive systems (gravity, natural circulation, accumulated pressure, etc.);
**(i)** Addition of severe accident mitigating equipment;
**(j)** Complete and standardized designs with pre-licensing (International Status and Prospects of Nuclear Power, 2008).

It is foreseen that before 2040, a new generation of nuclear power reactors, the so-called "Generation IV system," will be in the market once they have reached technical maturity and met sustainable development criteria, particularly those on waste management and preservation of energy resources.[ae] The so-called "Small Modular Reactors or SMRs" will be a realistic alternative solution for countries with no sufficient human and financial resources to build a nuclear power plant, or their grid capacity is not large enough to accommodate the connection of a large nuclear power reactor for electricity generation. So what is new in this

---

[ae] The first generation of nuclear power reactors was commercially available in the 1950s and 1960s. The second generation began in the 1970s in large commercial nuclear power plants and several of them still are operating today. Generation III was developed more recently in the 1990s with a number of evolutionary designs that offer significant advances in safety and economics. Several units have been built, primarily in East Asia. Advances to Generation III (the so-called "Generation III +") are underway resulting in several near-term deployable plants that are actively under development and are being considered for deployment in several countries. It is expected that the new nuclear power plants to be built between now and 2030 will likely be chosen from this type of reactor. Beyond 2030, the prospect for innovative advances through renewed research and development has stimulated interest worldwide in the Generation IV of nuclear power reactor systems (Morales Pedraza, 2012).

type of nuclear power reactor that promises an increase in nuclear energy use for electricity generation? The first significant difference between conventional nuclear power reactors and SMRs is that this new type of nuclear power reactor can be constructed in factories in prefabricated modules. That possibility expedites the construction of a single large nuclear power plant, including its construction underground, which improves containment and security, although it may hinder emergency access. In other words, SMRs are designed based on the modularization of their components, which means the structures, systems, and components are shop-fabricated, then shipped and assembled on-site, with the purpose of significantly reduce construction time and costs (Morales Pedraza, 2015, 2017).

One of the Generation IV systems' main features is that they will be much smaller in size and have low capital investment. That is a more flexible solution with much shorter building times and a lower potential risk due to smaller radioactive inventories and passive safety features (Schneider and Froggatt, 2007). Based on what has been said before, it is important to mention that the international community should be aware that each of the three previous generations of nuclear power reactors will coexist during the 21st century. These types of nuclear power reactors face specific technological challenges to be overcome on the path to sustainability. But all share the common goal of guaranteeing the highest level of safety (Morales Pedraza, 2012).

According to the PRIS—Power Reactor Information System (PRIS) (2021), during 2019, the electricity production using nuclear power at the world level increased by 3.7% with respect to 2018. The evolution in the use of nuclear energy for electricity generation at the world level during the period 2008–19 is shown in Fig. 1.37.

Based on the data shown in Fig. 1.37, the following can be stated: since 2012 has been registered a systematic increase in the use of nuclear energy for electricity generation at the world level. The growth recorded was 13.3% during the period 2012–19, rising from 2346.19 TWh in 2012 to 2657.16 TWh in 2019. Furthermore, it is expected that this trend will continue during at least the coming years.

In the North American region, the electricity generation capacity distribution by energy source in 2019 is shown in Fig. 1.38.

Considering the data included in Fig. 1.38, the following can be stated: in 2019, the nuclear energy capacity installed in the region represents 8.97% of the total energy capacity installed at the world level, behind natural gas with 43.91% and coal with 22.15%. In 201, and based on the IAEA information database (PRIS—Power Reactor Information System (PRIS), 2021), the total electricity generated in the North American region using all available energy sources reached 4,758,442 GWh. The power generated in the region using nuclear energy as an energy source reached, in the same year, 904,878 GWh, which represents 19% of the total.

In 2019, the US total net electricity generation was 4,118,051 GWh, of which 19.7% comes from nuclear energy (PRIS—Power Reactor Information System (PRIS), 2021). Annual electricity demand is projected to increase up to 5000 TWh in 2030, though in the short-term, it is depressed and has not exceeded the 2007 level. Annual per capita electricity consumption in 2017 was about 12,300 kWh. The total net summer capacity is about 1080 GW, less than one-tenth of which is nuclear (EIA Database, 2018).

In 2018, Canada generated 640,391 GWh, of which 14.9% comes from nuclear power plants (95,469 GWh) (PRIS—Power Reactor Information System (PRIS), 2021).

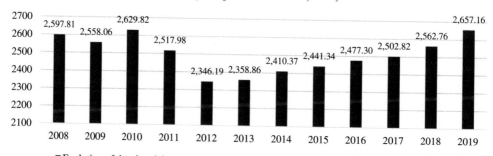

FIG. 1.37   Evolution of the electricity generation using nuclear energy during the period 2008–19 (TWh). *Source: April 2021. PRIS—Power Reactor Information System. www.iaea.org/pris, © IAEA and Author own work.*

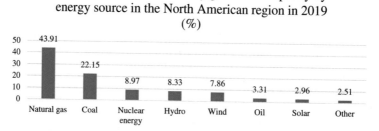

FIG. 1.38   Distribution of the electricity generation capacity by energy source in the North American region in 2019 (%). *Source: Statista 2019.*

The IAEA revised the projection of nuclear energy's contribution to the global energy balance for 2050. According to the IAEA projections, until 2050, the total electricity production using nuclear energy as an energy source will be between 2657.3 TWh produced in 2019 and 5762 TWh to be produced in 2050. Central and Eastern Asia will be the region with significant participation in nuclear energy in electricity production, between 39.5% and 41.5%, followed by Eastern Europe with between 15.7% and 13.3%. On the other hand, in the case of Latin America and the Caribbean, Africa, and Southeastern Asia, nuclear energy participation in electricity production will be very small.

Table 1.8 shows the latest IAEA estimates of nuclear electricity generating capacity per region up to 2050.

Table 1.8 shows that the greatest expansion of nuclear electricity generating capacity for the following 30 years is projected for Central and Eastern Asia. The increase will go from 565.8 TWh in 2019 to a minimum of 1158 TWh in 2050 or a maximum of 2393 TWh. On the other hand, Germany (2022), Switzerland (2034), and Belgium (2025) have concrete plans for

TABLE 1.8   IAEA's estimates of world nuclear electricity generating capacity and by region (TWh).

| Region | 2019 | 2030 | | 2040 | | 2050 | |
|---|---|---|---|---|---|---|---|
| | | Low | High | Low | High | Low | High |
| Northern America | 904.2 | 721 | 863 | 509 | 854 | 131.3 | 174.6 |
| Latin America and the Caribbean | 34 | 46 | 49 | 62 | 119 | 80 | 170 |
| Northern, Western, and Southern Europe | 723.8 | 601 | 724 | 408 | 718 | 348 | 562 |
| Eastern Europe | 358.2 | 428 | 504 | 434 | 711 | 459 | 764 |
| Africa | 13.6 | 14 | 32 | 27 | 83 | 52 | 119 |
| Western Asia | 2.0 | 59 | 71 | 85 | 146 | 111 | 190 |
| Southern Asia | 55.7 | 145 | 209 | 252 | 375 | 372 | 606 |
| Central and Eastern Asia | 565.8 | 859 | 1231 | 990 | 1904 | 1158 | 2393 |
| South-eastern Asia | – | – | – | 8 | 23 | 24 | 64 |
| Oceania | – | – | – | – | – | – | 16 |
| World Total | 2657.3 | 2872 | 3682 | 2774 | 4933 | 2929 | 5762 |

*Source: 2020. Energy, Electricity and Nuclear Power Estimates for the Period up to 2050. IAEA Reference Data Series No 1, 2020 Edition.*

shutting down all nuclear power reactors currently in operation. In France's case, the government has decided to close up to 17 nuclear power reactors now in service by 2025. Thus, although approximately 20 new countries are included in the 2050 projections, the global increase in the high forecast comes mainly from increases in the 31 countries already using nuclear energy for electricity production. The low forecast also includes approximately five new countries that might have their first nuclear power reactors in operation by 2050.

Several issues could affect the future implementation of nuclear power programs in a group of countries, including the USA and Canada. These issues are the following:

(1) Nuclear power has generated stronger political passions in comparison with any other energy alternatives source, with the exception, perhaps, of the use of coal for electricity generation;
(2) Due to the front-loaded cost structure of a nuclear power plant, high-interest rates, or uncertainty about interest rates, the financing of the construction of large nuclear power plants has become a limiting factor not only in the case of building a new nuclear power plant but for also the expansion in the use of this type of energy source for electricity generation in some countries that are already using nuclear energy for this purpose, such as Argentina, Mexico, Republic of Korea, UK, the USA, among others;
(3) Nuclear power's front-loaded cost structure also means that the cost of regulatory delays during construction is higher for nuclear power plants than for any other power plant. In countries where licensing processes were relatively untested in recent years, investors face potentially more costly regulatory risks with the construction of nuclear power plants than with the building of any other power plants;

**(4)** The nuclear industry is a global industry with good international cooperation, and hence the implications of an accident anywhere will have a strong negative impact on the nuclear industry worldwide;

**(5)** Nuclear terrorism may have a more far-reaching effect than comparable terrorism directed at other power plants;

**(6)** The nuclear power plant in itself is not a principal contributor to proliferation risk. However, proliferation worries can affect public and political acceptance of nuclear power;

**(7)** Among energy sources, high-level radioactive waste is unique to nuclear power. Therefore, the nuclear power industry might feel a disproportionately broad impact if significant problems are encountered in any of the repository programs that are most advanced in Finland, France, Sweden, and the USA (International Status and Prospects of Nuclear Power, 2008).

If the world meets even a fraction of the developing world's economic and social aspirations, then energy supplies must expand significantly. If the industrialized countries' increased economic and social development needs are considered, then the world energy supplies must expand even further (Morales Pedraza, 2012).

## Contribution of nuclear energy to the world and North America electricity generation

Table 1.9 includes information on the participation of nuclear energy in electricity generation by region in 2019.

Based on Table 1.9, the following can be stated: in 2019, the contribution of nuclear energy to world electricity generation varies considerably among regions. For instead, in Europe, nuclear-generated electricity accounts for 23.3% of the total electricity produced in that year. That is the highest contribution of nuclear energy to world electricity generation. In North America, it is 17.8% of the total, whereas in Africa and Latin America it is 1.9% and 1.6%, respectively. In Asia and the Pacific, nuclear energy accounts for 5.1% of world electricity generation. Based on the information included in Table 1.9, it is not difficult to conclude that the use of nuclear energy for electricity generation is concentrated in technologically advanced regions. For this reason, and at least during the coming years:

- Europe will remain the most significant nuclear energy user for electricity generation;
- North America was the second-largest user of nuclear energy for electricity production. It is expected that this situation will not change, at least during the coming years. However, there are no concrete plans for the expansion in the use of nuclear energy for electricity generation and heating in the region during the coming years, and this could change the position of nuclear energy within the energy mix in the region;
- With the most extensive regional population and fast-growing economies, Asia and the Pacific were, in 2019, the fourth-largest nuclear energy user for electricity production. However, it is important to single out that Asia and the Pacific is the region with the world most ambitious nuclear power plans to be implemented in the future, particularly

**TABLE 1.9** Level of participation of the different types of energies sources in electricity generation by region in 2019.

| Region | Fossil fuels | | Nuclear | | Renewables | | Total | |
|---|---|---|---|---|---|---|---|---|
| | Generation TWh | % | Generation TWh | % | Generation TWh | % | Generation TWh | % |
| North America, including Mexico | 3172 | 58.5 | 963.7 | 17.8 | 1253.8 | 23.1 | 5425.7 | 100 |
| Europe | 1518.5 | 38 | 928.5 | 23.3 | 1469.1 | 36.8 | 3993.3 | 100 |
| CIS | 965.8 | 67.5 | 211.2 | 14.8 | 251.7 | 17.6 | 1431 | |
| Asia and Pacific | 8998.5 | 70.9 | 647.3 | 5.1 | 2929.9 | 23.1 | 12,690.5 | 100 |
| Latin America | 405.6 | 30.5 | 24.6 | 1.9 | 898.8 | 67.6 | 1329.3 | 100 |
| Africa | 675.4 | 77.6 | 14.2 | 1.6 | 177.8 | 20.4 | 870.1 | 100 |
| Middle East | 1211.6 | 95.8 | 6.4 | 0.5 | 46.6 | 3.7 | 1264.7 | 100 |
| World total | 16,947.4 | 62.8 | 2796 | 10.4 | 7027.7 | 26 | 27,004.7 | 100 |

*Source: 2020. BP Statistical Review of World Energy, 69th ed. and Author own work.*

in China, and this could change the role of this type of energy source within the energy mix of the region;

- Latin America registered almost the same growth in its nuclear energy use for electricity generation since 1980. Its share in 2018 reached 1.6%, which is a little lower than the level registered during the past two decades;
- The Middle East is the region with the smallest nuclear share since 1980 and will continue in this position in the future. However, some countries in the area are developing a nuclear power program that could produce more electricity from nuclear power plants than in Africa in the coming years.

Fig. 1.39 and Table 1.10 include relevant information regarding the evolution of the energy capability factor during the period 2000–2019.

Based on the data included in Fig. 1.39 and Table 1.10, the following can be stated: the number of nuclear power reactors in operation during the period 2000–19 increased 7.4%, rising from 420 units in 2000 to 451 in 2019 (444 units in April 2021). However, the UCF (Utility Capability Factor) decreased by 5.7%, falling from 83.6% in 2000 to 77.9% in 2019. The aging of nuclear power reactors currently in operation at the world level is one of the main issues responsible for decreasing UCF.

In Fig. 1.40 and Table 1.11, the evolution of electricity supplied from nuclear power reactors connected to the electric grid during the period 2000–2019 is shown.

According to Fig. 1.40 and Table 1.11, the electricity supplied by nuclear power reactors connected to the electrical grid increased by 8.7% during 2000–19, rising from 2443.85 TWh in 2000 to 2657.16 TWh in 2019. Based on the US Energy Information Administration (AEO, 2020) report and PRIS—Power Reactor Information System (PRIS) (2021), the electricity generated by nuclear power plants in the USA, in 2019, reached 19.7%, but it is expected to decline to 12% in 2050; this means a decrease of 7.7%. Using the same report as reference, it can be stated that the electricity generated by renewables, in 2019, in the USA reached 19%. Therefore, it is expected that the participation of renewables in the US energy mix will increase up to 38% in 2050, double the electricity share in 2019.

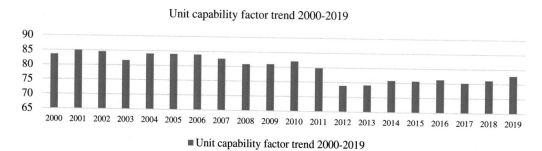

FIG. 1.39   Energy capability factor during the period 2000–19. *Source: April 2021. PRIS—Power Reactor Information System. www.iaea.org/pris, © IAEA and Author own work.*

TABLE 1.10  Energy capability factor during the period 1999–2019.

| Year | Number of commercially operated reactors with data | UCF [%] |
|------|------|------|
| 2000 | 420 | 83.6 |
| 2001 | 421 | 85.0 |
| 2002 | 429 | 84.6 |
| 2003 | 433 | 81.7 |
| 2004 | 435 | 84.0 |
| 2005 | 441 | 84.0 |
| 2006 | 442 | 83.9 |
| 2007 | 439 | 82.6 |
| 2008 | 439 | 80.8 |
| 2009 | 438 | 80.9 |
| 2010 | 441 | 82.0 |
| 2011 | 444 | 79.8 |
| 2012 | 436 | 73.9 |
| 2013 | 437 | 74.2 |
| 2014 | 435 | 75.8 |
| 2015 | 442 | 75.7 |
| 2016 | 447 | 76.3 |
| 2017 | 449 | 75.2 |
| 2018 | 454 | 76.2 |
| 2019 | 451 | 77.8 |

*Source: April 2021. PRIS—Power Reactor Information System. www.iaea.org/pris, © IAEA.*

# Limiting factors in the use of nuclear energy for electricity generation

Without a doubt, nuclear energy will be an important component of the energy balance in the Asian region in the coming years, followed by Europe, including Russia and North America. However, even in these regions, the use of this type of energy source will not play the same role due to some limiting factors. Among the limiting factors are, among others, the following:

**(a)** Management of the radioactive waste, particularly high–level nuclear waste;
**(b)** Proliferation security;
**(c)** Environment impact;
**(d)** Operational safety of nuclear power plants;
**(e)** Economic competitiveness;

1. General overview

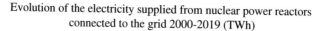

Evolution of the electricity supplied from nuclear power reactors
connected to the grid 2000-2019 (TWh)

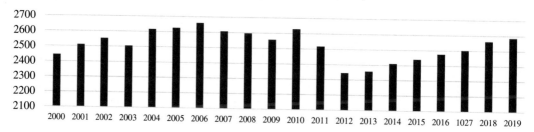

■Evolution of the electricity supplied from nuclear power reactors connected to the grid 2000-2019
(TWh)

**FIG. 1.40** Evolution of the electricity supplied from nuclear power reactors connected to the electrical grid during the period 2000–19. *Source: April 2021. PRIS—Power Reactor Information System. www.iaea.org/pris, © IAEA and Author own work.*

**(f)** High financial investment;
**(g)** Public acceptance;
**(h)** Human resources available, particularly high-qualified and with great experience, professionals, and technicians in the use of nuclear energy;
**(i)** Transport of uranium;
**(j)** Infrastructure available;
**(k)** Capacity of the electricity grids;
**(l)** Reactor technology available;
**(m)** Legal infrastructure:
**(n)** Public acceptance.

Some of the limiting factors mentioned above are briefly explained in the following paragraphs.

## Management of high-level radioactive waste

The management of radioactive waste, particularly high–level nuclear waste, is an unresolved problem for many countries' public opinion. The main irradiated fuel constituents discharged from LWRs are uranium, plutonium, actinides, and fission products. Uranium constitutes about 96% of the fuel unloaded from commercial nuclear power reactors. In light water reactors (LWRs), the spent fuel on discharge still contains 0.90% enriched in the fissile isotope U-235, whereas natural uranium contains only 0.7% of this isotope. Plutonium constitutes about 1% of discharged fuel weight; it is a fissile material that can be used as fuel in the present and future commercial nuclear power reactors. Minor actinides constitute about 0.1% of the weight of discharged fuel. They consist of about 50% neptunium, 47% americium, and 3% curium, which are very radiotoxic. Finally, fission products (iodine, technetium, neodymium, zirconium, molybdenum,

**TABLE 1.11** Evolution of the electricity supplied from nuclear power reactors connected to the electrical grid during the period 2000–19.

| Year | Number of operated reactors with data | Electricity supplied (TWh) |
|------|---------------------------------------|----------------------------|
| 2000 | 438 | 2443.85 |
| 2001 | 438 | 2511.09 |
| 2002 | 444 | 2553.18 |
| 2003 | 443 | 2504.78 |
| 2004 | 443 | 2616.24 |
| 2005 | 443 | 2626.34 |
| 2006 | 443 | 2660.85 |
| 2007 | 439 | 2608.18 |
| 2008 | 439 | 2597.81 |
| 2009 | 440 | 2558.06 |
| 2010 | 442 | 2629.82 |
| 2011 | 448 | 2517.98 |
| 2012 | 440 | 2346.19 |
| 2013 | 441 | 2358.86 |
| 2014 | 439 | 2410.37 |
| 2015 | 448 | 2441.34 |
| 2016 | 451 | 2477.30 |
| 2017 | 451 | 2502.82 |
| 2018 | 457 | 2562.76 |
| 2019 | 456 | 2657.16 |

*Source: April 2021. PRIS—Power Reactor Information System. www.iaea.org/pris, © IAEA.*

cerium, cesium, ruthenium, palladium, etc.) constitute about 2.9% of the weight of discharged fuel. They are considered the final waste form of nuclear power production at the present stage of knowledge and technological capacity unless a specific use is found for the non-radioactive platinum metals.

A typical 1000-MWe PWR unit operating at 75% load factor generates about 21 tons of spent fuel at a burn-up of 43 GWd/t; this contains about 20 tons of enriched uranium; 230 kg Pu; 23 kg minor actinides, and 750 kg fission products.

Without a doubt, the management of spent fuel should ensure that the biosphere is protected, and the public must be convinced of the effectiveness of the methods used for this specific purpose. Since the spent fuel contains very long-lived radionuclides, some protection is required for at least 100,000 years. There are two means to reach this goal. One of them is the following: society can wait for the natural decay of the radioactive elements by isolating them physically from the biosphere by installing successive barriers at a suitable depth in the ground. This strategy leads to deep geological disposal. The second one is the following: society can use nuclear reactions to transmute the very long-lived wastes into less radioactive or shorter-lived products. In several experts' opinions, deep geological repository disposal is the most appropriate solution today. Some countries are making progress on these sensitive issues, such as Finland, the USA, and Sweden. In the USA's case, President Trump recently approved the use of the Yucca Mountain site for the final disposal of high-level nuclear waste today temporary storage in all nuclear power plants in operation or closed in the country.

## Proliferation risk

According to Morales Pedraza (2012), the use of nuclear power for electricity generation entails potential proliferation risk, particularly for the possible misuse of specific nuclear technology, facilities, or materials used for this particular purpose. For this reason, the use of nuclear power for electricity generation should be expanded in the world only if the risk of proliferation from the operation of the commercial nuclear fuel cycle is zero. How to achieve this goal? First, the international community should strengthen the IAEA safeguards system's application to all states by forcing the adoption and implementation of the IAEA Additional Protocol. Second, the international community should adopt a multilateral approach to the nuclear fuel cycle. Third, the international community must take all necessary measures to prevent the acquisition of weapons-usable material, either by diversion (in the case of plutonium) or by misuse of nuclear fuel cycle facilities (including related facilities, such as research reactors or hot cells) now operating in different countries.

Adopting a multilateral approach to the nuclear fuel cycle should be done in a way that respects the right of any state to develop their nuclear fuel cycle but under full IAEA safeguards, including the Additional Protocol, without exception.

There are three issues of particular concern for the international community when any government considers the nuclear energy option as a real alternative to satisfy the foresee increase in electricity demand in the coming decades. These issues are, according to Morales Pedraza (2012), the following:

- The exists stocks of fissionable materials in the hands of several countries that are directly usable for the production of nuclear weapons;
- The number of nuclear facilities with inadequate physical protection and controls. Terrorist groups could use the lack of adequate physical protection of nuclear installations in several countries to have access to a certain amount of fissionable materials to use them for the production of a nuclear weapon;
- The transfer of sensitive nuclear technology, especially enrichment and reprocessing technology, to countries implementing a nuclear power program brings them closer to a nuclear weapons capability.

The proliferation risk due to the global growth in the use of nuclear energy for electricity production should be reduced to the minimum possible if the nuclear option is going to be considered by the international community as a realistic alternative for electricity production in several countries during the coming years (Morales Pedraza, 2012). The USA and Canada's experience in this critical issue for the international community should be used as a reference for other countries using nuclear energy for electricity generation or are going to start using this type of energy source for this specific purpose in the future.

## Environmental impact of the use of nuclear energy for electricity generation in the North American region

The public opinion of many countries has perceived nuclear power as very dangerous for the environment and the health of people, heightened by the three main accidents registered in the Three Mile Island (USA), Chernobyl (Ukraine), and Fukushima (Japan) nuclear power plants. Other accidents at fuel cycle facilities in the USA, Russia, and Japan in the past also have had a negative impact on the public opinion of several countries and increase their fear of the use of nuclear energy for electricity generation. As a result of these nuclear accidents, several countries have decided not to expand the use of nuclear energy for electricity generation; others have decided to phase out their nuclear power program; while others have prohibited the use of nuclear energy for electricity generation. For this reason, the design of new nuclear power reactors with stringent safety requirements is an indispensable condition to spread the use of nuclear energy for electricity generation during the coming years.

Nuclear energy produces very few emissions of $CO_2$ to the atmosphere. If the whole production cycle is taken into consideration (from the beginning of the construction of the nuclear power plant to their exploitation), then the production of one kWh of nuclear origin electricity supposes less than six grams of $CO_2$ emission to the atmosphere, mainly associated to the construction of the nuclear power plant and the transport of fuel. On the other hand, according to the type of technology used, a combined cycle gas power plant generates 430 g of $CO_2$ and a coal power plant between 800 g and 1050 g of $CO_2$. Based on these facts, the use of nuclear energy for electricity generation is one of the cleanest types of energy available in the world compared with any of the fossil fuel power plants currently in operation (Morales Pedraza, 2012).

## Economic competitiveness and financial investment

Nuclear power costs are competitive with other electricity generation forms, except where there is direct access to low-cost fossil fuels. Fuel costs for nuclear power plants are a minor proportion of total generating costs, though capital investment is greater than those for coal, natural gas, and oil-fired power plants. It is important to stress that even an increase in the uranium price does not make nuclear energy more expensive for electricity production compared with fossil fuel power plants. The reason is the following: nuclear power is hardly sensitive to fluctuations in the price of uranium, so that price shocks and market volatilities, as experienced recently, influence the generation price marginally.

In assessing the cost competitiveness of the use of nuclear energy for electricity generation, decommissioning and waste disposal costs should be taken into account. According to some expert's calculations, decommissioning costs are about 10–20% of a nuclear power plant's initial capital cost. Because of this type of cost, not included in the construction costs of any other electricity-generating power plants, the nuclear energy option could be more expensive than the use of any other energy source for electricity generation. However, if the social, health, and environmental costs of using fossil fuels fired power plants for electricity generation are also considered, then the use of nuclear energy for electricity generation is outstanding (Morales Pedraza, 2012).

The increased cost due to delay in constructing a nuclear power plant is another critical element that needs to be considered when the overall cost associated with the use of nuclear energy for electricity generation is analyzed. For all of the above reasons, nuclear power's future competitiveness will depend substantially on the additional costs, which may accrue to oil, coal, and gas generating power plants. However, it is uncertain how the real costs of meeting targets for reducing sulfur dioxide and greenhouse gas emissions will be attributed to fossil fuel power plants that will be included in the generation cost of conventional power plants (Morales Pedraza, 2012).

## Public acceptance of the use of nuclear energy for electricity generation

Public opinion is a crucial element that any government needs to consider when planning the use of nuclear energy for electricity generation. The future use of nuclear power for electricity generation in a particular country will depend on the public's perception of the importance of their purpose and the risks involved. The concern of the society arises because, although the probability of a nuclear accident is very low, the risks to have it can have significant consequences on the environment and on the health of the population that lives in areas not only close to the plant but in other areas far from the site as well. The Chernobyl nuclear accident is an example of an accident that affected not only the environment surrounded the plant and the health of the population located near the site, but the environment and the health of the population located far from the nuclear power plant site, including people living in other countries as well. This concern is not based on an objective perception but in intuitive judgment, due to the initial history on how nuclear energy was created in the past, the problems associated with the disposal of nuclear waste, the nuclear accidents that occurred in some nuclear facilities, and the optimistic initial expectations related to the use of nuclear energy for electricity generation, among others (Morales Pedraza, 2012).

Different studies carried out so far on the use of nuclear energy for electricity generation in several countries indicate that there is a lack of information in the community on this vital subject, as well as about the price of the different energy options available for electricity production and their impact on the environment and population. As a result, the different nuclear actors, particularly the scientific community, technologists, operators, regulators, political leaders, and legislators, have not been able, in some countries, to convince the public opinion on which of the energy options available for electricity generation are the most convenient for the country. For this reason, it is extremely important to make additional efforts to provide reliable and impartial information to the public about the use of the different types of energy sources available in the country for electricity generation. The purpose of these efforts is to assist public opinion in a particular country to define their position regarding using the different energy options available and identify the most appropriate for the country (Morales Pedraza, 2012).

The increase in the construction costs of a nuclear power plant and the unforeseen delays in the building of this type of power plant due to new safety and security requirements demanded to the operators after the nuclear accidents of the Three Miles Island, Chernobyl, and Fukushima Daiichi has stopped the construction of new nuclear power plants in the USA, Canada, Spain, Sweden, Japan, Germany, among others. As a result, many countries' public opinion is against the use of nuclear energy for electricity generation and, as a consequence of this position, politicians are also rejecting any support in developing the nuclear power sector in the USA, Canada, Germany, and in several countries within the EU, among others. In some of these countries, the government adopted a nuclear phase-out policy to shut down all nuclear power reactors in operation and stop any future construction of new units such as Germany, Switzerland, and Belgium. However, and despite several nuclear accidents of minor consequences that occurred in other countries, the opposition of the public opinion to the use of nuclear energy for electricity production in Asia, particularly in India, the Republic of Korea, China as well as in Russia, are not so strong to stop the ongoing programs for the construction of new nuclear power reactors.

It is important to single out that people are ready to support the nuclear energy option in many countries if and only if problems related to the final disposal of high-level nuclear waste and the safety of the nuclear power reactors are adequately solved. In the USA's particular case, people are unlikely to support nuclear power expansion without substantial improvements in costs and technology.

The carbon-free character of nuclear power is one of the main arguments used by those who support nuclear energy use for electricity generation and heating, particularly in Europe. However, in the USA's case, nuclear power's cost-free character does not appear to be one of the main elements that motivate the public opinion to prefer the nuclear option for electricity generation, compared with other energy sources available in the country.

In some developed countries, there is an impression that nuclear power is especially unpopular among available energy sources. In these countries, it is further assumed that nuclear power will not flourish unless this unpopularity can be overcome, even if using this type of energy for electricity generation has solid grounds to support it. The question to be answered is why public opinion in several countries is against the use of nuclear energy for electricity generation? Some of the factors behind the loss of public confidence in nuclear energy use for electricity generation in some developed countries were caused directly by the industry itself. The construction times and costs of many nuclear power plants were far higher than projected. The performance of many nuclear power plants was disappointing. The nuclear accidents at Three Mile Island, Chernobyl, and Fukushima nuclear power plants also exacerbated growing mistrust of the nuclear industry and its often vocal supporters within governments. This mistrust had its origin, at least partly, in the secretiveness of nuclear spokespeople within the nuclear industry in many countries during the occurrence of a nuclear accident and the lack of clarity and transparency in the information provided to the public. The suspicion that the nuclear industry and its supporters were able, for example, to put undue pressure on regulators further damaged their public credibility. Critics of the nuclear industry often had no apparent vested interest to do so, while the industry's responses increasingly came to be discounted – 'they would say that, would not they?' The passion which has surrounded the nuclear debate in recent years is, to a considerable degree, a legacy of these factors (Grimstom and Beck, 2002).

At the same time, perceptions of the availability of renewable energy sources for electricity generation were changing. When global fossil fuel supplies were under apparent threat (notably in the 1950s and again in the 1970s until the beginning of the 1980s), nuclear power programs were introduced in many countries with relatively little objections, at least by today's standards. The discovery of vast reserves of natural gas and oil, coupled with low prices and the development of the highly efficient combined cycle gas turbine by the mid-1980s, reduced the apparent need for nuclear power in many developed countries. However, some developing countries, notably India and China, did not share this perception, and, for this reason, both countries continue with the development of crucial nuclear power programs (Morales Pedraza, 2012).

Many developed and developing countries have anti-nuclear movements, some relatively small and without significant influence on policy decision-makers, while others have a high impact at the policy level. Environmental pressure groups are already consolidated within the industrialized world and increasingly establish themselves in the developing world.

Several specific explanations have been suggested for the apparent special unease felt about the use of nuclear power for electricity generation in many countries. They include:

- Links to the military, both real (the development of shared facilities for civil and military uses) and perceptual;
- Secrecy, coupled sometimes with an apparent unwillingness to give 'straight answers' (in part, perhaps, because of links to military nuclear operations in some countries, and in part because of commercial issues);
- The historical arrogance of many in the nuclear industry, dismissing opposition, however well-founded or sincerely held, as 'irrational';
- The apparent vested interest of many nuclear advocates, to be contrasted with the apparent altruism of opponents who, for example, are often not funded to take part in public inquiries;
- The perceived potential for large and uncontainable accidents and other environmental and health effects, notably those associated with the management of radioactive waste;
- The oversells of nuclear technology, especially in its early days, in particular concerning its economics, leading to a degree of disillusionment and distrust;
- General disillusionment with science and technology and with the 'experts know best' attitude of mind that was more prevalent in the years immediately after the Second World War;
- The broader decline of 'deference' toward 'authority,' including, for example, politicians and regulatory bodies (Grimstom and Beck, 2002).

Other explanations to the public opposition to the introduction of a nuclear power program or to expand the use of nuclear energy for peaceful purposes are the following:

- Radioactivity cannot be smelled, felt, or seen; this means has no color and is not cold or warm;
- Radioactivity can only be detected using special equipment, available only in special nuclear facilities and institutions.

Despite concerns against the use of nuclear energy for electricity production coming from the anti-nuclear establishment, as well as civil society organizations in several countries, a consensus is emerging that current and future energy needs in several countries can be satisfied without increasing the emission of $CO_2$ to the atmosphere if the use of nuclear power is included in the energy balance of these countries. The use of nuclear energy for electricity generation should be one of the types of energy to be included in any energy plan and energy balance in a group of countries, including developed and developing countries alike. However, to consider the nuclear option as a real alternative, it is crucial to overcome the following four challenges:

- Capital costs;
- Operational safety;
- Proliferation risk;
- Nuclear wastes.

These challenges will escalate if many new nuclear power plants are built in a growing number of countries. However, the effort to overcome these challenges is justified only if nuclear power can significantly reduce global warming and supply electricity systematically and economically.

## Pros and cons associated with the use of nuclear energy for electricity generation in the North American region

Since the beginning of nuclear energy use for electricity generation, two different groups of positions have been identified. One group favored using nuclear power for electricity generation, while the other was strongly against it. Those who supported nuclear energy for electricity generation in the past are quite different from those who promote it today. At the beginning of the 1950s, nuclear energy for electricity production was considered competitive, safe, and clean regarding fossil fuel use for the same purpose. In 2010, this position was forgotten, and the defense of the nuclear option orbits around the urgent need to stop greenhouse emissions and climate change (Fernández Vázquez and Pardo Guerra, 2005).

### The pros

There are several strong arguments in favor of the use of nuclear energy for electricity generation. These arguments are, according to Morales Pedraza (2012), the following:

- It brings technological development in advanced areas from the technical point of view in comparison with any other form of energy;
- It is a proven technology that can satisfy large-scale energy demands during the coming years;
- It provides a continuous supply of energy. Other available technologies such as hydro, solar, and wind energy depend on weather conditions, which are very difficult to predict;
- There are no supply problems, at least in the medium and long terms, regarding nuclear fuel. Global stocks of uranium are more than enough to satisfy any future increase in the world energy demand. But, what is very important, proven reserves of uranium are not located in politically sensitive regions of the world;
- At this moment, the international cost of nuclear fuel is acceptable. It can be afforded by those countries with nuclear power programs and those thinking to introduce this type of program in the future.

However, it is important to be aware of the following: the opponents of the nuclear option forget that a series of nuclear energy characteristics requires being highly cautious when dealing with atoms. These are the following:

- First, it generates dangerous waste that is difficult to isolate, cannot be reprocessed by nature's cycles, and lasts for several thousand years, posing a tremendous threat to the environment and human health if not well handled. Even though the conventional sources and methods used in electricity production generate residues that have to be managed, none of them pose as many risks as nuclear energy, nor do they require such a long-term management program.
- Second, nuclear energy is not entirely and absolutely secure, as demonstrated by the accidents in Three Mile Island, Chernobyl, and Fukushima nuclear power plants. Though no energy source is inherently safe, an oil spill is not the same as a radioactive spill.
- Third, the construction, dismantling, and decommissioning of nuclear facilities are extremely expensive. Finally, and above all, nuclear energy is intrinsically linked to the shadow of nuclear proliferation, which humanity has sought to eliminate without success since the 1950s (Fernández Vázquez and Pardo Guerra, 2005).

The international community should be aware that by the first decades of the 21st century, all forms of primary energy for electricity production will be needed if sustainable development is to be achieved. In this context, we have the moral obligation to utilize those energy resources that lead to the lowest possible negative environmental impact (Current Issues in Nuclear Energy: Nuclear Power and the Environment, 2002).

One of the available energy sources that do not emit greenhouse gas (carbon dioxide, methane, nitrous oxide, among others), or any gas-causing acid rain or photochemical air pollution (sulfur dioxide and nitrogen oxides), when it is used for electricity generation, is nuclear energy. When used for electricity generation, this type of energy does not also emit to the atmosphere any carcinogenic, teratogenic, or mutagenic metals (As, Hg, Pb, Cd, etc.). The utilization of nuclear energy does not release gases or particles that cause urban smog or deplete the ozone layer. At the same time, nuclear power is the only energy technology that treats, manages, and contains its waste completely and is segregated from the public and the environment. It does not require large areas for resettling huge populations because it is a highly concentrated energy source. Hence, its environmental impact on land, forests, and waters is minimal. The use of nuclear energy for electricity production avoids some 10% of additional $CO_2$ emissions to the atmosphere, considering all economical sectors and about one-third in the power sector. However, it is important to stress that nuclear power alone cannot solve the environmental load created by the emissions of greenhouse gases, but without the use of nuclear power, no other solution for this crucial problem exists within a reasonable time and the state of the art of energy generation technologies (Current Issues in Nuclear Energy: Nuclear Power and the Environment, 2002).

## The cons

The following are, among others, the main cons of the use of nuclear energy for electricity generation:

- The negative impact and the long-term consequences of a nuclear accident[af];
- High initial capital investment;
- Long construction time;
- Delay in the construction of large nuclear power plants;
- Strict safety regulations;
- Need of high-qualified and trained human resources with experience in the operation of this type of power plant;
- The need to consider dismantling and decommissioning costs;
- Management of the spent nuclear fuel and its costs;
- Proliferation risk (Morales Pedraza, 2012).

The use of nuclear energy source for electricity generation requires being highly cautious when dealing with this energy option because it is not an accident-free industry and, in case of a severe accident, the consequences for the environment and the population could be not only very high but could have a long-term negative effect. Although it can be considered a very rare event, the Chernobyl nuclear accident is an example of the negative consequences for the environment and the population with a long-term negative effect of a nuclear accident.

## Looking forward in the use of nuclear energy for electricity generation in the North American region

According to BP Statistical Review of World Energy (2019), the energy mix in the North American region (including Mexico) was structured, in 2018, in the following manner:

- Fossil fuels: 59.4% of the total for the region;
- Renewables: 22.9% of the total;
- Nuclear energy: 17.7% of the total.

It is expected that, in 2025, the energy mix in the North American region (including Mexico) will be structured in the following manner:

- Fossil fuels: 55% of the total, a decrease of 4.4% with respect to 2018;
- Renewable: 29% of the total, an increase of 6.1% with respect to 2018;
- Nuclear energy: 16% of the total, a decrease of 1.7% with respect to 2018 (see Fig. 1.41).

[af] Nuclear power has one of the lowest levels of fatalities per unit of energy generated compared to other energy sources. Coal, oil, natural gas, and hydroelectricity each have caused more fatalities per unit of energy due to air pollution and accidents (Markandya and Wilkinson, 2007). Since its commercialization in the 1970s, nuclear power has prevented about 1.84 million air pollution-related deaths and the emission of about 64 billion tons of carbon dioxide equivalent that would have otherwise resulted from the burning of fossil fuels (Kharecha and Hansen, 2013). Accidents in nuclear power plants include the Chernobyl disaster in the Soviet Union in 1986, the Fukushima Daiichi nuclear disaster in Japan in 2011, and the more contained Three Mile Island accident in the United States in 1979. There have also been some nuclear submarine accidents.

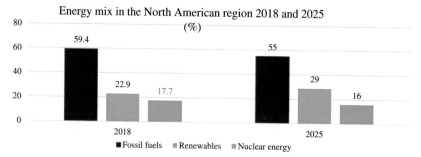

**FIG. 1.41**    Energy mix in the North American region 2018 and 2025. *Source: 2019. BP Statistical Review of World Energy, 68th ed. and Author own work.*

Based on the abovementioned data, the following can be stated: North America's share of energy generation from renewable and nuclear energy sources will grow from 40.6% in 2018 to 45% in 2025; this means an increase of 4.4%. During the period considered, fossil fuels and nuclear energy participation in the region's energy mix will be lower in 2025 than in 2018. Renewables are the only one that is expected to increase in the same period. However, it is important to stress that this projection assumes the Clean Power Plan (CPP) is upheld and will take effect in the USA before 2022. A recent agreement among Canada, Mexico, and the USA established a goal of 50% of electricity generation from clean energy sources by 2025, something that seems somehow difficult to achieve.

According to Martin (2016), electricity generation in the USA currently represents more than 80% of the North American region's electricity generation (in 2018 was 81.9% and in 2019 reached 86.5%). It is also projected that the extension of certain tax credits, significant cost reductions, and recognition of future CPP or similar requirements will substantially increase in the use of renewable energy sources for electricity generation between 2018 and 2025. On the other hand, the US coal-fired generation is expected to decline by 13% between 2015 and 2025, while natural gas-fired generation is projected to increases by 4%.

In Canada's case, power generation already reached, in 2018, a total of 66.2% using renewable energy, mainly because of Canada's extensive hydroelectric capacity installed in that year (81 GW). Despite its already huge hydroelectricity capacity installed, Canada plans to further increase this capacity by 2025, in addition to increasing wind and solar capacity by that year. US EIA projects an important reduction in coal use for electricity generation in the country between 2015 and 2025, consistent with the Canadian government's plans to gradually phase out existing coal-fired power plants in this decade. However, it is essential to single out that the combined share of renewables and nuclear energy in Canada's total electricity generation is expected to fall to 75% by 2025 because of increases in the use of natural gas for this specific purpose and the closure of nuclear power reactors currently in operation. Overall, Canada's electricity generation represented, in 2018, about 18.1% of the North American total electricity generation.

According to World Energy Outlook 2018, the share of electricity generation from nuclear power worldwide, the second-largest source of low-carbon electricity today after hydropower, "stays at around 10%, but the geography changes as the generation in China overtakes the USA and the EU before 2030." Some 66.7% of today's nuclear fleet in advanced

Net growth in world nuclear generation during the period 2016-40

FIG. 1.42   Net growth in world nuclear generation during the period 2016–40. *Source: 2017. World Energy Outlook 2017. International Energy Agency and Author own work.*

economies has more than 30 years old. Decisions to extend, or shut down the oldest nuclear power plants, will have significant implications for energy security, investment and, $CO_2$ emissions.

The World Energy Outlook New Policies Scenario expects about US$1.1 trillion of investment in nuclear power by 2040, leading to an increase in nuclear power production of around 46%. While significant, the nuclear share of power generation declines to 10%, and the increased output is less than half of what is assumed in the Sustainable Development Scenario. Furthermore, the net growth in world nuclear generation during the period 2016–40 is quite concentrated, with 91.5% of the net production increase accounted for by two countries: China with 889 TWh or 72.1% of the total, and India with 239 TWh or 19.4% of the total[ag] (see Fig. 1.42). The USA and Canada have no concrete plans to expand their nuclear power programs in the coming years.

Despite what has been said about the use of nuclear energy for electricity generation, this type of energy source is under serious consideration in several countries without a nuclear power program. Thirty-one states are already using nuclear energy for electricity generation. Fifty-nine countries not currently using this type of energy source are considering introducing a nuclear energy program in the coming years. These countries are, according to the World Nuclear Association (WNA) (2019), the following:

(a) **In Europe**: Albania, Serbia, Croatia, Portugal, Norway, Poland, Belarus Estonia, Latvia, Lithuania, Ireland, and Turkey;
(b) **In the Middle East and North Africa**: Gulf States including UAE, Saudi Arabia, Qatar, and Kuwait, Yemen, Israel, Syria, Jordan, Egypt, Tunisia, Libya, Algeria, Morocco, and Sudan;
(c) **In West, Central and Southern Africa**: Nigeria, Ghana, Senegal, Kenya, Uganda, Tanzania, Zambia, Rwanda, Ethiopia, and Namibia;
(d) **In Central and South America**: Chile, Ecuador, Bolivia, Peru, and Paraguay;
(e) **In Central and Southern Asia**: Azerbaijan, Georgia, Kazakhstan, Mongolia, Sri Lanka, Uzbekistan, and Bangladesh;

[ag]The net growth in the rest of the world is 105 TWh (8.5% of the total).

**(f) In South-Eastern Asia and Oceania**: Indonesia, Vietnam, Thailand, Laos, Cambodia, Singapore, Myanmar, Malaysia, and Australia;
**(g) In East Asia:** North Korea.

Besides, it is essential to single out the following: The institutional arrangements to be created to support the introduction of a nuclear power program varies from country to country. Usually, governments in the industrialized world are heavily involved in planning the introduction of different energy types in the energy mix, including nuclear energy, and, in some cases, financing the construction of nuclear power plants. On the other hand, in developing countries, governments are generally involved in all phases of introducing nuclear power programs from planning until the nuclear power plant's operation, including financing this type of program.

In countries where nuclear energy for electricity generation is going to be introduced for the first time, there is always a lack of nuclear engineers and other scientists, professionals, and technicians duly prepared and trained. For this reason, how the construction of nuclear power plants is carried out is often on a turnkey basis. In this case, the nuclear power reactors' supplier assumes all technical and commercial risks in delivering a functioning plant on time and within the budget approved, or as an alternative, set up a consortium to build, own, and operate the plant.

## References

Anon., 2018. 2018 IHS Markit Wind O&M Benchmarking in North America: Aging Turbines, Rising Costs—A Study of O&M Costs for North American Wind Power Plants. HIS Markit.

Admin, 2019. Pros and Cons of Renewable Energy. https://www.prosancons.com/energy/pros-and-cons-of-renewable-energy/.

AEO, 2020. Annual Energy Outlook 2020 With Projections to 2050. U.S. Energy Information Administration.

Annual Energy Outlook 2013, 2013. Levelized Cost of New Generation Resources. US EIA.

Anuta, H., Pablo, R., Michael, T., 2018. Renewable Power Generation Costs in 2018. International Renewable Energy Agency (IRENA), Abu Dhabi.

Anon., 2019. Bioenergy. IRENA.

Blodgett, L., 2014. Profiling Geothermal Energy in North America. Issue 9, vol. 118 Geothermal Energy Association.

BNEF, 2019. Clean Energy Investment Trends 2018. Bloomberg New Energy Finance (BNEF). https://www.carbon-brief.org/analysis-fossil-fuel-emissions-in-2018-increasing-at-fastest-rate-forseven-years.

Bord, R.J., Fisher, A., O'Connor, R.E., 1998. Public perceptions of global warming: United States and international perspectives. Clim. Res. 11 (1), 75–84.

Anon., 2019. BP Energy Outlook 2019. 2019 edition.

Anon., 2019. BP Statistical Review of World Energy, sixty-eighth ed.

Anon., 2020. BP Statistical Review of World Energy, sixty-ninth ed.

Anon., 2016. Canadian Hydropower Facts. Canadian Hydropower Association.

CanWEA, 2016. Canada Wind Energy Story. Retrieved January 27, 2017, from https://canwea.ca/wp-content/uploads/2017/01/canadas-wind-energy-story-2016-e-v3.pdf.

Cillizza, C., 2018. Donald Trump Buried a Climate Change Report Because 'I Don't Believe It'. CNN. Archived from the original on December 30, 2018; Retrieved December 30, 2018.

Anon., 2016. Clean Power Plan for Existing Power Plants. EPA. Archived from the original on March 25, 2016. Retrieved August 14, 2018.

CleanTechnica, 2018. Saudi Arabia's 1st Wind Farm Receives Strikingly Low Bid Prices. CleanTechnica. https://cleantechnica.com/2018/07/27/saudi-arabia-1st-wind-farm-receives strikingly-low-bid-prices/. (Accessed 7 March 2019).

Anon., 2018. Coal: America's Power. National Mining Association Fact Sheet.

Corscadden, K., Wile, A., Yiridoe, E., 2012. Social license and consultation criteria for community wind projects. Renew. Energy 44, 392–397. https://doi.org/10.1016/j.renene.2012.02.009.

Anon., 2019. Cumulative Capacity of Bioenergy in North America in 2015 and 2021, by Country 2019. Statista.

Anon., 2002. Current Issues in Nuclear Energy: Nuclear Power and the Environment. International Nuclear Societies Council (INSC); American Nuclear Society (ANS), Illinois. January 2002.

DiChristopher, T., 2018. Trump Administration Moves to Keep Failing Coal and Nuclear Plants Open, Citing National Security. CNBC.

Edvard, 2011. Renewables and the Impact on the Environment. Electrical Engineering Portal.

EIA, 2019a. Short-Term Energy Outlook (STEO). US Energy Information Administration (EIA), USA.

EIA, 2019b. International Energy Outlook 2019 (IEO 2019). US Energy Information Administration (EIA), USA.

EIA, 2019c. Natural Gas Imports and Exports. Energy Information Administration (EIA).

Anon., 2018. EIA Database.

Anon., 2019. EIA Energy From the Sun. US Energy Information Administration (EIA).

Anon., 2019. EIA Photovoltaics and Electricity. US Energy Information Administration (EIA).

Anon., 2019. EIA Solar Energy and the Environment. US Energy Information Administration (EIA).

Electricity Facts, 2019. Natural Resources Canada. 2019-08-09. Retrieved 2019-10-14.

Anon., 2019. Energy Fact Book 2019–2020 Canada. Natural Resources Canada.

Energy Policy-Natural Resources Canada, 2014. Energy Policy. Natural Resources Canada; Government of Canada.

F.T., 2019. Shell Aims to Become World's Largest Electricity Company. Financial Times (F.T.).

Federal Energy Regulatory Commission, 2005. Remedying Undue Discrimination Through Open Access Transmission Service and Standard Electricity Market Design—Order Terminating Proceeding (Docket No. RM0112-000). Federal Energy Regulatory Commission.

Feldman, D., Margolis, R., 2019. Solar Industry Update. National Renewable Energy Laboratory.

Fernández Vázquez, E., Pardo Guerra, J.P., 2005. Latin America Rethinks Nuclear Energy. IRC, Americas. September 13, 2005.

Flórez, P., Arturo, C., 2007. National Energy Plan. Context and Strategy 2006–2025. Minister of Mines and Energy, Colombia.

Anon., 2019. Global Energy Statistical Yearbook 2018. Enerdata.

Anon., 2020. Global Energy Statistical Yearbook 2019. Enerdata.

Global Energy Transformation: A Roadmap to 2050, 2019. Key Findings. World Energy Council; Turkish National Committee.

Grimstom, M.C., Beck, P., 2002. Double or Quits?: The Future of Civil Nuclear Energy.

GTM, 2019. Trends Shaping the Global Solar Market in 2019. Green Tech Media. https://www.greentechmedia.com/articles/read/trends-shaping-the-global-solar-market-in2019#gs.BMccMo5Q. (Accessed 7 March 2019).

GWEC, 2019. 51,3 GW of Global Wind Capacity Installed in 2018. Global Wind Energy Council (GWEC).

Hagen, B., Pijawka, D., 2015. Public perceptions and support of renewable energy in North America in the context of global climate change. Int. J. Disaster Risk Sci. https://doi.org/10.1007/s13753-015-0068-z. Springer.

Hamilton, M.S., 2013. Energy Policy Analysis: A Conceptual Framework. M.E. Sharpe, Inc., Armonk, NY.

Hansler, J., 2018. U.S., Syria Are Now Only Ones not in Climate Deal. CNN. Archived from the original on June 16, 2018. Retrieved May 4, 2018.

Hennick, M., 2018. Renewable Energy and the Environment. HahaSmart.solar.

IEA, 2018a. Key World Energy Statistics 2018. International Energy Agency (IEA), Paris.

IEA, 2018b. Coal 2018; Analysis and Forecasts to 2023. International Energy Agency (IEA), Paris.

Intergovernmental Panel on Climate Change, 2007. Natural Forcing of the Climate System. Archived from the original on 29 September 2007. Retrieved 29 September 2007.

Intergovernmental Panel on Climate Change, 2011. Special Report on Renewable Energy Sources and Climate Change Mitigation. Cambridge University Press, Cambridge; and New York, NY.

Intergovernmental Panel on Climate Change (IPCC), 2013. Climate Change 2013: The Physical Science Basis; Contribution of Working Group I to the Fifth Assessment Report of the Intergovernmental Panel on Climate Change. Cambridge University Press, Cambridge.

Anon., 1957. International Atomic Energy Agency Statute.

Anon., 2008. International Status and Prospects of Nuclear Power. Report by the Director-General, GOV/INF/2008/10-GC (52)/INF/6, August 12, 2008.

IRENA, 2018. Renewable Power Generation Costs in 2017. International Renewable Energy Agency (IRENA), Abu Dhabi.

IRENA, 2019a. Renewable Power Generation Costs in 2018. International Renewable Energy Agency, Abu Dhabi.

IRENA, 2019b. Global Energy Transformation: A Roadmap to 2050 (2019 edition). International Renewable Energy Agency, Abu Dhabi.

Joshi, P., 2018. The Real Pros and Cons of Renewable Energy Sources. Help Save Nature.

Kharecha, P.A., Hansen, J.E., 2013. Prevented mortality and greenhouse gas emissions from historical and projected nuclear power. Environ. Sci. Technol. 47 (9), 4889–4895. Bibcode:2013EnST...47.4889K https://doi.org/10.1021/es3051197. 23495839.

Lantz, E., Flowers, L., 2011. Social acceptance of wind energy projects: country report of United States. In: IEA Wind Task 28.

Laverty, G., 2015. Canada's Provincial Leaders Reach Agreement on Energy Strategy. SNL Energy Power Daily, pp. 1–3.

Markandya, A., Wilkinson, P., 2007. Electricity generation and health. Lancet 370 (9591), 979–990. https://doi.org/10.1016/S0140-6736(07)61253-7. 17876910.

Market Line 2018, 2018. Renewable Energy in North America.

Martin, L., 2016. Renewables Share of North America Electricity Mix Expected to Rise. US EIA.

Morales Pedraza, J., 2012. Nuclear Power: Current and Future Role in the World Electricity Generation. Nova Science Publishers, ISBN: 978-1-61728-504-2.

Morales Pedraza, J., 2015. The current status and perspectives for the use of small modular reactors for electricity generation. In: Advances in Energy Research. vol. 21. Nova Science Publishers (Chapter 1).

Morales Pedraza, J., 2017. Small Modular Reactors for Electricity Generation: An Economic and Technologically Sound Alternative. 2017 edition, Springer International Publishing, https://doi.org/10.1007/978-3-319-52216-6. ISBN: Print: 978-3-319-52215-9, Online: 978-3-319-52216-6, January 2017.

Motyka, M., 2020. 2020 Renewable Energy Industry Outlook. Deloitte.

North American Center for Transborder Studies (NACTS), 2011. Cooperation on Transborder Renewable Energy Development and Exchange. Arizona State University, Tempe, AZ.

Oestmann, J., 2017. American-Made Ethanol Should Be Part of Trump's Energy Policy. Morning Consult.

Phadke, R., 2010. Steel forests or smokestacks: the politics of visualization in the Cape Wind controversy. Environ. Polit. 19 (1), 1–20.

Philibert, C., 2005. The Present and Future Use of Solar Thermal Energy as a Primary Source of Energy. IEA. Archived (PDF) from the original on 12 December 2011.

Pineau, P.-O., 2013. Fragmented markets; Canadian electricity sectors' underperformance. In: Evolution of Global Markets: New Paradigms, New Challenges, New Approaches. Academic Press, pp. 363–392.

Planete Energies, 2014. The Two Types of Solar Energy.

Anon., April 2021. PRIS—Power Reactor Information System (PRIS). International Atomic Energy Agency (IAEA).

Rand, J., Hoen, B., 2017. Thirty years of North American wind energy acceptance research: what have we learned? Energy Res. Soc. Sci. https://doi.org/10.1016/j.erss.2017.05.019.

Rapier, R., 2019. Ethanol Industry in Free Fall Since President Trump's Inauguration. forbes.com.

Anon., 2019. Renewable Energy and Jobs Annual Review 2019. IRENA.

Anon., 2019. Renewable Energy Facts. Natural Resources, Canada. 2019-10-02. Retrieved 2019-10-14.

Anon., 2019. Renewable Energy Statistics 2019. IRENA.

Renewables in North America, 2019. The New Energy Era.

Resources for the Future, 2016. Harmonizing the Electricity Sectors Across North America: Recommendations and Action Items From Two RFF/US Department of Energy Workshops; Washington, D.C.

Rinkesh, 2019. Renewable Energy Pros and Cons. Conservative Energy Future.

Robles, J., 2017. La Política Energética de EEUU es Continuista, pese a Trump. Cambio 16.

Rogers, K., Davenport, C., 2019. Trump Saw Opportunity in Speech on Environment. Critics Saw a "1984" Moment. Archived from the original on July 10, 2019; Retrieved July 10, 2019—via NYTimes.com.

Russo, T., 2017. The Trump Effects on U.S. Hydropower. Russo on Energy.

Schneider, M., Froggatt, A., 2007. The World Nuclear Industry Status Report 2007. Greens-EFA Group in the European Parliament, Brussels. November 2007.

Schneider, M., Froggatt, A., 2018. The World Nuclear Industry Status Report 2018.

Schneider, M., Froggatt, A., 2019. The World Nuclear Industry Status Report 2019.

Schroeder, R., 2018. Trump Administration to Prop Up Coal, Nuclear Power Facilities. Market Watch.

Siegel, J., 2018. Forgotten Hydropower Plots a Comeback in Trump Era. Washington Examiner.

Anon., 2011. Solar Energy Perspectives: Executive Summary. International Energy Agency. Archived from the original (PDF) on 3 December 2011.

Somerville, R., 2007. Historical Overview of Climate Change Science; Intergovernmental Panel on Climate Change. Retrieved 29 September 2007.

Sovacool, B.K., 2009. Rejecting renewables: the socio-technical impediments to renewable electricity in the United States. Energy Policy 37 (11), 4500–4513. https://doi.org/10.1016/j.enpol.2009.05.073.

Tanti, T., 2018. The key trends that will shape renewable energy in 2018 and beyond. In: World Economic Forum Annual Meeting.

Taylor, J., 2016. Trump's Energy Policy: 10 Big Changes. Forbes.

Anon., 2008. The National Repository of Yucca Mountain. U.S. Department of Energy.

Union of Concerned Scientists, 2013. Environmental Impacts of Renewable Energy Technologies.

US DoE, 2015. Wind Vision: A New Era for Wind Power in the United States. U.S. Department of Energy, Washington, DC.

Vidangos, N., Griffith, L., Flores-Espino, F., McCall, J., 2016. Electricity in North America. U.S. Department of Energy; Office of Energy Policy and Systems Analysis.

Anon., 2019. Wikipedia Energy Policy.

Anon., 2019. Wikipedia Energy Policy of Canada 2019.

Anon., 2020. Wikipedia Photovoltaic System.

Anon., 2019. Wikipedia Solar Energy 2019.

Wiser, R., Bolinger, M., 2016. 2015 Wind Technologies Market Report. Lawrence Berkeley National Laboratory, Berkeley, CA.

Wiser, R., Jenni, K., Seel, J., Baker, E., Hand, M., Lantz, E., Smith, A., 2016. Expert elicitation survey on future wind energy costs. Nature Energy 1, 16135.

Anon., 2019. World Energy Issues Monitor 2019. World Energy Council.

Anon., 2018. World Energy Outlook 2018. International Energy Agency. Executive Summary.

World Hydropower Congress, 2019. Regional Focus: North America.

World Nuclear Association (WNA), 2019. Renewable Energy and Electricity.

Wüstenhagen, R., Wolsink, M., Burer, M.J., 2007. Social acceptance of renewable Energy innovation: an introduction to the concept. Energy Policy 35, 2683–2691.

# The use of hydropower for electricity generation

## Introduction

The energy contained in sunlight vaporizes water from the oceans, lakes, and rivers and later deposits it in rain or snow on the land. Hydroelectric power plants convert the kinetic energy contained in falling water into electricity. The energy contained in flowing water is ultimately derived from the Sun and, for this reason, is continually being renewed (see Fig. 2.1). For this reason, hydropower is the most commonly used renewables energy source for electricity generation all over the world, particularly in Canada, within the North American region. The participation of hydropower plants in Canada's energy mix in 2019 reached 62%. In the USA, it reached only 6.6%.

People have a long history of using the force of water flowing in streams and rivers to produce mechanical energy. "Thousands of years ago, people used hydropower to turn paddle wheels on rivers to grind grain. Before steam power and electricity were available in the USA and other countries, grain and lumber mills were powered directly with hydropower" (EIA Hydropower Explained, 2020).

Without a doubt, hydropower is a mostly low-tech, usually familiar renewable energy resource that generally makes waves only of the watery sort. The giant hydropower projects of the 20th century have become part of the North American region's landscape. Nevertheless, relatively few more giant hydropower plants are likely ever to get built in the region in the future, particularly in the USA. The US population is not in favor of the construction of this type of energy project.

It is well known that the Trump administration does not support the use of renewable energy sources for electricity generation. However, despite the Trump administration's position on renewables, the role of wind and solar energies within the North American energy mix has been growing every year. The capacity installed of both types of energy sources in the USA also follows the same trend. At the same time, the concern of the population about the negative impact on the environment of large hydropower plant construction and operation has been growing, particularly in the USA. It is expected that his trend will continue without

*Non-Conventional Energy in North America*
https://doi.org/10.1016/B978-0-12-823440-2.00010-X

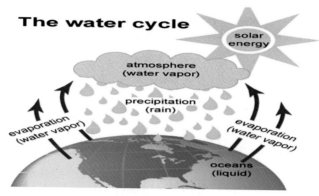

**FIG. 2.1**   Water cycle. *Source: Adapted from National Energy Education Development Project (public domain).*

change, at least during the coming years. For this reason, it is unlikely that new big hydropower plants be constructed in the region during the coming years, particularly in the USA.

Nevertheless, in the Americas, hydropower still is the continent's largest renewable energy source used for electricity generation. The USA and Canada are among the top producers of hydroelectricity in the world, particularly in the case of Canada (Hydroelectric Power in North America: What Are the Facts?, 2020).

According to Fig. 2.1, the water cycle has three steps. These steps are the following:

- Solar energy heats water on the surface of rivers, lakes, and oceans. This action of the Sun evaporates the surface water of rivers, lakes, and oceans;
- Water vapor condenses into clouds and later falls as precipitation in the form of rain and snow;
- Precipitation collects in streams and rivers, flowing into oceans and lakes, where it evaporates and begins the cycle again (EIA Hydropower Explained, 2020).

It is important to single out that "the amount of precipitation that drains into rivers and streams in a geographic area determines the amount of water available for producing electricity" using hydropower plants. "Seasonal variations in precipitation and long-term changes in precipitation patterns, such as droughts, can have large effects on the availability of hydropower production" (EIA Hydropower Explained, 2020).

The unique source for electricity generation by hydropower plants is water. "The volume of the water flow and the change in elevation, or fall, often referred to as the "head," from one point to another, determine the amount of available energy in moving water. In general, the greater the water flow and the higher the head, the more electricity a hydropower plant can produce" (see Figs. 2.2–2.4) (EIA Hydropower Explained, 2020).

For this reason, hydropower plants are usually located on or near a water source, such as rivers, lakes, or oceans. This specific characteristic of a hydropower plant sometimes makes it difficult to be selected for electricity generation. The reason is straightforward: because, in some cases, cities and towns where this electricity is needed are located far away from the sites where the hydropower plant must be constructed, increasing the transmission losses. That makes it less efficient to use a hydropower plant for electricity generation than other renewable energy sources, such as solar and wind energies.

## Hydroelectric Dam

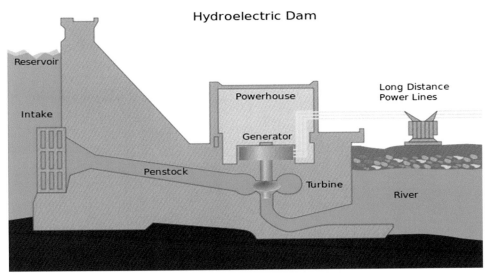

FIG. 2.2 Hydroelectric dam. *Source: https://commons.wikimedia.org/wiki/File:Hydroelectric_dam.svg#mw-jump-to-license.*

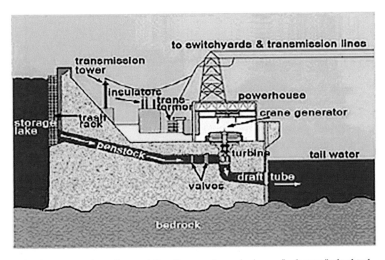

FIG. 2.3 Inside a hydropower plant. *Source: https://www.usbr.gov/uc/power/hydropwr/index.html.*

According to Figs. 2.2–2.4, the following can be stated regarding the operation of a hydropower plant: in this type of power plant, water flows through a penstock, then pushes against and turns blades in a turbine to spin a generator to produce electricity that is supplied to houses, towns, villages, cities, industries, and other installations and facilities. As can be seen, the only fuel used in a hydropower plant for electricity generation is water.

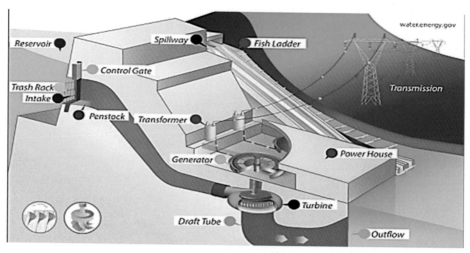

**FIG. 2.4** Hydroelectric power plant. *Source: https://www.energy.gov/podcasts/direct-current-energygov-podcast/episode-7-water-wattage.*

## Types of hydropower plants

Hydropower plants can be classified into three groups. These groups are the following:

- Low, medium, and high-head hydropower plants.
- Run-of-the-river hydropower plants[a];
- Pumped-storage hydropower plants.

Hydropower plants, based on the head and storage capacity availability, can be categorized as follows:

- Low-head hydropower plants (between 2 m and 20 m);
- Medium-head hydropower plants (between 20 m and 150 m);
- High-head hydropower plants (+ 150 m).

High-head hydropower plants are the most common and large hydropower plants currently in operation at the world level. This type of hydropower plant generally utilizes a dam to store water at an increased elevation. The use of a dam (see Fig. 2.5) to store water associated with a hydropower plant also allows collecting water during rainy seasons and releasing it during dry seasons or when the electricity is not needed, or when the electricity demand is higher. This possibility results in consistent and reliable electricity production, able to meet demand when needed. High-head hydropower plants with storage facilities are very valuable to electric utilities because they can be quickly adjusted to match the electrical demand during peak hours.

---

[a] This type of hydropower plant are usually built on rivers with steady natural flows or regulated flows discharged from upstream reservoirs, have little or no storage capacity, and power is generated using the river flow and water head. Run-of-the-river hydropower plants are less appropriate for rivers with large seasonal fluctuations.

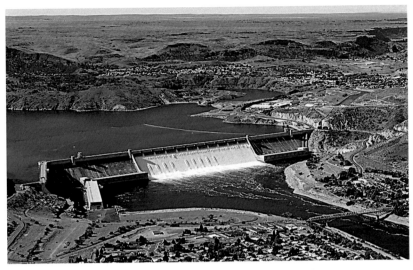

FIG. 2.5    The largest hydropower plant in the US: Grand Coulee. *Source: Photo courtesy of US Bureau of Reclamation.*

Low-head hydropower plants are those that utilize either a low dam or weir to channel water or use the "run of the river." Run-of-river hydropower plants is a facility that channel flowing water from a river through a canal or penstock to spin a turbine. Typically, a run-of-river hydropower plant has little or no storage facility to store the water. However, it provides a continuous electricity supply (baseload), with some operation flexibility for daily fluctuations in demand through water flow regulated by the facility. Nevertheless, it is essential to know that the "run of the river" generating hydropower output varies with seasonal water streams in a river.

The pumped-storage hydropower plants use excess electrical system capacity, generally available at night, to pump water from one reservoir at a low site to another at a higher elevation. Then, during periods of peak electricity demand, water from the higher reservoir is released through turbines to the lower reservoir in order to produce the electricity demanded. Although most of the pumped-storage hydropower plants are not net producers of electricity, because it actually consumes more electricity to pump the water up than is recovered when released, they are a valuable component of the electricity supply system. This type of power plant's importance is its ability to store electricity in the form of water for its use later when electrical peak demands are higher. Pumped-storage hydropower plants are even more critical if intermittent power sources, such as solar parks or wind farms, are hooked into the system (Morales Pedraza, 2013).

In the form of a pumped-storage hydropower plant, energy storage is a recognized and mature technology that has continuously changed to suit the needs of switching power systems in many countries all over the world. Pumped-storage hydropower plants currently account for over 94% of installed global energy storage capacity and over 96% of the energy stored in grid-scale applications. Besides, a pumped-storage hydropower plant can help accommodate renewable energy sources such as wind and solar energies within many countries' energy mix to provide the electrical grid's necessary stability. The flexibility it can provide through its storage and ancillary grid services is increasingly important in ensuring power supply meets demand across multiple timescales.

Multiple studies have identified the vast potential for pumped-storage hydropower plant sites worldwide. There is growing research on retrofitting pumped-storage hydropower plant sites at disused mines, underground caverns, and non-powered dams, among other locations. As a result of a rebirth of interest in pumped-storage hydropower plants worldwide, it is estimated that global pumped-storage hydropower plant capacity will grow up to 78 GW by 2030. This increase is considerably higher than any other form of energy storage technology (IRENA Global Renewables Outlook 2020, 2020).

However, it is important to stress that market regulations and policy frameworks do not yet incentivize the use of the technology mentioned above for electricity generation. The flexibility and storage services provided by pumped-storage hydropower plants "are not appropriately valued. As a result, the lack of a strong business case has hindered private sector investment in pumped-storage hydropower development, as evidenced by the low growth rate in 2019. In that year, only 304 MW of new pumped-storage hydropower capacity was added, as some projects in China missed their deadlines" (IRENA Global Renewables Outlook 2020, 2020).

According to the IRENA Renewable Energy Statistics 2020 report, in 2019, the USA had 19.2 GW pumped storage hydropower plant capacity installed in the country, representing 15.9% of the world total (99.1% at the regional level). According to the same report, Canada has 174 MW pumped-storage hydropower plant capacity installed in the country, representing only 0.001% of the world total (0.9% at the regional level). It is important to note that this capacity has not increased in both countries in the last several years. It is expected that this situation will not change in the near future. Besides, there are no plans to construct new pumped storage hydropower capacity in any of the countries mentioned above for the coming years.

## Classification of hydropower plants according to their capacity

According to the capacity of the hydropower plants, they can be classified as follows[b]:

(a) **Large conventional hydropower plants:** These facilities have a generation capacity of more than 300 MW;

(b) **Medium conventional hydropower plants:** These facilities have a generation capacity from 100 MW to 300 MW (having a dam and a reservoir) and, in the case of run-of-river facilities, between 10 MW and 100 MW. Medium conventional hydropower plants can sometimes be economically viable for rural electrification, particularly in those rural areas that have adequate hydropower technical potential;

(c) **Small conventional hydropower plants[c]:** These facilities have a generation capacity from 1 MW to 10 MW. Small conventional hydropower plants can sometimes also be

---

[b] It is important to stress that there is no worldwide consensus on definitions regarding size and categories of hydropower plants (Egré and Milewski, 2002).

[c] In general, small hydropower plants can use existing infrastructure such as dams or irrigation channels for electricity generation. This type of hydropower plants are located close to villages to avoid expensive high-voltage distribution equipment. They can use pumps as turbines and motors as generators for a turbine/generator set, and have a high level of local content both in terms of materials and work force during the construction period and local materials for the civil works (Kumar et al., 2011). The transmission losses of this type of hydropower plants are very low.

economically viable for rural electrification, particularly in those rural areas that have adequate hydropower technical potential;

**(d) Mini-hydropower plants:** These facilities have a generation capacity from 100 kW to 1 MW;

**(e) Micro-hydropower plants:** These facilities have a generation capacity of less than 100 kW.

## Hydropower electricity generation and capacity installed in the North American region

Hydropower plants for electricity generation at the world level, despite slowed capacity growth compared to previous years, increased by 2.2 in 2018 compared to 2017 and 2.5% in 2019 compared to 2018. It is the world's largest electricity source with multiple non-power benefits, unique storage, and flexible services. Hydropower plants can also support integrating other types of renewable energy sources such as wind and solar energies into the country's energy mix. For this reason, hydropower plants can play an integral role in the recovery effort and the clean energy transition at the world level as well as at regional and national levels.

On the other hand, hydropower plants can safely supply clean water for agriculture, homes, towns, villages, and businesses and help to mitigate the impacts of extreme weather events such as floods and drought. "To maximize the contribution hydropower can make to the world economy, policy-makers should recognize the urgency for a bold and ambitious green recovery plan as part of the global response to COVID-19, involving significant new investment by public and private sectors" (IRENA Global Renewables Outlook 2020, 2020).

During the coming years, it is expected only a 1.3% increase per year in the use of hydropower plants for electricity generation at the world level below the increase rate of the last years. This percentage is lower than the one registered in 2018 and 2019 and much lower than the one registered over the past 20 years. In the foreseen scenario, China remains the largest source of growth in the use of hydropower plants for electricity generation. Still, this increase is becoming more broadly based because of significant recording increases in other regions such as Asia, Latin America, and Africa. Hydropower is also the dominant source of renewable electricity generation in the North American region. It will continue to play an essential role in achieving both the Sustainability Development Goals and the targets set out in the Paris agreement on climate change.

According to the IRENA's Global Renewables Outlook 2020 and IRENA Renewables Energy Statistics 2020 report, there is now around 1308 GW (1,307,994 MW) of installed hydropower plant capacity globally, increasing about 1% compared with 2018. In 2019, the new hydropower plant capacity installed at the world level reached almost 13 GW (12,975 MW).[d]

---

[d] The following are the countries that led the way in 2019 concerning annual investment, net capacity additions and electricity production using hydropower plants: Brazil, China, Lao, Bhutan, and Talikistan. However, the countries with the highest capacity installed and the highest electricity generation using hydropower plants at the end of 2019 were the following: China (28% of the total), Brazil (9% of the total), Canada (7% of the total), the USA (7% of the total), and Russia and India (4% of the total).

The electricity generated by hydropower plants at the world level in 2019 reached 4267 TWh, representing an increase of 2.2% compared with 2018 (see Fig. 2.6).

Based on the report mentioned above, hydropower capacity at the world level will need to grow by around 60% by 2050. That means that 850 GW new capacity should be installed over the next 30 years to help limit the growth in global temperature below 2 °C above pre-industrial levels, as agreed in the Paris agreement on climate change. "Such growth would help generate some 600,000 skilled jobs over the coming decade and would require an esti-mated investment of US$1.7 trillion" (Facts About Hydropower, 2020).

To reach the 2050 target, the yearly average growth in hydropower plant capacity would need to reach an estimated 2% a year on average. Between 2015 and 2019, hydropower in-stalled capacity increased by 8%, rising from 1211 GW in 2015 to 1308 GW in 2019.[e] The aver-age year-on-year growth in installed capacity was 2.1%. In 2019, the growth rate was around 1% in comparison with 2018. However, it is important to single out that the annual growth in hydropower installed capacity can vary considerably depending on when major hydropower power plant projects, which take years to develop, become operational. "Notwithstanding, this underlines the need for investment in hydropower to increase significantly over the next decade and more" (IRENA Global Renewables Outlook 2020, 2020).

However, to satisfy the foresee electricity demand during the coming years, additional hydropower plant capacity should be built worldwide. This goal also can be reached by:

- Upgrading and modernizing some 600 GW of existing capacity in hydropower plants older than 30 years;

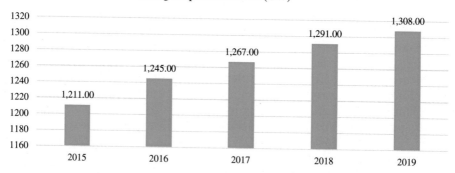

FIG. 2.6　Hydropower installed capacity growth at the world level during the period 2015–19. *Source: IRENA Global Renewables Outlook 2020.*

[e] The use hydropower to replace burning coal for electricity generation could reduce between 3.5 and 4 billion metric tons of additional greenhouse gases to be emitted annually, and global emissions from fossil fuels and industry would be around 10% higher. "In addition, using hydropower instead of coal avoids the production of around 150 million tons of air polluting particulates, 60 million tons of sulfur dioxide, and 8 million tons of nitrogen oxide – avoiding many health, environment and climate impacts" (IRENA Global Renewables Outlook 2020, 2020).

- Tapping tens of thousands of non-powered dams that are not used now for any particular activities;
- Doubling the existing pumped-storage hydropower plant capacity to support the use of more variable renewable energy for electricity generation such as wind and solar energies.

In the North American region, the evolution of the hydropower plant capacity installed during the period 2010–19 is included in Fig. 2.7.

According to Fig. 2.7, the following can be stated: the hydropower plant capacity installed in the North American region during the period 2010–19 increased by 4.4%, rising from 176,101 MW in 2010 to 183,822 MW in 2019. However, it is important to single out that hydropower plant installed capacity in the region increased only 0.9% during the period 2010–14. After that year, the hydropower plant installed capacity increased by 3.4%. It is expected that the trend reported after 2014 will continue without change during the coming years.

If only pumped-storage hydropower plant capacity is considered, then the evolution of this type of power plant capacity in the North American region during the period 2010–19 is shown in Fig. 2.8.

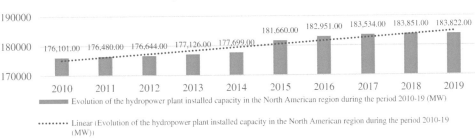

FIG. 2.7   Evolution of the hydropower plant capacity installed in the North American region during the period 2010–19. *Source: 2020. IRENA Renewable Energy Statistics 2020 and Author own calculations.*

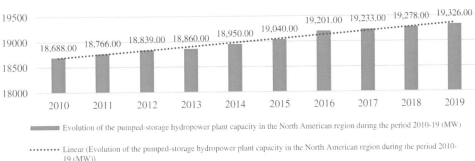

FIG. 2.8   Evolution of the pumped-storage hydropower plant capacity in the North American region during the period 2010–19. *Source: 2020. IRENA Renewable Energy Statistics 2020 and Author own calculations.*

According to Fig. 2.8, the pumped-storage hydropower plant capacity in the North American region increased by 3.4% during the period 2010–19, rising from 18,688 MW in 2010 to 19,326 MW in 2019. This trend is expected to continue in the future.

The evolution of the electricity generation by hydropower plants in the North American region during the period 2010–18 is included in Fig. 2.9.

Based on the data included in Fig. 2.9, the following can be stated: the electricity generated by hydropower plants in the North American region increased by 9.6% during the period 2010–18, rising from 637,794 GWh in 2010 to 698,754 GWh in 2018. The period under consideration can be divided into three parts. In the first part covering the years 2010 and 2011, the electricity generated by hydropower plants increased 12.9%, rising from 637,794 GWh in 2010 to 720,358 GWh, the highest peak registered in this period. The second part covering the period 2011–2015, the electricity generated by hydropower plants decreased by 9.3%, falling from 720,358 GWh in 2011 to 653,422 GWh in 2015. The third part covering the period 2015–2018, the electricity generated by hydropower plants increased by 6.9%, rising from 653,422 GWh in 2015 to 698,754 GWh in 2018. Due to the behavior of hydropower electricity generation in recent years, it is difficult to predict the trend of electricity generation using hydropower plants in the region for the coming years.

If only pumped-storage hydropower plants are considered, then the electricity generation's evolution by this type of power plant in the North America region during the period 2011–18 can be seen in Fig. 2.10.

According to Fig. 2.10, the electricity generation by pumped-storage hydropower plants in the North American region during 2011–18 decreased by 5.8%, falling from 22,939 GWh in 2011 to 21,614 GWh in 2018. Due to ups and downs in hydroelectricity generation registered in the region and the USA during the last years, it is difficult to predict the electricity generation's behavior during the coming years. In Canada, however, this trend is expected to change during the coming years due to the entry in operation of five pumped-storage hydropower plants with a total capacity of 2400 MW. The new pumped-storage hydropower plant capacity that will be installed in the county is almost 14-fold the current hydropower capacity installed.

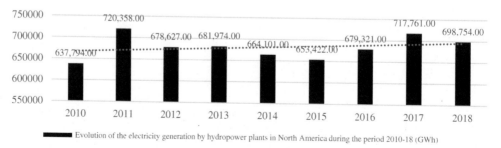

FIG. 2.9  Evolution of the electricity generation by hydropower plants in the North American region during the period 2010–18. *Source: 2020. IRENA Renewable Energy Statistics 2020 and Author own calculations.*

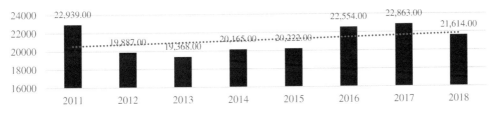

FIG. 2.10   Evolution of the electricity generation by pumped-storage hydropower plants in the North American region during the period 2011–18. *Source: 2020. IRENA Renewable Energy Statistics 2020 and Author own calculations.*

Although growth in the use of hydropower plants for electricity generation in the North American region remains modest compared to other areas of the world, there is an increased focus on pumped-storage projects within the North American region, particularly in the USA. Without a doubt, hydropower's role in the clean energy mix at the world level will increase during the coming years as more intermittent renewable energy sources such as wind and solar energies are deployed, particularly in the USA. This deployment is spread unevenly across the whole country (see Fig. 2.11). "Some regions, like the Pacific Northwest, generate a significant amount of their electricity from hydropower plants, while others have only limited hydro resources available" (Hydroelectric Power Resources Form Regional Clusters, 2011).[f]

In Canada, major storage projects are under construction in Keeyask generating power in Manitoba, Site C in British Columbia, Muskrat Falls in Newfoundland and Labrador, and Romaine-4 in Quebec. These new constructions could increase the current pumped-storage capacity in the country, which is now very low (174 MW in 2019).

[f] Without a doubt, hydropower has been the leading source of renewable energy across the world, and in the North American region as well. Hydropower plant capacity was built up in North America between 1920 and 1970 when thousands of dams were built. There are over 82,000 large dams in the USA alone (Chen et al., 2016; US Army Corps of Engineers, 2016). In addition, over 2 million small low-head dams fragment USA rivers (Fencl et al., 2015), and their cumulative impacts are largely unknown, since they have escaped careful environmental assessment. However, only 2400 of the nation's 82,000 existing dams are used to generate power. Installing turbines in existing dams presents a promising and cost-effective power source.It is important to stress that, in the last ten years, the Department of Energy (DoE) has spent US$1.2 billion on research and development for other renewable sources like wind, solar, and geothermal, but only US$10 million on hydropower. During the last decades, big dams stopped being built in developed countries, "because the best sites for dams were already developed and environmental and social concerns made the costs unacceptable. Nowadays, more dams are being removed in North America and Europe than are being built" (Moran et al., 2018). In Canada, there is more than 500 hydropower plants located in every province and territories of the country.

Distribution of conventional hydroelectric plants in the lower 48 states

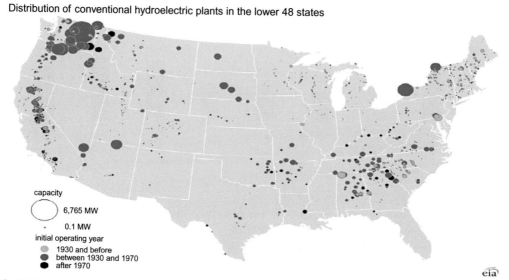

**FIG. 2.11**    Hydroelectric generators in and around the USA. *Source: US EIA, 2011. Today in Energy. Hydroelectric Power Resources From Regional Clusters, Derived From Energy Velocity. https://search.usa.gov/search?utf8=%E2%9C%93&affiliate=eia.doe.gov&sort_by=&query=Today+in+Energy+hydroelectric+power+resources+from+regional+clusters&commit=-Search.*

Summing up, the following can be stated, according to the Canadian Hydropower Association (Canadian Hydropower Association Water Power Canada 2020, 2020), the IRENA's Global Renewables Outlook 2020 report, and IRENA Renewable Energy Statistics 2020:

- In the USA, total hydropower plant capacity, including pumped-storage hydropower plant capacity, remained at 103 GW in 2019 (102,769 MW), ranking 43 at the world level. While most recent growth comes from small hydropower plants, there is still 50 GW of untapped hydropower plant potential, including 30 GW of pumped-storage hydropower plant capacity in the country.
- In Canada, hydropower remains the dominant electricity supply source, representing 61% of total electricity generation and 55% of the total installed generation capacity. There are more than 500 hydropower facilities in the country. The full hydropower plant capacity installed in Canada in 2019 reached 81,053 GW. It is expected that the hydropower installed capacity soon exceeding 85,000 MW. "Canada is the second-largest generator of hydroelectricity in the world after China" (Canadian Hydropower Association Water Power Canada 2020, 2020);
- Together, Canadian and USA hydropower resources represent 93.5% of the hydropower plant capacity installed in the region, if Greenland and Mexico are included;
- A total of 183,822 MW of hydropower plant capacity are installed in Canada and the USA in 2019;
- A total of 698,754 GWh of hydroelectricity was generated in Canada (381,750 GWh) and the USA (317,004 GWh) in 2019;

- A total of more than 250,000 MW of potential capacity is still available in Canada and the USA;
- A total of more than 50,000,000 households are powered annually in Canada and the USA using the electricity generated by hydropower plants;
- A total of 350,000,000 tons per year of greenhouse gas emissions are avoided due to the use of hydropower plants in Canada and the USA.

## Investment in hydropower plants in the North American region

Without a doubt, for hydropower plants, location is critical to success in a construction investment project. There are many good reasons to invest in hydropower plants[g] at the world level and the North American region in particular. Among these reasons are carbon tariffs, rising energy consumption worldwide and, particularly in the North American region, and the current high level of environmental pollution registered in many countries all over the world due to the use of fossil fuels for electricity generation and heating. It is important to recall that the USA is one of the three world's most pollutant countries due to its high use of coal for electricity generation and heating. However, most people are concerned with the negative impact on the environment of solar PV parks and wind farms, their instability, whether dependency, the price of the electricity generated by these two types of renewable energy sources, and the investment return. On this last issue, it is important to know that hydropower is a renewable energy source that offers a good investment return, despite its strong location-dependency.

In the North American region, the total investment in renewable energy sources in 2018 reached US$55 billion (18% of the total), US$9 billion in Canada (almost 3% at the world level and 16.4% at the regional level), and US$46 billion in the USA (15.1% at the world level and 83.6% at the regional level). The level of the investment in renewables in the North American region, in 2018, ranking third at the world level, behind Asia and the Pacific with US$154 billion, and Europe with US$57 billion (IEA, 2019).

Another critical element that should be in future investors' minds in constructing power plants is the plant's lifespan. If the lifespan of hydropower plants is compared to solar parks and wind farms lifespan, then without a doubt, hydropower plants have, by far, the most extended lifespan among these three types of energy sources. Hydropower plants have a service life of 100 years, depending on the dam's physical characteristics and the soil accumulation in the reservoirs. In other words, in terms of a power plant's service life, hydropower

---

[g] Many dams are currently under construction all over the world, mainly in China, Africa (dams are built in Ethiopia with Chinese support), Brazil, and India. For this reason, these countries will cover a growing proportion of their primary energy needs with hydropower in the near future. Of all these countries, China has the highest growth of hydropower consumption, as a share of primary energy consumption. In China, several dams are being built at the same time. The Yangtzeekiang Dam (with an incredible generator capacity of 22.5 GW) is the largest hydropower plant in the world, and can supply around 60 million inhabitants with electricity, which is an enormous technical achievement. This is particularly impressive because the electricity generated is available at any time of the day. Solar and wind energies, on the other hand, are strongly dependent on the weather (Boegelsack, 2019).

has the edge over different energy sources, including nuclear power plants, which have, in some specific nuclear power reactors, around 60 years of life services.

On the other hand, hydropower has a very high energy-return-on-energy-invested (EROI).[h] Hydropower EROI is about 0.5 without storage and between 0.3 and 0.4 with energy storage. By comparison, nuclear power plants have an EROI between 0.4 and 0.8, sometimes even less. Coal use, on the other hand, has an EROI of about 0.3. Coal and nuclear energy have the advantage of not having to store the power, and energy is continuously available. That improves the EROI significantly (Boegelsack, 2019).

According to the IEA World Energy Investment 2019 report, "energy investment remained at US$1.85 trillion in 2018 while a rise in fossil fuel supply investment offset lower power and stable efficiency spend (see Fig. 2.12). More spending in upstream oil and gas and coal supply was offset by lower spending on fossil-fuel-based and renewable power. Investment in energy efficiency was relatively stable."

According to Fig. 2.12, the following can be stated: since 2015, the North American region has been investing in the energy efficiency of around US$50 billion, behind Asia and the Pacific, and Europe. By far, the Asian and the Pacific region is the area of the world registered with the major investment in energy efficiency in the last years. This trend is expected to continue in the coming years.

It is important to stress that the power sector was again, in 2018, the largest sector for investment at the world level for the third year in a row. This trend also reflects the growing importance of electricity, whose demand growth in 2018 was nearly twice as fast as overall energy demand.[i] According to the IEA World Energy Investment 2019 report, in the past

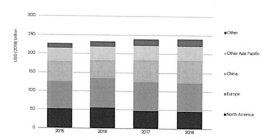

FIG. 2.12   Global investment in energy efficiency by region. *Source: IEA, 2019. World Energy Investment 2019. https://www.iea.org/reports/world-energy-investment-2019.*

---

[h] EROI is the ratio of the amount of usable energy (the energy) delivered from a particular energy resource to the amount of energy used to obtain that energy resource. Arithmetically the EROEI can be defined as:

$$EROI = Energy\ delivered\ /\ Energy\ required\ to\ deliver\ that\ energy$$

When the EROI of an energy source is less than or equal to one, that energy source becomes a net "energy sink", and can no longer be used as an energy source, but depending on the system might be useful for energy storage (Wikipedia Energy Return on Investment, 2020).

[i] There is a strong link between income levels and energy investment. Nearly 90% of energy investment in 2018 was concentrated in high- and upper-middle income countries and regions. High-income countries, with just over 15% of the global population, accounted for over 40% of energy investment in 2018. However, investment in this group is down somewhat from five years ago, largely due to lower spending in Europe and Japan, but rose in 2018 with stronger spending in fuel supply and the power sector predominantly in the USA.

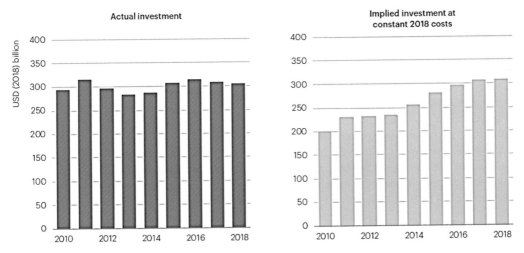

**FIG. 2.13** Evolution of the investment in renewable energy sources during the period 2010–18. *Source: IEA, 2019. World Energy Investment 2019. https://www.iea.org/reports/world-energy-investment-2019.*

years, energy investment trends indicate the need for courageous decisions in order to make the energy system more sustainable at the world level (see Fig. 2.13).

The government's role in supporting energy investment in the energy sector is critical to reducing investors' risks, particularly in the emerging energy sectors that urgently need more capital to get the world on the right track. Energy investment in the USA, accounted for most growth in energy supply investment in recent years, "has been catching up with China, mainly due to oil and gas supply and electricity networks, while China's power sector spending has fallen"[j] (IEA, 2019).

According to Fig. 2.13, the level of investment in renewable energy sources during the whole period 2010–18 was above US$250 billion, and since 2015, above US$300 billion. At constant 2018 costs, the investment in renewable energy sources grew up almost every year within the period 2010–18, particularly since 2014. However, during the period 2017–18, the investment level in renewables was practically the same. This trend reflects the fact that "in recent years, there has been a broad shift in favor of projects with shorter construction times that limit capital at risk" (IEA, 2019).

The USA energy investment, including the investment in the power sector, fossil fuel supply, energy efficiency, and renewables for transport and heating, reached, in 2018, a total of US$351 billion; this represents an increase of 9.3% with respect to the level registered in 2016 (US$321 billion).

[j] The USA has been responsible for most of the growth in energy supply investment in recent years, with increases in both oil and natural gas, supported by more spending on shale gas, and in the power sector. "While oil and gas spend has moderated somewhat in the past three years (even as it grew strongly from 2017 to 2018), that for electricity networks rose. Compared to 2015, investment in renewable power and gas power remained relatively stable, but at high levels. Meanwhile, investment in energy efficiency has declined over the period" (IEA, 2019).

Annual global investment in hydropower plants reached, in 2018, a total of US$50 billion, according to the IEA World Energy Investment 2019 report. While at similar levels to previous years, it is still well "short of the estimated US$100 billion a year required" to help meet the energy-related components of the IEA's Sustainable Development Scenario, according to the report mentioned above. Under this scenario, which is compatible with the Paris agreement's objectives, around an additional 800 GW of hydropower plant capacity would need to be commissioned by 2040 and 850 GW of newly installed hydropower plant capacity requiring an investment of up to US$1.7 trillion by 2050.

Finally, it is important to single out that while public energy research and development spending rose modestly in 2018, led by the USA and China, most countries are not spending too much on energy research. The reason for this situation is the following: hydropower is the world's primary renewable energy resource used for electricity generation in many countries all over the world, but questions have been raised about its consistency under forecast climate change. According to different experts' opinions, large dams seem to be something that a country should not try to build if it cares about sustainability and has no condition to avoid affecting the environment and population. "To move toward sustainability, future hydropower development needs to give more attention to how climate change may affect hydropower production and make greater efforts to reduce the environmental and social costs borne by people near the dams." Besides, "those harmed by the dams need to be adequately compensated, the number of people that must be resettled should be reduced, and most importantly, innovative technologies that reduce all of these negative outcomes should be developed, especially instream turbines and other forms of renewable energy" (Moran et al., 2018).

## Hydroelectricity costs in the North American region

Without a doubt, hydropower plants for electricity generation are an excellent way to increase energy production without mortgaging the country's future to another country's fossil fuel supplies. Over 600 hydropower plants are presently under construction, and over 3000 are planned for the near future at the world level. Most of these hydropower plants are or will be located in Asia and Latin America. If new hydropower plants under construction and planned are considered, then the current hydropower plant capacity at the world level will be doubled (Conca, 2014).

The electricity generated by hydropower plants is clean, and it is considered the most mature, reliable, and cost-effective renewable power generation technology available globally in 2020. "Hydropower operations usually have flexibility in their design as well and can be structured to meet baseload and peak demands." Without a doubt, hydropower is the most flexible renewable energy source available at the world level. "It is also capable of responding to demand fluctuations in a short time sequence" (Meyers, 2016).

On the other hand, it is important to stress that hydropower is the only large-scale and cost-efficient energy storage technology available today, among all energy sources. Hydropower is still the only type of renewable energy source offering technological, efficient, and economically viable large-scale storage despite new and encouraging technological developments in renewable energy storage technologies.

"Currently, more than 25 countries in the world depend on hydropower for 90% of their electricity supply (99.3% in the case of Norway), and 12 countries are 100% reliant on hydropower plants for electricity generation. Hydropower also produces a large share of electricity in 65 countries and is used for electricity generation at some level by more than 150 countries" (Meyers, 2016).

In 2018, Canada raking fourth (81,053 MW), and the USA, ranking third (102,769 MW), are among the countries having the largest hydropower generation capacity at the world level. In the same year, Canada ranked third at the world level in the use of hydropower plants for electricity generation with 381,750 GWh, and the USA ranked fourth with 317,004 GWh.

There are two major cost components for hydropower projects. According to the IRENA report entitled "A cost analysis of hydropower" (IRENA, 2012), these primary components are the following:

- The civil works associated with the construction of a hydropower plant, including any infrastructure development required to access the site, such as roads and bridges, and the project development costs;
- The cost related to electro-mechanical equipment to be installed in the hydropower plant.[k] This cost tends to vary significantly less than the civil engineering costs. The electro-mechanical equipment is a mature, well-defined technology whose costs are not significantly influenced by the site characteristics. As a result, the variation in the installed costs per kW for a given hydropower plant is almost exclusively determined by the local site considerations and the civil work involved.

According to the IRENA report mentioned above, project development costs include several components. These components are "planning and feasibility assessments, environmental impact analysis, licensing, fish and wildlife/biodiversity mitigation measures, development of recreation amenities, historical and archaeological mitigation, and water quality monitoring and mitigation" (IRENA, 2012). The civil works costs can be broadly grouped into the following categories:

- Dam and reservoir construction associated with the building of the hydropower plant;
- Tunneling and canal construction related to the building of the hydropower plant;
- Powerhouse construction associated with the construction of the hydropower plant;
- Site access infrastructure, such as roads, bridges, etc.;
- Electrical grid connection between the national electrical grid and the hydropower plant;
- Engineering, procurement, and construction associated with the building of a hydropower plant;
- Developer/owner costs (including planning, feasibility, authorization, etc.) (IRENA, 2012).

To construct hydropower plants in sites located far away from existing transmission networks, the construction of transmission lines can contribute significantly to the total costs associated with building the hydropower plant. Accessing remote sites may also require the

[k] This type of equipment includes the turbines, generators, transformers, cabling, and control systems required for the correct operation of the hydropower plant.

construction of roads and other infrastructures at the site. These activities will inevitably increase the total construction cost of the hydropower power plant. According to the IRENA mentioned above report, average investment costs for large hydropower plants with storage facilities range from US$1050 per kW to US$7650 per kW. In the specific case of small hydropower plants, the range is slightly higher, between US$1300 per kW and US$8000 per kW. It is important to single out that adding additional capacity at existing hydropower plants in operations or existing dams that do not have a hydropower plant associated with it can be significantly cheaper and cost as little as US$500 per kW.

At the world level, according to the IRENA report entitled "Renewable Power Generation Cost in 2018," the global weighted-average cost of electricity was US$0.47, a decrease of 11% concerning 2017, but 29% higher than in 2010. The global-weighted average LCOE of hydropower projects was the same or lower than the cheapest fossil fuel-fired option cost in all but two years between 2010 and 2018.

The evolution of the total installed costs by hydropower plants and the global weighted average during the period 2010–18 is shown in Fig. 2.14.

According to Fig. 2.14, installed costs for hydropower have increased 21.2% during the period 2010–18, rising from US$1232 per kW in 2010 to US$1492 per kW in 2018. Total installed costs span a wide range, reflecting the very site-specific nature of hydropower projects, but generally fall within the US$1000 and US$2500 per kW range. In 2018, total installed costs fell from their recent highs registered in 2016 and 2017.

In the USA, the average cost of the electricity generated by hydropower plants is US$0.85 per kWh. That is about 50% of the electricity cost produced by nuclear power plants, 40% of the cost of the electricity generated by fossil fuel power plants, and 25% of the cost of the electricity generated by natural gas power plants (Meyers, 2016). In the case of Canada, electricity energy use varies significantly from province to province. Provinces with plentiful and cheap electricity from large-scale electricity projects like British Columbia and Quebec tend

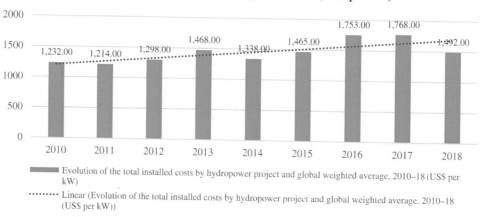

FIG. 2.14 Evolution of the total installed costs by hydropower plants and the global weighted average during the period 2010–18. *Source: 2018. IRENA Renewable Power Generation Cost Report 2018.*

to use more power per person than those that rely on other energy to do things like heating their homes and water. The Canadian electricity price in 2020 was around US$0.17 per kWh as average (US$0.11 per kWh in 2018). In the specific case of the electricity generated by hydropower plants, the prices move between US$0.54 per kWh to US$0.17 per kWh for monthly consumption of 1000 kWh, excluding taxes.

Proper hydropower plant maintenance is crucial in successfully running and operating the plant, as with all power plants. According to Halliday Hydropower International Renueve from Rivers report entitled "Operational and Maintenance Cost of Hydropower Plant (2020)", one of the relevant tasks associated with the operations and maintenance of a hydropower plant is to ensure "that there is not a buildup of debris on the intake screen, especially after bad weather, such as storms and floods." Other maintenance costs would include grease lubrication once a month. Every two to three years, hydraulic fluids should be changed. "If a drive belt is fitted, this would typically need replacing every three years. According to the manufacturers' claims, the generator bearings should last up to 15 years before they need to be changed, although they should be checked and serviced every two years." Several factors have an impact on the longevity of the generator bearings. These factors are their "quality, the regularity of inspections, and maintenance." Other maintenance costs "are labor costs, electrical costs for connecting to the electricity distribution grid, and planning and project management fees."

It is important to single out that hydropower plants' operation and maintenance costs vary depending on the site, size, head, and type of hydro turbine design installed.

The IRENA report entitled "Renewable Power Generation Cost 2018" confirmed that the average operational and maintenance cost associated with hydropower plants is slightly less than 2% of total installed costs per year, with a variation of between 1% and 3% of this total. Larger hydropower plants have operational and maintenance expenses below the 2% average, while smaller hydropower plants approach 3% or are higher than the average operational and maintenance costs.

The construction of hydropower plants can be a worthwhile investment and can build up long-term cash flows. Hydropower plants' typical life expectancy of around 100 years is a long-term solution to creating power more sustainably.

## Hydropower plant construction in the North American region

The North American region is one of the three areas in the world with the highest number of hydropower plants in operation. In the USA, there are around 2400 hydropower plants, while in Canada, more than 500. However, North America is currently the region with fewer hydropower plants under construction and planned at the world level. Most of the hydropower plants under construction are located in Canada, several of them on the USA border.

Canada exports nearly 9% of the electricity it generates to the USA. A total of 34 active major international transmission lines are connecting Canada to the USA.

In 2019, according to the "International Hydropower Association (IHA) country file Canada 2019", work progressed on major hydropower projects under construction across three Canadian provinces. In British Columbia, BC Hydro's 1100 MW Site C hydropower plant on the Peace River commenced construction in July 2015 and is anticipated to be completed by 2024. The hydropower plant will generate about 5300 GWh per year. The period of the building of the hydropower plant mentioned above is nine years. The Keeyask Hydropower Limited Partnership develops the 695 MW Keeyask hydropower plant on the Nelson River in Manitoba. The construction of the hydropower plant commenced in July 2014, with completion scheduled for 2021. It is expected to generate about 4400 GWh. The period of construction is seven years. Work continues in Hydro-Québec's La Romaine 4 hydropower plant, scheduled to go into service in 2021 at a capacity of 245 MW. Finally, it is important to stress that "Alberta's legislature approved the construction of the proposed Canyon Creek Pumped Hydro Energy Storage Project. The project will have a storage capacity of 75 MW for 37 hours of full-capacity operation" (International Hydropower Association (IHA) Country File Canada, 2019).

According to the IRENA report entitled "A cost analysis of hydropower 2012", "hydropower is a capital-intensive technology with long lead times for development and construction due to the significant feasibility, planning, design, and civil engineering works" associated with the construction of this type of power plant.

On the other hand, according to the International Energy Agency (IEA) World Energy Investment 2019 report, if all hydropower plant construction stages are considered, then the normal construction time could be between four and seven years. The mentioned time includes the period covering from the beginning of the project's design until the power plant is ready to start operation.

It is important to stress that constructing a very small hydropower plant is disproportionately expensive than building a bigger one. The reason is straightforward: hydropower projects of any size must include a substantial fixed-cost element at the design and approving stages and slightly lesser during the installation stage. That is why it is generally advisable to get an economically viable hydropower plant, the maximum power output to be at least 25 kW, preferably 50 kW. Smaller hydropower plants can make sense, particularly at sites with higher heads or other intangible benefits, such as sustainability, environmental protection, or publicity, which are valued as much as the return on investment.

## The efficiency of the hydropower plants in the North American region

In general terms, efficiency is the output of a process compared to the input. In the context of a power plant, there are three types of efficiencies. These are the following:

- **Economic efficiency.** A power plant's economic efficiency is essentially the specific cost of producing useful output. The type of efficiency is the main driver behind shaping process plant design and operation;
- **Operational efficiency.** A power plant's operational efficiency is generally called the "capacity factor" or sometimes the "load factor." This type of efficiency measures the actual output from a process compared to the maximum potential output;

- **Energy efficiency**.[1] A power plant's energy efficiency refers to the ratio between output and input. In general, this type of efficiency is also called "energy conversion efficiency" when considering mixed inputs and outputs.

The use of hydropower plants for electricity generation "is by far the most efficient and clean method of large-scale electric power generation. The conversion process captures kinetic energy and converts it directly into electric energy. There are no inefficient intermediate thermodynamic or chemical processes and no heat losses" (Electropaedia Hydroelectric Power, n.d.).

According to the paper mentioned above, it is important to single out that "the conversion efficiency of a hydroelectric power plant depends mainly on the type of water turbine employed and can be as high as 95% for large installations." In the case of smaller hydropower plants, "with output powers, less than 5 MW may have efficiencies between 80% and 85%. It is, however, difficult to extract power from low flow rates" (Electropaedia Hydroelectric Power, n.d.).

The best fossil fuel power plants have an operational efficiency of about 50%. In the case of a nuclear power plant, its operational efficiency could be higher than 90%.

The following are, according to Electropaedia Hydroelectric Power (n.d.), the turbine types that can be installed in a hydropower plant:

- **Impulse turbines**. "Impulse turbines require tangential water flow on one side of the turbine runner (rotor) and must, therefore, operate when only partially submerged. They are best suited to applications with a high head, but a low volume flow rate such as fast-flowing shallow watercourses though it is used in a wide range of situations with heads from as low as 15 meters up to almost 2000 meters" (Electropaedia Hydroelectric Power, n.d.) A Pelton turbine is a type of impulse turbine that can reach an efficiency factor of up to 95%;
- **Reaction turbines**. "Reaction turbines are designed to operate with the turbine runner fully submerged or enclosed in a casing to contain the water pressure. They are suitable for lower heads of water of 500 meters or less, and they are the most commonly used high power turbines." A Francis turbine is a type of reaction turbine that, on a large scale used in dams, is "capable of delivering over 500 MW of power from a head of water of around 100 meters with efficiencies of up to 95%." A Propeller and Kaplan turbines are a reaction turbine "designed to work fully submerged. It is similar in form to a ship's propeller and is the most suitable design for low headwater sources with a high flow rate, such as those in slow-running rivers. Designs are optimized for a particular flow rate, and efficiencies drop off rapidly if the flow rate falls below the design rating. The Kaplan turbine version has variable pitch vanes to enable it to work efficiently over a range of flow rates" (Electropaedia Hydroelectric Power, n.d.).

---

[1] Energy efficiency is closely related to the 'heat rate', which is the quantity of heat required to produce a unit of useful output. Therefore, a lower heat rate is more efficient and gives a higher percentage energy efficiency. The relationship most frequently used to describe heat rate and efficiency in respect of electrical power generation is:

$$\text{Energy efficiency} = 3600 / \text{heat rate}$$

## Types of incidents in hydropower plants registered in the North American region

According to Wikipedia Dam Failure (2020), "a dam failure or dam burst is a catastrophic type of failure characterized by the sudden, rapid, and uncontrolled release of impounded water or the likelihood of such an uncontrolled release."

There are two sources of dam failure. One of them is the aging of the construction materials; the other is the accumulation of sediment behind the dam impoundment. In the first case, they are susceptible to failure, sometimes resulting in numerous fatalities and significant property loss. For example, heavy rains from a single tropical storm in 1994 caused more than 230 dams to fail in Georgia, USA (Stamey, 1996). In 2016, Oroville Dam Spillway began to fail in California after heavy rains, resulting in the evacuation of 190,000 people from their homes. More famously, in 1976, the Teton Dam in Idaho failed, with resulting losses exceeding US$2 billion (in 2017 US dollars). Many USA dams constructed during the period 1930–50 have significant potential for failure. These dams have already passed their 50-year lifespan, and around 85% of them will reach that milestone by 2020 (Maclin and Sicchio, 1999).

One crucial element that should be in those governments' minds that support hydropower plants' construction is that "repairing a small dam can be up to three times the cost of removing it" (Born et al., 1998). The high cost of repairing a small dam is an important reason for the growing trend to remove dams today in the USA. More than 60 dams per year have been removed in the USA, a trend that began in 2006 and continues until today.[m] Varying by the amount of sediment load on the river, sedimentation problems occur faster than the loss of structural integrity (Morris and Fan, 1997). Before 1960, sedimentation rates were not consistently factored into dam design criteria; thus, many dams are expected to fill at rates exceeding design expectations (Kondolf et al., 2014). "Today, engineers typically design reservoirs to incorporate a 100-year sediment storage pool" (Moran et al., 2018).

However, these calculations often fail to consider changes in other land use, such as road construction, which can increase sediment yield by two orders of magnitude, and extreme events due to climate change. These unforeseen events will likely increase sediment transport toward reservoirs, reducing the lifespan of the hydropower plants. The tendency to overlook factors that could increase sediment loads continues until today, particularly in tropical countries. "Some river basins are being targeted for hydropower development given their potential to produce energy but with little consideration to reducing the environmental and social consequences of such energy development" (Moran et al., 2018).

It is essential to single out the following: often, large dams are promoted by governments with the idea that the local population will benefit from the construction of hydropower plants. However, the evidence suggests otherwise. A recent study using a database of 220 dam-related conflicts found that, in dams surrounded by controversies and tension, repression, criminalization, violent targeting of activists, and assassinations were common (Del Bene et al., 2018).

---

[m] Since 2006, "the contribution of hydropower to the USA' electrical supply has steadily declined to 6.1% of energy consumption, and other energy sources, such as nuclear, gas, coal, solar, and wind, began to replace it" (Moran et al., 2018).

According to this study, millions of people worldwide are affected by dam construction either because they are permanently resettled due to the filling of the reservoirs or because their livelihoods get disrupted with the construction and operation of the dam (Scudder, 2012). People who are displaced often get an undervalued price for their land or buildings that do not consider the social, cultural, and religious value of their land or the way that people make their livelihoods on the land or the stretch of river (Hanna et al., 2016; Heming et al., 2001; Castro-Diaz et al. (2018)). Besides, after resettlement, people often lose their social networks, jobs, and other social prosperity types, which has, without a doubt, economic, cultural, social, and significant health consequences (Scudder, 2012; Wang et al., 2013).

On the other hand, communities that are not displaced, like those downstream, generally do not get any compensation. However, the effects of a hydropower plant's construction and operation on their livelihoods are just as great as the effects on those who require resettlement (Richter et al., 2010; Castro-Diaz et al., 2018). This problem seems to be even more significant, considering that most people affected by the hydropower plant's construction and operation are the poorest and more vulnerable in their societies. They are often indigenous and traditional communities. Appropriate compensation mechanisms should consider that men and women are impacted differently by a hydropower plant's construction and operation. For this reason, this compensation mechanism must ensure that the most vulnerable people are recompensed in an appropriate manner (Castro-Diaz et al., 2018).

The use of compensation mechanisms that are not always in the form of monetary compensation is an important innovation that needs to be considered for future energy development plans associated with hydropower plants' construction and operation. Until today, no appropriate attention has been given to compensation forms that strengthen communities and individuals affected by a hydropower plant's construction and operation. That can be done by investing in understanding the history and the social organization of these communities and working with them to sustain the integrity of their social, economic, and political relationships. Regrettably, the contrary has been more common: resettling people without concern for any of these issues, and sometimes, even seeming to purposely break up any pre-existing social organization as a way of preventing to act after the construction of the dam is built on lobbying for adequate compensation (Leturcq, 2018).

The use of a hydropower plant with a dam associated with it for electricity generation has an important advantage compared to other renewable energy sources, like wind and solar energies. This advantage is that it can be dispatched quickly at any time, enabling utilities to balance load variations on the electric distribution system (Resch et al., 2008). However, this type of power plant also has a significant advantage compared to the same renewable energy sources mentioned above.

Summing up, it can be stated that the leading causes of dam failures (see Fig. 2.15), according to Wikipedia Dam Failure (2020) and Disaster Survival Resources Dam Failure 2020 (2020), are the following:

- Sub-standard construction materials or techniques used in the construction of the dam;
- Spillway design error;
- Lowering of dam crest height, which reduces spillway flow;
- Geological instability caused by changes to water levels during filling or poor surveying;
- Inadequate maintenance, especially of outlet pipes;

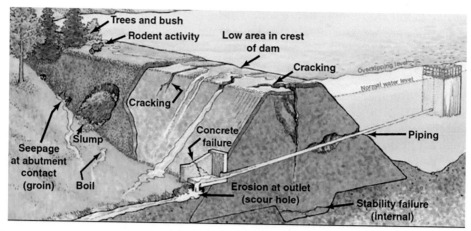

FIG. 2.15  Causes of a dam failure. *Source: https://www.fema.gov/about/website-information.*

- Extreme inflow;
- Human, computer, or design error;
- Internal erosion or piping, especially in earthen dams;
- Overtopping caused by floods that exceed the capacity of the dam;
- Deliberate acts of sabotage;
- Movement and/or failure of the foundation supporting the dam;
- Settlement and cracking of concrete or embankment dams;
- Natural disasters such as an earthquake.

According to Wikipedia Dam Failure (2020), a list of the main dam failures reported in the North American region until May 2020 is shown below (see Table 2.1).

Based on the information included in Table 2.1, the following can be stated: a total of 42 incidents caused by dam failure were registered until May 2020 in the North American region. Out of this total, the USA reported 38 dam failures for 90.5% of the total. Canada reported, in the same period, only four dam failures for 9.5% of the total. Based on these figures, it can be concluded that the construction of a dam in Canada is more secure and suffered fewer damages than in the USA.

## Advantages and disadvantages in the use of hydropower plants in the North American region

Without a doubt, hydroelectricity has general advantages over using other renewable energy sources for electricity generation in the North American region. These advantages are the following:

- Hydropower is continually renewable thanks to the recurring nature of the water cycle;
- Hydropower causes no pollution on cities, environment, and population;
- Hydropower is one of the cheapest sources of electricity generation available worldwide.

**TABLE 2.1**    List of dam failures in the North American region.

| Dam/incident | Date | Location | Country | Fatalities | Details |
|---|---|---|---|---|---|
| Hogs Back dam | 1829-04-03 | Ottawa | Upper Canada | – | Inexperience with cold-weather engineering allowed a small leak in the wall to form on March 28 and the dam slump on April 2. The following day, on April 3, the dam failed and washed away down the Rideau River. A new dam of a different design was built atop the foundation of the original later that same year (Rideau Canal, 2020) |
| Mill River dam | 1874 | Williamsburg, Massachusetts | United States | 139 | Lax regulations and cost-cutting led to an insufficient design, which fell apart when the reservoir was full. A total of 600 million gallons of water were released, wiping out four towns and making national headlines. This dam break led to increased regulation of dam construction. |
| South Fork dam/Johnstown flood | 1889-05-31 | Johnstown, Pennsylvania | United States | 2208 | Blamed on poor maintenance by owners, who lowered crest by a meter or more (Coleman, 2018), the court deemed it an "Act of God." Followed exceptionally heavy rainfall was reported. A total of 1600 homes were destroyed. |
| Walnut Grove dam | 1890 | Wickenburg, Arizona | United States | 100 | Heavy snow and rain following public calls by the dam's chief engineer to strengthen the earthen structure |
| Austin dam | 1900-04-07 | Austin, Texas | United States | 8 | In an extreme current, two 250-ft sections of the dam slid about 20 m downstream intact. The town was left without electrical power for months. |
| Hauser dam | 1908-04-14 | Helena, Montana | United States | 0 | Massive flooding coupled with poor foundation quality. Workers managed to warn people downstream. |
| Broken Down dam | 1908-09-24 | Fergus Falls, Minnesota | United States | 0 | Design flaw; dam built on water springs. Four downstream dams and bridges were destroyed; a fourth dam was opened and saved. No fatalities. |

*Continued*

TABLE 2.1  List of dam failure in the North American region—cont'd

| Dam/incident | Date | Location | Country | Fatalities | Details |
|---|---|---|---|---|---|
| Austin dam | 1911-09-11 | Austin, Pennsylvania | United States | 78 | Poor design, use of dynamite to remedy structural problems. Destroyed paper mill and much of the town of Austin. Replacement failed in 1942. |
| Lower Otay dam | 1916 | San Diego County, California | United States | 14 | Over-topped from flooding following heavy rains. Locally, blame was placed on Charles Hatfield, who had been contracted by The City of San Diego for his rainmaking efforts. Court cases following the dam's failure resulted in neither liabilities being passed to Mr. Hatfield nor the original payment, as both of the court's decisions ruled the event an "Act of God." |
| Sweetwater dam | 1916-01-27 | San Diego County, California | United States | 0 | Over-topped from flooding, spillway inadequate, the water rose over a meter higher than the dam and waterfalled over its surface. The dam had been raised after a similar earlier overtopping. Partial failure. |
| Lake Toxaway dam | 1916-08-13 | Transylvania County, North Carolina | United States | 0 | Heavy rains and lack of water-level controls caused the dam to give way. The private lake was destroyed; the resort area failed. The dam was later rebuilt in the 1960s |
| St. Francis dam | 1928-03-12 | Santa Clarita, California | United States | + 451 | Geological instability of the canyon wall. The designer inspected it hours before it failed. |
| Castlewood dam | 1933 | Franktown, Colorado | United States | 2 | Bad design and maintenance, with proximate cause of heavy rain. The dam failed at 1am on August 3, 1933, with dam waters just 15 miles from the city of Denver. Warnings to the city by 4am allowed most people to move out of the floodwaters' way (Cherry Creek Flood in 1933, 2015; Castlewood Canyon State Park, 2007; Disaster Nearly Drowns Denver in 1933, 2019). |
| Baldwin Hills Reservoir | 1963-12-14 | Los Angeles | United States | 5 | Subsidence is caused by over-exploitation of a local oil field. A total of 277 homes were destroyed. |

**TABLE 2.1** List of dam failure in the North American region—cont'd

| Dam/incident | Date | Location | Country | Fatalities | Details |
|---|---|---|---|---|---|
| Spaulding Pond dam (Mohegan Park) | 1963-03-06 | Norwich | United States | 6 | More than US$6 million estimated damages. |
| Swift dam | 1964-06-10 | Montana | United States | 28 | Failed in heavy rains. Another nearby dam did likewise. |
| Buffalo Creek Flood | 1972-02-26 | West Virginia | United States | 125 | The unstable loose constructed dam created by a local coal mining company collapsed in heavy rain. A total of 1121 persons were injured, 507 houses were destroyed, over 4000 left homeless. |
| Canyon Lake dam | 1972-06-09 | South Dakota | United States | 238 | Flooding, dam outlets clogged with debris. 3057 injuries, over 1335 homes, and 5000 automobiles destroyed. |
| Teton dam | 1976-06-05 | Idaho | United States | 11 | Geological problems, including unsuitable bedrock, seismic activity, and caves. USGS, before completion: "Since such a flood could be anticipated, we might consider a series of strategically-placed motion-picture cameras to document the process." Water leakage eroded the earthen wall and led to dam failure. A total of 13,000 heads of cattle died. |
| Laurel Run dam | 1977-07-19 | Johnstown | United States | 40 | Heavy rainfall and flooding over-topped the dam. Six other dams failed the same day, killing five people. |
| Kelly Barnes dam | 1977-11-06 | Georgia | United States | 39 | Unknown, possibly design error as the dam was raised several times by owners to improve power generation. |
| Lawn Lake dam | 1982-07-15 | Rocky Mountain National Park | United States | 3 | Outlet pipe erosion; dam under-maintained due to location. |
| Upriver dam | 1986-05-20 | Spokane | United States | 0 | Lightning struck the power system; turbines shut down. The water rose behind the dam while trying to restart. Backup power systems failed, could not raise spillway gates in time. The dam overtopped (rebuilt). |

*Continued*

TABLE 2.1    List of dam failure in the North American region—cont'd

| Dam/incident | Date | Location | Country | Fatalities | Details |
|---|---|---|---|---|---|
| Meadow Pond dam | 1996-03-13 | New Hampshire | United States | 1 | Design and construction deficiencies failed in heavy icing conditions. |
| Saguenay Flood | 1996-07-19 | Quebec | Canada | 10 | Problems started after two weeks of constant rain, which severely engorged soils, rivers, and reservoirs. Post-flood inquiries discovered that the network of dikes and dams protecting the city was poorly maintained. |
| Martin County coal slurry spill | 2000-10-11 | Martin County | United States | 0 | Failure of a coal slurry impoundment. The water supply for over 27,000 residents was contaminated. The spill was 30 times larger than the Exxon Valdez oil spill and one of the worst environmental disasters ever in the southeastern USA. |
| Silver Lake dam | 2003-05-14 | Michigan | United States | 0 | Heavy rains caused the earthen Fuse plug dam and bank to wash away. A total of 1800 people evacuated. The flood caused the failure of the downstream Tourist Park dam. |
| Hope Mills dam | 2003-05-26 | North Carolina | United States | 0 | In heavy rains, the floodgate was held shut by water pressure. A total of 1600 people evacuated. |
| Big Bay dam | 2004-03-12 | Mississippi | United States | 0 | A small hole in the dam grew, spouted higher, and eventually led to failure. A total of 104 buildings were damaged or destroyed. |
| Taum Sauk reservoir | 2005-12-14 | Lesterville, Missouri | United States | 0 | Computer/operator error; gauges intended to mark dam full were not respected; dam continued to fill. Minor leakages had also weakened the wall through piping. The dam of the lower reservoir withstood the onslaught of the flood. |
| Ka Loko dam | 2006-03-14 | Kauai, Hawaii | United States | 7 | Heavy rain and flooding. Several possible specific factors include poor maintenance, lack of inspection, and illegal modifications (Hawaii Reporter, 2013). |

**TABLE 2.1**  List of dam failure in the North American region—cont'd

| Dam/incident | Date | Location | Country | Fatalities | Details |
|---|---|---|---|---|---|
| Lake Delton | 2008-06-09 | Lake Delton | United States | 0 | Failure in June 2008 Midwest floods; nearby highway washed out, creating a new channel that drained the lake. |
| Kingston Fossil Plant coal fly ash slurry spill | 2008-12-22 | Roane County | United States | 0 | Failure of a fly ash slurry pond. |
| Hope Mills dam | 2010-06-16 | North Carolina | United States | 0 | A sinkhole caused dam failure. The second failure of the dam will be replaced. |
| Testalinda dam | 2010-06-13 | Oliver | Canada | 0 | Heavy rain, low maintenance. Destroyed at least five homes. Buried Highway 97. |
| Delhi dam | 2010-07-24 | Iowa | United States | 0 | Heavy rain, flooding, malfunctioning spillway, and structural problems. Around 8000 people had to be evacuated. Replacement uncertain due to lake-dredging debt. |
| Mount Polley tailings dam failure | 2014-08-04 | British Columbia | Canada | 0 | The Tailings dam collapsed due to negligent operation; the reservoir was overfilled beyond design parameters despite repeated warnings of the danger (CBC News, 2017a,b; Macleans.ca, 2014). |
| Maple Lake | 2017-10-05 | Paw, Michigan | United States | 0 | A heavy rainstorm caused a dam section to crumble because of the weight of a pond above, which happened around 5 am. |
| Sanford dam, Patricia Lake | 2018-09-15 | Boiling Spring Lakes, North Carolina | United States | 0 | Overtopping after over 36″ rainfall during landfall of Hurricane Florence. |
| Spencer dam failure | 2019-03-14 | Near Spencer, Nebraska | United States | – | |
| Edenville dam | 2020-05-19 | Edenville, Michigan | United States | None reported | Heavy rains overtopped and breached the dam. |
| Sanford dam | 2020-5-19 | Sanford, Michigan | United States | None reported | The Edenville Dam's failure immediately upstream caused a large inflow into Sanford Lake, which overtopped the dam, flowing "all over and around" its structure (which remained intact). |

*Source: 2020. Wikipedia Dam Failure.*

Undoubtedly, hydropower plants' use for electricity generation has benefits and consequences for the environment and population that lives in the area where it will be built, so it is necessary to reach a balance to maintain sustainability.

According to Morales Pedraza (2013, 2015) and Parshley (2018), hydropower plants for electricity generation have certain advantages and disadvantages over other energy sources. These advantages are, among others, the following:

A-**Advantages**

- **No fuel cost**: the use of hydropower plants for electricity generation does not require any fuel to operate like most other energy sources, particularly in the case of conventional energy sources;

- **Low operation and maintenance costs**: the labor costs associated with a hydropower plant's operation are usually low since the plants are automated and have limited personnel on-site to operate them. The maintenance cost of a hydropower plant is even lower than in the case of other power plants. It is slightly less than 2% of total installed costs per year, with a variation of between 1% and 3% of total installed costs per year, depending on the size plant;

- **Lower cost of electricity**: the cost of electricity produced by a hydropower plant is quite low. For this reason, it is very attractive building a hydropower plant, but this is not possible in all the places where electricity is needed. Sometimes, they have to be located far from cities that need them. In those cases, the transmission loss could be high.

- **The recovery period of the investment**: the recovery period of the investment in the construction of a medium-size hydropower plant is estimated to be between four and seven years or between five and eight years;

- **No emissions of greenhouse gases**: hydropower plant during its operation does not emit greenhouse gases and pollute the air because its operation does not use any fossil fuel type. As a result, a hydropower plant's construction is very attractive as a cheap electricity source that does not produce carbon dioxide and methane, although under certain conditions. According to some expert's opinions, hydropower's most devastating effect might be that it is not emission-free, contrary to popular belief. There has been a lot of discussion about greenhouse gas emissions from reservoirs from submerged vegetation, particularly methane. "As trapped material decays in reservoirs, methane bubbles are released; tropical locations tend to have more vegetation and, therefore, higher methane emissions. These bubbles occur in natural reservoirs as well, but their rate increases when water passes through turbines" (Parshley, 2018);

- **Energy storage**: energy storage is possible in certain types of hydropower plants, such as hydropower pumped-storage plants. This type of hydropower plant is the ideal plant for the consumption of the electricity generated by solar parks and wind farms when this electricity is not needed and is not convenient to send to the electrical network. The hydropower pumped-storage plants consume the power generated by solar parks and winds farms by pumping water from a low reservoir to a high reservoir, and when there is a need for power in the electrical grid that solar parks and winds farms cannot generate, then the hydropower plant releases the water from the high reservoir for the production of the electricity demanded;

- **Size diversity**: hydropower plants can be built of various sizes from 1 MW to 10,000 MW, making them very versatile. Europe's governments promote the construction of small hydropower plants because they cause less ecological effects than large-scale hydropower plants. There is a strong movement of people against the construction of larger hydropower plants in most developed countries due to the negative impact on the environment and population. Micro-hydropower plants are also possible to build in many countries with a minimum adverse effect on the environment and population, particularly for the supply of electricity to small towns and villages located far from the electrical grid;
- **Reliability**: the use of hydropower plants for electricity generation is much more reliable than the use of wind farms and solar parks for the same purpose, albeit less than the use of oil, natural gas, coal, and nuclear power plants as a baseload electricity source. The use of hydropower plants for electricity generation is more or less predictable well in advance. However, it can decrease in the summer months when it rains little or during dry seasons;
- **High load factor**: the load factor for solar parks vary from 15% to 30%, and in the case of wind farms could be as higher as 47%, which is quite low in comparison with the load factor of power plants using fuel fossils and nuclear energy, which may have a load factor between 70% and 90% or more in some specific cases. Hydropower plants, on the other hand, has a load factor between 30% and 60% or more in some particular cases;
- **Long life**: Hydropower plants have a very long-life service of around 100 years, which is much longer than even nuclear power plants' life service. The long-life service of a hydropower plant implies that the cost of the life cycle of this kind of power plant becomes very low in the long term;
- **Hydroelectric resources are widely distributed throughout the world**: the potential of hydroelectricity exists in around 150 countries, and about 70% of the economically feasible potential remains to be developed, mainly in developing countries;
- **No fuel is burned**: fuel does not burn in a hydropower plant, so the environmental contamination is minimal. The use of hydropower plants for electricity generation is by far the cleanest way to produce electricity;
- **Support for the development of other renewable energy sources**: hydropower is an important element of support for the use of other renewable energy sources such as wind and solar energies for electricity generation, particularly in the case of hydropower pumped-storage plants;
- **Promotes energy security and price stability**: the price of the electricity generated by hydropower plants is very stable because it is not using any type of fossil fuel for electricity generation, and for this reason, it is independent of fluctuations in the fuel price;
- **Contributes to the storage of freshwater**: the reservoirs associated with a hydropower plant also offer the opportunity to improve the management of freshwater, allowing flood control, facilitating the irrigation of the land for food production, and river transport goods;
- **Hydropower plants are incredibly flexible in their operation**: hydropower plants allow smooth frequency control and rapid response in situations of emergency or high electricity demand;

- **Impulse force**: hydropower plants, especially those of higher capacity, are a driving force for regional development.

**B-Disadvantages**

However, there are also certain disadvantages to the use of hydropower plants for electricity generation. The main disadvantages are, according to Morales Pedraza (2013, 2015), Robbins (2007), Patrick and Chansen (1998), Sentürk (1994), Urban and Mitchell (2011), and Parshley (2018), the following:

- **Damage to the ecosystem and land loss could be beneficial for other purposes**: hydropower plants that use dams for their operation will submerge large land areas due to a reservoir's requirement. The large reservoirs required for hydropower plants' operation cause the immersion of vast areas upstream of the dams, destroying towns, villages, forests, swamps, and lowlands. Land loss is often exacerbated by the fragmentation of the habitat in the surrounding areas caused by the reservoir (Robbins, 2007);

- **Hydropower plants can be detrimental to the surrounding aquatic ecosystems, both upstream and downstream of the power plant site**: hydropower generation changes the downstream river's environment. The water leaving a turbine generally contains very little suspended sediment, causing scrubbing of riverbeds and riverbanks' loss. Because turbine gates often open intermittently, rapid or even daily fluctuations in river flow are observed;

- **Sedimentation and lack of flow**: when water flows, it can carry heavier particles downstream. That has a negative effect on dams and, subsequently, on hydropower plants, particularly in rivers or in catchment areas with high sedimentation. Sedimentation can fill a reservoir and reduce its ability to control flooding, in addition to causing additional horizontal pressure on the upstream part of the dam (Patrick and Chansen, 1998; Sentürk, 1994);

- **Changes in the amount of river water flow will correlate with the amount of energy produced by a dam**: lower river water flows will reduce the amount of water stored in a reservoir and reduce the amount of water used to generate electricity. The result of decreased river water flow may be an energy shortage in highly dependent areas on the electricity produced by a hydropower plant. The risk of water flow shortage may increase as a result of climate change (Urban and Mitchell, 2011);

- **Methane emissions**: these emissions, in the case of reservoirs associated with hydropower plants, are because the plant material in the flooded areas decomposes in an anaerobic environment and forms methane, a greenhouse gas. When the reservoir is large compared to the generation capacity (less than $100\,W$ per $m^2$ of the surface) and the felling of forests in the area was not undertaken before the construction of the reservoir begins, the emission of greenhouse gases from the reservoir may be higher than of a thermal power plant. "In 2016, researchers at Washington State University conducted a comprehensive meta-analysis, looking at 100 studies of emissions from over 250 reservoirs, and found that each square meter of reservoir surface emitted 25% more methane than previously recognized. In some cases, greenhouse gas emissions from hydropower were higher than a comparable fossil fuel power plant" (Parshley, 2018);

- **Relocation of the population living in the reservoir construction area**: a significant disadvantage of a reservoir associated with hydropower plants is the need to relocating the community who live where the repository will be built;
- **Risks of failure**: because large conventional reservoirs associated with hydropower plants retain large volumes of water, a failure due to poor water containment construction of the reservoir, the occurrence of natural disasters, or sabotage actions can be catastrophic for downstream settlements and the infrastructure associated with the power plant. Dam failures have been some of the largest human-made disasters in history. Banqiao's dam failure in southern China led directly to the death of 26,000 people, and 145,000 others died later by epidemics. Millions of people were left homeless. Furthermore, creating a dam in a geologically wrong location can cause extensive damage, such as the one caused by the 1963 disaster at the Vajont Dam in Italy, where nearly 2500 people died. In the North American region, a total of 42 dam failures have been reported, 90.5% of them in the USA;
- **High investment costs**: particularly in the case of large hydropower plants;
- **Dependency of the hydrology** (precipitation).

## The future of hydropower in the North American region

The coronavirus pandemic has underlined the hydropower sector's resilience and its critical role in delivering clean, reliable, and affordable energy, especially during a crisis. For this reason, a bold and ambitious green recovery plan involving significant investment in sustainable hydropower and other renewable energy sources for electricity generation and heating will be needed as part of the policy response at the world, regional, and national levels.

A summary of the achievement in the use of hydropower plants for electricity generation at the world level can be found in the following paragraphs, according to the Hydropower Status Report (2020) and IRENA Renewable Energy Statistics 2020:

- Electricity generation from hydropower plants achieved a record of 4306 TWh in 2019, the single most significant contribution from a renewable energy source in humankind history;
- Almost 13 GW of new hydropower plant capacity was built and put into operation in 2019. However, this new hydropower plant capacity is lower than the new hydropower plant capacity added in 2018 (22.4 GW);
- The total global hydropower installed capacity reached 1,307,994 GW in 2019. This total represents an increase of 1% concerning the level achieved in 2018, but is 1.2% down on the five-year annual average of 2.1%, and well below an estimated 2% yearly growth required to meet the Paris agreement climate change targets;
- Fifty countries added new hydropower capacity in 2019. The states with the highest individual increases in hydropower plant installed capacity were Brazil with 4.63 GW (4629 MW), and China with 4.14 GW (4142 MW);
- India overtook Japan (50,008 MW) as the fifth largest world hydropower producer, with its total installed capacity now standing at 50,225 MW;

- The 11,233 MW Belo Monte project in Brazil became fully operational in 2019. Other significant projects include the 1285 MW Xayaburi project in Laos, followed by the 990 MW Wunonglong and 920 MW Dahuaqiao projects in China.
- There was a decrease in the growth of pumped-storage hydropower plant installed capacity in 2019 due to project delays in China. However, growing interest in new pumped-storage hydropower projects has been observed across the world. For this reason, it can be expected an increase in pumped-storage hydropower plant capacity during the coming years.

In the specific case of the North American region, the following are, among others, the primary trend associated with the future use of hydropower plants for electricity generation:

- Hydroelectric development has been immense in Canada but less relevant in the USA. In Canada, the Canadian Shield rivers provide many sites, especially in Quebec and Ontario, to construct hydropower plants for electricity generation. These rivers are linked to the Great Lakes, which, in turn, tie into a power grid developed from Appalachian rivers. The north-central and north-eastern areas are thus excellent sites for the construction of hydropower plants for electricity generation. In addition, the snow-fed rivers from the high Cordilleras were impounded (as at the Grand Coulee, Hoover, Glen Canyon, Fort Peck, and Garrison dams), providing an immense amount of power that can be used for electricity generation.
- In Canada, hydropower plants generate most of the electricity produced (381,750 GWh in 2018), and this situation will not change during the coming decades. The hydropower technical potential in Canada has been estimated to be around 155,000 MW, almost double the current hydropower plant capacity installed in 2019 (81,053 MW). The number of jobs associated with the Canadian hydropower sector in 2019 is around 130,000.
- In the USA, the situation is different than in Canada. According to USA EIA (EIA Electricity Explained, 2020), in 2019, there were three significant energy sources for electricity generation in the country. These three energy sources were fossil fuels (coal, natural gas, and petroleum), nuclear power, and renewable energy sources (hydro, solar, wind, biomass, geothermal, tide, among others). Most electricity in the country is generated with steam turbines using fossil fuels. In 2019, 62% of the electricity generated in the country was produced by fossil fuels (38% by natural gas, 23% by coal, and 1% by oil), 20% by nuclear energy, and 17% by renewable energy sources (7.3% by wind energy, 6.6% by hydro, 1.8% by solar energy, 1.4% by biomass, and 0.4% by geothermal energy). In 2019, hydroelectricity represented only a small percentage of the total electricity generated in the USA (6.6%). However, considering only the electricity generated by all renewable energy sources, hydropower plants produced, in 2019, 42.7% of the total. In the USA, hydropower generates enough power "to serve the needs of 28 million residential customers. That is equal to all the homes in Wisconsin, Michigan, Minnesota, Indiana, Iowa, Ohio, Missouri, Nebraska, Kansas, North and South Dakota, Kentucky, and Tennessee" (Meyers, 2016).

Without a doubt, hydropower is an incredible energy source, tracing its usage back to the watermills that have been a staple of civilizations since ancient times. Today, hydropower is the largest source of renewable energy used to produce electricity in the world, with an installed capacity of 1,307,994 MW in 2019, generating a total of 4,267,085 GWh in 2018 (IRENA Renewable Energy Statistics 2020, 2020).

When considering new hydropower projects in the North American region, two critical elements should be in the USA and Canadian governments' minds. These elements are a) climate change and b) environment and social damage. Climate change is projected to significantly impact hydropower technology development in the future, as the large repository of water necessary for large hydropower plants' operation is increasingly under threat for environmental reasons. Climate change has fewer and fewer right places to build environmentally, economically, and socially acceptable large hydropower plants (Unwin, 2019).

Besides, it is crucial to be aware that the construction of large hydropower plants often involves considerable environmental damage and social displacement. The reservoir's construction floods a vast land area, forcing the entire population that lives in this area to move to another place. Building a large hydropower plant with a dam destroys jobs, breaks social relationships, and has adverse environmental and social effects. If the site selected is far away from large cities, then the electricity transmission losses could be very high.

## Hydropower plant installed capacity in Canada

In Canada, hydropower uses for electricity generation remain, in 2019, the dominant source of electricity supply, representing 62% of total electricity generation in the country and 55% of the total installed generation capacity. Based on this data, it can be stated that Canada is one of the few countries in the world that generate a significant amount of its electricity from hydropower plants.

The evolution of the hydropower plant installed capacity in Canada during the period 2010–19 is shown in Fig. 2.16.

According to Fig. 2.16, the following can be stated: the hydropower plant installed capacity in Canada increased by almost 8% during the period 2010–19, rising from 75,078 MW in 2010 to 81,053 MW in 2019. It is important to single out that during the period 2012–14, the

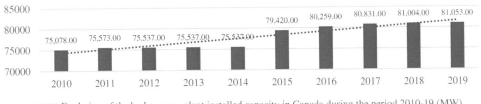

Evolution of the hydropower plants installed capacity in Canada during the period 2010-19 (MW)

**FIG. 2.16** The evolution of the hydropower plant installed capacity in Canada during the period 2010–19. *Source: 2020. IRENA Renewable Energy Statistics 2020.*

hydropower plant installed capacity in the country kept at the same level. After 2014, the hydropower plant installed capacity increased by 7.3%. It is expected that the hydropower plant installed capacity in Canada will continue to grow during the coming years because the Canadian government considers this type of clean energy source as the primary energy source for electricity generation within its energy mix. Canada is ranking four at the world level according to the hydropower plant installed capacity in 2019, behind China with 356,403 MW, Brazil with 109,092 MW, and the USA with 102,769 MW (IRENA Renewable Energy Statistics 2020, 2020).

In the specific case of pumped-storage hydropower plant installed capacity in Canada, since 2012, this type of power plant's capacity has stayed at 174 MW. However, it is expected that this trend will change during the coming years due to the entry into operation of five pumped-storage hydropower plants now under construction with a total capacity of 2400 MW. As a result, the new pumped-storage hydropower plant capacity to be installed in the county is almost 14 times the current hydropower capacity installed.

Besides, it is important to stress that the Canadian hydropower plants supply approximately 1% of all US electricity consumed in the country, particularly in New York, New England, the Midwest, and the Pacific Northwest, according to the Canadian Hydropower Association.

It is also important to single out that since 2005, the Canadian renewable electricity generation has grown more using hydropower as a source of energy (growth of 40,000 GWh) than from wind and solar energies together (30,000 GWh). That means that electricity generation's growth using hydropower plants is 33.3% higher than the electricity generation using wind farms and solar parks together. In 2019, the electricity generated by hydropower plants, wind farms, solar parks, biomass power plants, and geothermal power plants represented 62% of Canada's total electricity production that year (Region North and Central America, 2020).[n]

Several factors, including the near-completion of several new major hydropower power plants and the regulated phase-out of coal-fired power plants, drive the country's continued growth in hydroelectricity production and installed capacity.

"The most significant hydropower projects have a total installed capacity of almost 3000 MW and include the following facilities:

- Site C in British Columbia (1100 MW, in-service by 2025);
- Muskrat Falls in Newfoundland and Labrador (824 MW, in service by 2020);
- Keeyask in Manitoba (695 MW, in service in 2020);
- La Romaine Complex Unit 4 in Quebec (245 MW in service in 2021)" (Region North and Central America, 2020).

Pumped-storage hydropower projects in Canada's development include five hydropower plants with a total installed capacity of up to 2400 MW. These power plants are the following:

- TC Energy Pumped-Storage Hydropower Plant with a capacity of 1000 MW in Ontario;
- Marmora Pumped-Storage Hydropower Plant with a capacity of 400 MW in Ontario;

[n] According to the BP Statistical Review of World Energy 2020 report, the electricity generated by all renewable energy sources in 2019 in Canada represented 65.3% of the total.

- Brazeau Pumped-Storage Hydropower Plant with a capacity between 300 MW and 900 MW in Alberta;
- Canyon Creek Pumped-Storage Hydropower Plant with a capacity of 75 MW in Alberta;
- Moon Lake Pumped-Storage Hydropower Plant with a capacity of 25 MW in British Columbia (Region North and Central America, 2020).

Several new transmission projects that would interconnect Canadian provinces with surplus hydroelectricity with those provinces phasing out coal-fired power plants (and territories seeking cleaner alternatives to diesel fuel in remote areas) continue to be studied by the competent Canadian authorities and energy industry representatives. Additional transmission lines capacity to enable electricity imports from the USA and electricity exports to the USA are also under development or construction. It is forecast that hydropower plants' electricity production will have grown by a further 9% by 2030 from the level registered in 2019.

## Hydropower electricity generation in Canada

The participation of renewable energy sources in the country's electricity generation, including hydro, wind, solar, biomass, and geothermal energies, represented approximately 65.3% of Canada's total electricity production in 2019 (around 660.4 TWh) (BP Statistical Review of World Energy 2020 report). The whole electricity generation by hydropower plants, in 2019, represents 88.6% of the total electricity produced by all renewable energy sources (431.3 TWh).

Based on the data included in Fig. 2.17, the following can be stated: the electricity generation in Canada by hydropower plants increased by 8.6% during the period 2010–18, rising from 351,461 GWh in 2010 to 381,750 GWh in 2018. In 2019, hydropower plants in Canada generated 382 TWh (BP Statistical Review of World Energy 2020, 2020). However, it is important to single out that the electricity generated by pumped-storage hydropower plants

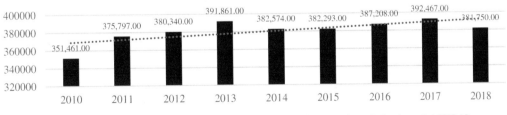

**FIG. 2.17** The evolution of the hydropower electricity generation in Canada during the period 2010–18. *Source: 2020. IRENA Renewable Energy Statistics 2020.*

remained stable throughout the whole period at 111 GWh. Therefore, it is expected that electricity generation using hydropower plants will continue the trend shown in Fig. 2.20 during the coming years. Furthermore, in the case of pumped-storage hydropower plants, it is expected that the electricity generated by this type of power plant in the future will increase as a result of the entry into operation of five new facilities with a capacity of 2400 MW.

The oscillation in the level of electricity generated by hydropower plants shown in Fig. 2.17 is due to the increase or decrease in the level of rain registered each year.

## The current and future role of hydropower in Canada

Without a doubt, Canada needs a rapid transition from the use of coal and oil to the use of more renewable energy sources for electricity generation to meet its climate change commitments within the Paris agreement on climate change. "It is also clear that the shift will require a great deal more low-carbon electricity to power everything from electric cars, to public transit, to residential and commercial building heating, to industrial processes" (Wilt, 2017).

According to some experts' calculations, the Canadian energy sector's full decarbonization will essentially double the country's electricity demand. This increase should be met by the increased use of renewable energy sources for this purpose, particularly the use of hydropower plants.

The Government of Canada's report entitled "Mid-Century Long-Term Low-Greenhouse Gas Development Strategy," published in 2016, cited projections of an increase between 113% and 295% in the total electricity generation during the period 2013–50. According to the three sources cited in the report, a significant increase in new hydropower plant capacity between 36,000 MW and 130,000 MW will be registered within the period mentioned above. In Canada, where most of the hydropower plants were commissioned in the 1970s, the energy transition will be carried out from the smallest fix to complete refurbishments of existing hydropower plants in operation as well as the construction of new hydropower plants.

Canada had, in 2019, a total of 81,053 MW in hydropower plant capacity installed, generating a total of 381,750 GWh in 2018, making it the third-largest hydro producer in the world, after China with an electricity generation of 1,232,100 GWh, and Brazil with 388,971 GWh. For this reason, Canada has a much cleaner electrical grid than most other countries in the world.

Most of the scenarios that can be considered up to 2050 or beyond indicate that there would be a need to increase the current level of hydropower plant installed capacity in the country significantly. By 2050, hydropower plant capacity will be expected to double the current hydropower plant capacity that exists in the country today if Canada wishes to fulfill its commitments with the Paris agreement on climate change.

## Hydropower plant installed capacity in the United States

Hydroelectric power in the USA is, as of 2019, the largest renewable energy source in both generation and nominal capacity, with 102,769 MW capacity installed generating, in 2018, a total of 317,004 GWh or 7.4% of the world hydropower generation (IRENA Renewable Energy Statistics 2020, 2020). In the USA, hydropower plants can be found in 50 states. According to National Hydropower Association sources, some states in the Pacific Northwest generate most of their electricity from hydropower plants. However, the USA's federal hydropower infrastructure is aging, and while some investment has been made in the past, it is simply not enough. Federal funding for capital investment, operation, and maintenance appropriations, including necessary maintenance of the existing hydropower infrastructure, has been severely constrained for years. This lack of appropriate resources to support the hydropower plants' correct functioning ignores the fact that the US federal government owns 48% of the country's total installed hydroelectric generation capacity.

Besides, several recent studies have noted potential increases in federal hydropower through efficiency gains and non-federal power development at federal sites. For instance, Corps and Reclamation staff estimated that approximately 2.4 GW of new hydropower capacity could be built in the future, representing a 7% increase on top of existing federal hydropower capacity in operation. This new potential may be possible through additional hydropower construction or the refurbishment of existing hydropower plants (Miller, 2018).

In the USA, there are three types of hydropower plants. Some conventional hydropower plants are incorporated into dammed reservoirs, providing them with a relatively reliable water supply. Other conventional hydropower plants are run-of-river, meaning they have no reservoir, and their generating output is subject to changes in river flows. Finally, unlike conventional hydropower plants, pumped-storage hydropower plants can function as energy storage. They can reverse their turbines, using the grid's power to move water from a reservoir located at a low level into a reservoir located at a high level. The water is then stored there and will wait to generate electricity at a later time when it is needed.[o]

In the USA, total hydropower capacity, including pumped-storage facilities, remained at 102,769 MW in 2019. While most recent hydropower growth comes from small hydropower projects, "the Department of Energy estimates there is nearly 50 GW of untapped hydropower potential, including 30 GW of pumped-storage facilities and 5 GW of development at non-powered dams. The existing hydropower plants continue to play a critical role, providing carbon-free flexibility and reliability as fossil fuels are replaced with intermittent renewables, particularly coal-fired power plants. Most states are setting carbon-free goals over the next 30 years and relying on hydropower to help achieve these" (Region North and Central America, 2020).

The evolution of hydropower plant installed capacity in the USA during the period 2010–19 is shown in Fig. 2.18.

According to Fig. 2.18, the USA's hydropower plant installed capacity during the period 2010–19 increased by 1.7%, rising from 101,023 MW in 2010 to 102,769 MW in 2019. Therefore, it is expected that the hydropower plant installed capacity in the USA will continue to increase during the coming years, following the trend shown in Fig. 2.18.

In the specific case of pumped-storage hydropower plant capacity in the USA, the evolution of this capacity during the period 2010–19 is shown in Fig. 2.19.

---

[o] It is important to mention that the Bath County pumped-storage hydropower plant is the largest such facility in the world. Other hydropower power plants of this type include Raccoon Mountain pumped-storage hydropower plant, Bear Swamp pumped-storage hydroelectric power plant and Ludington pumped-storage hydropower plant on Lake Michigan and previously the largest in the world (Wikipedia Hydroelectric Power in the United States, 2020).

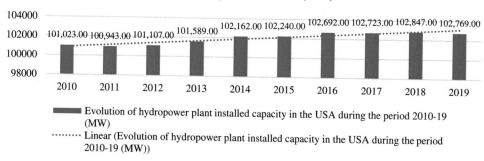

FIG. 2.18   The evolution of the hydropower plant capacity in the USA during the period 2010–19. *Source: 2020. IRENA Renewable Energy Statistics 2020.*

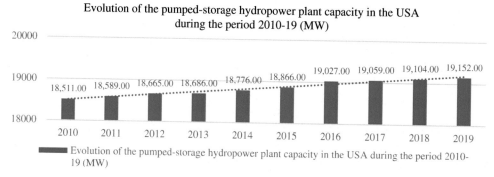

FIG. 2.19   The evolution of the pumped-storage hydropower plant installed capacity in the USA during the period 2010–19. *Source: 2020. IRENA Renewable Energy Statistics 2020.*

Based on the data included in Fig. 2.19, the following can be stated: pumped-storage hydropower plant installed capacity in the USA during the period 2010–19 increased by 3.5%, rising from 18,511 MW in 2010 to 19,152 MW in 2019. It is expected that this trend will continue during the coming years.

## Hydropower electricity generation in the United States

The first industrial use of hydropower to generate electricity in the USA was in 1880 to power 16 brush-arc lamps at the Wolverine Chair Factory in Grand Rapids, Michigan. The first USA hydroelectric power plant to sell electricity opened on the Fox River near Appleton, Wisconsin, on September 30, 1882. Without a doubt, hydropower was one of the first energy sources used for electricity generation in the USA.

According to the USA EIA source, until 2019, hydropower was the largest annual USA renewable electricity generation source. It accounted for about 6.6% of the whole USA utility-scale electricity generation and 42.7% of entire utility-scale renewable electricity generation. However, "hydroelectricity's share of total USA electricity generation has decreased over time, mainly because of increases in electricity generation from other sources" (EIA Hydropower Explained, 2020).

The evolution of the electricity generation by hydropower plants in the USA during the period 2010–18 is shown in Fig. 2.20.

According to Fig. 2.20, the following can be stated: electricity generation by hydropower plants in the USA increased by 10.7% during the period 2010–18, rising from 286,333 GWh in 2010 to 317,004 GWh in 2018. Therefore, it is anticipated that the general trend shown in Fig. 2.20 will continue without change in the future.

In the specific case of electricity generation using pumped-storage hydropower in the USA, this production's evolution during the period 2010–18 is shown in Fig. 2.21.

According to Fig. 2.21, the following can be stated: the evolution of the electricity generation by pumped-storage hydropower plants in the USA during the period 2010–18 decreased by 12.5%, falling from 24,067 GWh in 2010 to 21,053 GWh in 2018. Therefore, it is expected that the general trend shown in Fig. 2.21 will continue without change, at least during the coming years.

In the USA, there are about 1460 conventional and 40 pumped-storage hydropower plants in operation. The oldest operating USA hydropower facility is the Whiting plant in Whiting, Wisconsin, which started operation in 1891 and has a total generation capacity of about 4 MW. Most USA hydroelectricity is produced at large dams on major rivers, and most of these hydroelectric dams were built before the mid-1970s by federal government agencies. The largest USA hydropower plant is the Grand Coulee hydro dam on the Columbia River on the Washington and Oregon border with 7070 MW total generation capacity (EIA Hydropower Explained, 2020).

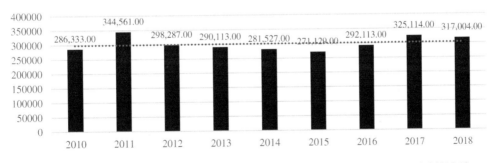

Evolution of electricity generation by hydropower plants in the USA during the period 2010-18 (GWh)

■ Evolution of electricity generation by hydropower plants in the USA during the period 2010-18 (GWh)
⋯⋯ Linear (Evolution of electricity generation by hydropower plants in the USA during the period 2010-18 (GWh))

**FIG. 2.20** The evolution of the electricity generation by hydropower plants in the USA during the period 2010–18. *Source: 2020. IRENA Renewable Energy Statistics 2020.*

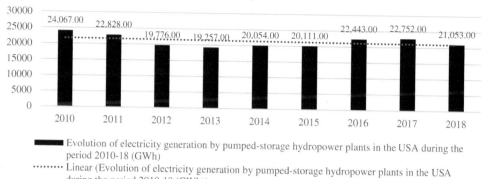

FIG. 2.21   The evolution of the electricity generation by pumped-storage hydropower plants in the USA during the period 2010–18. *Source: 2020. IRENA Renewable Energy Statistics 2020.*

Without a doubt, the USA has a long history of building low-head or small dams (two million of them were constructed in the country).[P] However, according to Fencl et al. (2015), the claim of the minimal impact of low-head or small dams is mostly untested. By their abundance, small dams can substantially impact flowing aquatic ecosystems (Premalatha et al., 2014). "Small hydro dams possess the same characteristics as large dams, with the only difference being their size" (Moran et al., 2018).

Summing up, the following can be stated: the Hydropower Vision (2016) report highlights the enormous potential of untapped hydropower resources across the USA, finding that hydropower could grow from around 101 GW of combined generating and storage capacity to nearly 150 GW by 2050. More of this growth (more than 50%) will be registered by 2030. "Growth under this scenario would result from a combination of 13 GW of new hydropower generation capacity (upgrades to existing plants, adding power at existing dams and canals, and limited development of new stream-reaches), and 36 GW of new pumped-storage capacity. Between 2017 and 2050, hydropower could save US$209 billion in avoided damages from greenhouse gas emissions," including US$185 billion in savings from the existing hydropower fleet being operated through 2050, "US$58 billion from avoided healthcare costs and economic damages due to air pollution, and 30 trillion gallons of water, equivalent to roughly 45 million Olympic-size swimming pools" (Hydropower Vision, 2016). According to the mentioned report, over 195,000 hydropower-related jobs spread across the country in 2050. During the period 1950–2015, hydropower plants provided a cumulative 10% of USA electricity and 85% of cumulative USA renewable power generation.

In the near term, USA hydropower generation growth until 2030 is estimated as 9.2 GW, driven "primarily from upgrades of existing hydropower plants (5.6 GW) and powering non-powered dams (3.6 GW). Long-term growth of 3.4 GW between 2030 and 2050 includes

---

[P] There are over 82,000 large dams in the USA alone as it was mentioned before.

1.7 GW" (Hydropower Vision, 2016) of new stream-reach development, for a total of 12.6 GW of new growth by 2050. It is essential to single out that the "potential exists to increase new stream-reach development beyond this level" (Hydropower Vision, 2016). However, "this development is unlikely to occur without significant, transformational innovation in technology and development approaches that can lower costs and meet environmental sustainability requirements" (Hydropower Vision, 2016).

On the other hand, pumped-storage hydropower plant capacity will likely increase in the near term (up to 2030) by 16.2 GW. In the long term (up to 2050), pumped-storage hydropower plant capacity is projected to increase by 19.3 GW. Based on what has been said before, it is expected to have, by 2050, a total of 35.5 GW of new pumped-storage hydropower plant capacity installed in the USA. This growth is driven primarily by developed growth in other renewable energy sources, such as wind and solar energies, "and by the inherent flexibility of pumped-storage and its ability to provide needed operating reserves and other essential grid reliability services" (Hydropower Vision, 2016). With increased pumped-storage hydropower deployment in the USA, it is projected that this type of power plant "will provide more operating reserves (52%) than any other energy technology by 2050" (Hydropower Vision, 2016).

## The current and future role of hydropower in the United States

Existing hydropower facilities have high value within the USA energy sector, providing low-cost, low-carbon, renewable energy as well as flexible grid support services. According to the DoE Office of Energy Efficiency and Renewable Energy (n.d.) in its report entitled "A New Vision for United States Hydropower," the use of hydropower plants for electricity generation in the USA "has significant near-term potential to increase its contribution to the nation's clean generation portfolio via economically and environmentally sustainable growth through optimized use of existing infrastructure. In the USA, significant potential exists for constructing new pumped-storage hydropower plants to give the electrical grid more flexibility. Simultaneously, installing more pumped-storage hydropower plants will allow an increased integration of unstable generation resources, such as wind farms and solar parks. Meeting the long-term potential for growth at potential sites that are not developed for hydropower is contingent upon a continued commitment to innovative technologies and strategies to increase economic competitiveness while meeting the need for environmental sustainability."

While hydropower development has, in some cases, adverse effects on river systems and the species that depend upon them, in general, the use of hydropower plants for electricity generation offers more advantages than disadvantages. In addition, the development of hydro technology continues to reduce the negative environmental impact caused by building a dam. Accordingly, the Hydropower Vision (2016) report shows increasing expectations for new hydropower development under which environmental gains are maintained, and the trend of improvement continues. "Sustainable hydropower fits into the water-energy system by ensuring that the ability to meet energy needs is balanced with the functions of other water management missions in the present as well as into the years ahead. In some cases, dam removal and site restoration may be part of meeting the sustainability objective" (Hydropower Vision, 2016).

The future development of the hydropower sector in the USA rests on three foundational principles. These principles are, according to the Hydropower Vision (2016), the following:

- Optimization;
- Growth;
- Sustainability.

Seven key factors characterize the critical role that hydropower has in the USA power sector. These key factors are, according to Hydropower Vision (2016), the following:

- Hydropower has been a cornerstone of the USA electrical grid and the most relevant within all renewable energy sources used in the country for electricity generation, providing 42.7% of the total electricity generated by this type of energy source.[q] Renewable energy sources generated 17.3% of the total electricity produced in the USA in 2019 (BP Statistical Review of World Energy 2020, 2020). The use of hydropower plants for electricity generation provides low-cost, low-carbon, renewable, and flexible power within the country for more than a century;
- Existing hydropower plants in the USA have high value based on their ability to provide flexible electricity generation when most needed, and energy services, ancillary grid services, multi-purpose water management, and social and economic benefits, including avoidance of greenhouses gas emissions;
- Hydropower technology can grow and contribute to additional electricity production in the future USA energy mix, including significant near-term potential for economically and environmentally sustainable growth, and despite the new USA administration energy policy that supports the use of fossil fuels for electricity generation, particularly coal. The growth mentioned above can be achieved by optimizing existing hydropower infrastructure through plant upgrades and adding new generation capabilities to non-powered dams and water conveyances, such as irrigation canals;
- Long-term hydropower growth potential, particularly at undeveloped sites within the USA, will rely on developing innovative and economically competitive hydropower technologies that are not yet ready to be used. The long-term hydropower growth potential will also depend on the extent to which new hydropower projects can be developed at lower costs and with improved environmental sustainability strategies;
- The USA has significant resource potential for new pumped-storage hydropower plants "as a continued storage technology, enabling grid flexibility and greater integration of variable generation resources, such as wind and solar energies" (Hydropower Vision, 2016);
- Hydropower technical design innovations, advanced project implementation strategies, optimized regulatory processes, and the application of sustainability principles will be important in determining the future role that hydropower technology can have within the USA energy mix (Hydropower Vision, 2016).

---

[q] According to the BP Statistical Review of World Energy 2020 report, hydropower plants generated, in 2019, a total of 35.6% of the total electricity generated by all renewable energy sources in the country.

The economic and societal benefits of both existing and potential new hydropower plants in the USA are substantial. They include job creation, cost savings in avoided mortality and economic damages from air pollutants, and prevented greenhouse gas emissions. It is projected that the USA hydropower plant capacity could grow from its current 102,769 MW registered in 2019 to nearly 150 GW of combined electricity generating and storage capacity by 2050; this means an increase of 45.9% for the whole period (Hydropower Vision, 2016).

Finally, it is important to stress that the USA has been responsible for most of the growth in energy supply investment during the period 2010–19, "with increases in both oil and gas, supported by more spending on shale, and in the power sector. While oil and gas spending has moderated somewhat in the past three years (even as it grew strongly from 2017 to 2018), that for electricity networks rose. Compared to 2015, investment in renewable power and gas power remained relatively stable but at high levels. Meanwhile, investment in energy efficiency has declined over the period" (IEA, 2019).

# References

Boegelsack, T., 2019. Investment Opportunities and Energy Return of Hydropower and Geothermal Energy; Bögelsack Energy & Fuels Review.

Born, S.M., Genskow, K.D., Filbert, T.L., Hernandez-Mora, N., Keefer, M.L., White, K.A., 1998. Socioeconomic and institutional dimensions of dam removals: the Wisconsin experience. Environ. Manag. 22 (3), 359–370.

BP Statistical Review of World Energy 2020, 2020. British Petroleum, sixty-ninth ed.

Anon., 2020. Canadian Hydropower Association Water Power Canada 2020.

Castlewood Canyon State Park, 2007. A Brief History. Colorado Parks and Wildlife, State of Colorado.

Castro-Diaz, L., Lopez, M.C., Moran, E., 2018. Gender-differentiated impacts of the Belo Monte hydroelectric dam on downstream fishers in the Brazilian Amazon. Hum. Ecol. 46, 411–422. Google Scholar.

CBC News, 2017a. Former Tailings Pond Engineers for Mount Polley Say They Made Warnings. CBC News. Retrieved 2017-12-28.

CBC News, 2017b. Mount Polley Mine Tailings Breach Followed Years of Government Warnings. CBC News. Retrieved 2017-12-28.

Chen, J., Shi, H., Sivakumar, B., Peart, M.R., 2016. Population, water, food, energy, and dams. Renew. Sustain. Energy Rev. 56, 18–28. Google Scholar.

Anon., 2015. Cherry Creek Flood in 1933. Denver Public Library; Western History and Genealogy Division.

Coleman, N.M., 2018. Johnstown's Flood of 1889—Power Over Truth and The Science Behind the Disaster. Springer International AG, ISBN: 978-3-319-95215-4.

Conca, J., 2014. The Hidden Cost of Hydroelectric Power. Forbes.

Del Bene, D., Scheidel, A., Temper, L., 2018. More dams, more violence? A global analysis on resistances and repression around conflictive dams through co-produced knowledge. Sustain. Sci. 13, 617–633. Google Scholar.

Disaster Nearly Drowns Denver in 1933, 2019. Ion Colorado. February 1, 2019.

Anon., 2020. Disaster Survival Resources Dam Failure 2020.

DoE Office of Energy Efficiency and Renewable Energy, n.d. A New Vision for United States Hydropower. Water Power Technologies Office.

Egré, D., Milewski, J.C., 2002. The diversity of hydropower projects. Energy Policy 30 (14), 1225–1230. Research Gates.

Anon., 2020. EIA Electricity Explained. EIA.

Anon., 2020. EIA Hydropower Explained. EIA.

Electropaedia Hydroelectric Power, n.d. Woodbank Communications Ltd, U.K. https://www.mpoweruk.com/hydro_power.htm.

Anon., 2020. Facts About Hydropower. International Hydropower Associations (IHA).

Fencl, J.S., Mather, M.E., Costigan, K.H., Daniels, M.D., 2015. How big of an effect do small dams have? Using geomorphological footprints to quantify spatial impact of low-head dams and identify patterns of across-dam variation. PLoS One 10, e0141210. Google Scholar.

Hanna, P., Vanclay, F., Langdon, E.J., Arts, J., 2016. The importance of cultural aspects in impact assessment and project development: reflections from a case study of a hydroelectric dam in Brazil. Impact Assess. Proj. Apprais. 34, 306–318. Google Scholar.

Hawaii Reporter, 2013. Kauai Dam Breach Killed 7 People Five Years Ago, But Criminal Charges Against Dam Owner Still Pending. Hawaii Reporter. http://www.hawaiireporter.com/kauai-dam-breach-killed-7-five-years-ago-but-criminal-charges-still-pending/123;. Retrieved July 24, 2013.

Heming, L., Waley, P., Rees, P., 2001. Reservoir resettlement in China: past experience and the three gorges dam. Geogr. J. 167, 195–212. CrossRef; Google Scholar.

Anon., 2020. Hydroelectric Power in North America: What Are the Facts? Energy Matters.

Anon., 2011. Hydroelectric Power Resources Form Regional Clusters. USA Energy Information Administration (EIA).

Hydropower Status Report, 2020. Sector Trends and Insights 2020.

Hydropower Vision, 2016. A New Chapter for America 1st Renewable Electricity Source. USA Department of Energy, USA.

IEA, 2019. World Energy Investment 2019. IEA, Paris. https://www.iea.org/reports/world-energy-investment-2019. 2020.

Anon., 2019. International Hydropower Association (IHA) Country File Canada.

IRENA, 2012. A cost analysis of hydropower. In: Road Transport: The Cost of Renewable Solutions.

Anon., 2020. IRENA Global Renewables Outlook 2020.

Anon., 2020. IRENA Renewable Energy Statistics 2020.

Anon., 2018. IRENA Renewable Power Generation Cost Report 2018.

Kondolf, G.M., 2014. Sediment Handling at the Intake of the Hydropower Plants: A Toolbox for Decision Making. Norwegian University of Science and Technology (NTNU).

Kumar, A., Shei, T., Ahenkora, A., Caceres Rodriguez, R., Devernay, J.M., Freitas, M., Hall, D., Killingtveit, A., Liu, Z., 2011. Hydropower. In: Edenhofer, O., Pichs-Mdruga, R., Sokona, Y., Seyboth, K., Matschoss, P., Kadner, S., Stechow, C. (Eds.), IPCC Special Report on Renewable Energy Sources and Climate Change Mitigation. Cambridge University Press, Cambridge, and New York.

Leturcq, G., 2018. Dams in Brazil: Social and Demographic Impacts. Springer International Publishing, Cham. Google Scholar.

Macleans.ca, 2014. First Nations Chief: Warning About B.C. Tailings Pond 'Ignored'. Macleans.ca. 2014-08-05; Retrieved December 28, 2017.

Maclin, E., Sicchio, M., 1999. Dam Removal Success Stories. Restoring Rivers Through Selective Removal of Dams That Don't Make Sense. American Rivers, Friends of the Earth, & Trout Unlimited, Washington, DC. Google Scholar.

Meyers, G., 2016. Hydropower Costs/Renewable Energy Hydroelectricity Costs vs. Other Renewable & Fossil Costs. PlanetSave.

Anon., 2016. Mid-Century Long-Term Low-Greenhouse Gas Development Strategy. Government of Canada.

Miller, R., 2018. A look at innovative financing strategies for federal hydropower infrastructure. In: Investing in Hydropower. HDR.

Morales Pedraza, J., 2013. Energy in Latin America and the Caribbean: The Current and Future Role of Renewable and Nuclear Energy Sources in the Regional Electricity Generation. Nova Publishers.

Morales Pedraza, J., 2015. Electrical Energy Generation in Europe. The Current Situation and Perspectives in the Use of Renewable Energy Sources and Nuclear Power for Regional Electricity Generation. Springer, ISBN: 978-3-319-16082-5.

Moran, E.F., Lopez, M.C., Moore, N., Müller, N., Hyndman, D.W., 2018. Sustainable hydropower in the 21st century. PNAS 115 (47), 11891–11898. first published November 5, 2018 https://doi.org/10.1073/pnas.1809426115.

Morris, G.L., Fan, J., 1997. Reservoir Sedimentation Handbook: Design and Management of Dams, Reservoirs and Watershed for Sustainable Use. McGraw-Hill, New York, NY. 805 pp.

Anon., 2020. Operational and Maintenance Cost of Hydropower Plant. Hallidays Hydropower International Renueve From Rivers.

Parshley, L., 2018. The cost and benefits of hydropower. Undark Magazine, Smithsonian Magazine.

Patrick, J., Chansen, H., 1998. Teaching Case Studies in Reservoir Siltation and Catchment Erosion. TEMPUS Publications, Great Britain.

Premalatha, M., Abbasi, T., Abbasi, S.A., 2014. A critical view on the eco-friendliness of small hydroelectric installations. Sci. Total Environ. 481, 638–643.

Anon., 2020. Region North and Central America. International Hydropower Association (IHA).

Resch, G., Held, A., Faber, T., Panzer, C., Toro, F., Haas, R., 2008. Potentials and prospects for renewable energies at global scale. Energy Policy 36, 4048–4056.

Richter, B.D., 2010. Lost in development's shadow: the downstream human consequences of dams. Water Altern. 3, 14–42. Google Scholar.

Rideau Canal, 2020. Tales of the Rideau: Washed Away, The Hogs Back Dam. www.rideau-info.com. Retrieved February 2, 2020.

Robbins, P., 2007. Hydropower; Encyclopedia of Environment and Society. University of Wisconsin, Madison, USA.

Scudder, T., 2012. The Future of Large Dams: Dealing with Social, Environmental, Institutional and Political Costs. Routledge, London. Google Scholar.

Sentürk, F., 1994. Hydraulics of Dams and Reservoirs (reference. ed.). Water Resources Publications, Highlands Ranch, CO, ISBN: 0-918334-80-2, p. 375.

Stamey, T.C., 1996. Summary of Data-Collection Activities and Effects of Flooding From Tropical Storm Alberto in Parts of Georgia, Alabama, and Florida. July 1994 (U.S. Geological Survey, Reston, VA), Series No. 96-228, Google Scholar.

Unwin, J., 2019. The Future of Hydropower Energy. Power Technology.

Urban, F., Mitchell, T., 2011. Climate Change, Disasters and Electricity Generation. Overseas Development Institute and Institute of Development Studies, London.

US Army Corps of Engineers, 2016. National Inventory of Dams. Available at nid.usace.army.mil/. (Accessed 12 October 2018), Google Scholar.

Wang, P., Wolf, S.A., Lassoie, J.P., Dong, S., 2013. Compensation policy for displacement caused by dam construction in China: an institutional analysis. Geoforum 48, 1–9. Google Scholar.

Anon., 2020. Wikipedia Dam Failure.

Anon., 2020. Wikipedia Energy Return on Investment.

Anon., 2020. Wikipedia Hydroelectric Power in the United States.

Wilt, J., 2017. What's the Future of Hydroelectric Power in Canada? The Narwhal.

# 3

# Solar energy for electricity generation

## Introduction

Solar energy is a reliable source for providing clean energy to many countries worldwide, particularly those located in sunny areas. Although solar energy is still a costly energy source used for electricity generation in many countries compared to the use of other energy sources for the same purpose, particularly natural gas and oil, it is a fact that solar systems' cost has been decreasing significantly during the last decade. It is expected that this trend will continue for years to come making the use of solar energy more attractive for electricity generation for several countries all over the world. Experts predict that the price of solar electricity production will be comparable to other energy sources in the next decade (Delucchi and Jacobson, 2013). However, despite the falling of solar power cost, the cost associated with this energy source is a costly option for many developing countries.

In 2019, the first year when the expansion of the use of renewable energy sources for electricity generation is seen to slow down, total solar power capacity expansion was also well below average, and non-renewable capacity expansion was also remarkably low. According to the IRENA Renewable Capacity Statistics 2020 report, Asia once again achieved over 50% of new renewable facilities installed in 2019 despite a slightly slower pace. On the other hand, Europe and North America saw an upturn in their renewable capacity expansion during 2019. It is important to single out that wind and solar power accounted for 90% of the world's newly added renewable capacity in 2019. In part, this extremely high share reflects low hydropower growth as several large projects missed expected completion deadlines.

According to IRENA 2020 Insights on Renewables (IRENA, 2020a), "the share of renewable energy sources in world power capacity expansion continued its upward trend and reached 72% in 2019. Similarly, the renewable share of total power generation capacity rose from 33.3% in 2018 to 34.7% in 2019"; this means an increase of 1.4%. "At the regional level, in 2019, non-renewable capacity expansion continued to follow long-term trends, with net growth in Asia, the Middle East, and Africa, but net decommissioning in Europe and North America and little change in other regions. However, renewables still accounted for at least 70% of total capacity expansion in almost all regions in 2019. The two exceptions were Africa and the Middle East, where renewables accounted for only 52% and 26%, respectively, of net additions" (IRENA, 2020a).

Two leading solar technologies are available for electricity generation in the world. They can be classified as passive or active according to how they capture, convert, and distribute solar energy. Active technologies include the use of solar PV panels and thermal solar collectors to collect energy from the Sun. Among the passive technologies, there are different techniques framed in bioclimatic architecture: the orientation of buildings toward the Sun, the selection of materials with a good thermal mass or that have properties for light scattering, as well as the design of spaces through ventilation natural (Wikipedia Solar Energy, 2019).

## Types of solar energy

It is important to single out that solar energy is used in many countries in all regions of the world, not only for electricity generation but also for heating and desalinating water. There are two types of solar power facilities:

- Solar photovoltaic (PV);
- Concentrated solar power (CSP).

### Solar photovoltaic

Solar PV, also called "solar cells", are electronic devices that transform the sunlight into electricity using special materials with photovoltaic effects (see Fig. 3.1). In other words, a solar PV system converts the Sun's radiation, in the form of light, into usable electricity and heating. It comprises two main components: a) the solar array; and b) the balance of system components (BOS).

**FIG. 3.1**   Solar arrays at Desert Sunlight California, USA. *Source: US Department of the Interior First Solar Desert Sunlight Solar Farm, California, USA. https://commons.wikimedia.org/wiki/File:02-09  15_First_Solar_Desert_Sunlight_ Solar_Farm_(15863210084).jpg.*

Various aspects can categorize a solar PV system. These aspects are the following:

- Grid-connected vs. stand-alone systems;
- Building-integrated vs. rack-mounted systems;
- Residential vs. utility systems;
- Distributed vs. centralized systems;
- Rooftop vs. ground-mounted systems;
- Tracking vs. fixed-tilt systems;
- New constructed vs. retrofitted systems.

Other distinctions may include the following aspects:

- Systems with microinverters vs. central inverter;
- Systems using crystalline silicon vs. thin-film technology;
- Systems with modules from Chinese vs. European and US manufacturers.

Undoubtedly, the most developed source of solar energy today is solar PV. According to reports from the environmental organization Greenpeace, solar PV could supply electricity to two-thirds of the world population by 2030 (Teske, 2008). Thanks to the technological advances achieved in this type of energy source, sophistication, and economy of scale, the cost of the electricity generated by solar PV parks has been reduced significantly in recent years and has increased its efficiency.[a]

For the above reason, today, the average cost of electricity generation from solar PV parks is competitive compared with other non-renewable energy sources in a growing number of countries located in different geographic regions. Other solar technologies, such as thermal solar energy, reduce their costs considerably. However, their use for electricity generation is still low in several regions of the world, particularly in the North American region. The prices of solar panels used in solar PV parks have dropped by a factor of ten during the last decade. This price reduction has made electricity generated by solar PV parks more competitive than other energy sources. With the help of wind energy, it opened the path to a global transition from the use of conventional energy sources for electricity generation and heating to increased use of all renewable energy sources. That transition is indispensable to mitigate global warming. However, it is important to know that the use of solar PV as one of the main sources for electricity generation, heating, and desalination requires energy storage systems or global distribution by high-voltage direct-current power lines causing additional cost.

A solar PV park (see Fig. 3.1) employs solar modules, each comprising several solar cells, which generate electrical power on a commercial scale or arranged in smaller installations to be used for mini-grids or personal use. A solar PV system for residential (see Fig. 3.2), commercial, or industrial energy supply, in addition to the solar array, has several components often summarized as the balance of system (BOS). That system balances the power-generating subsystem of the solar array with the power-using side of the AC-household devices and the utility grid (see Fig. 3.3). BOS-components include (see Fig. 3.4):

---

[a] However, it is a fact that solar PV efficiency is still low compared to other energy sources.

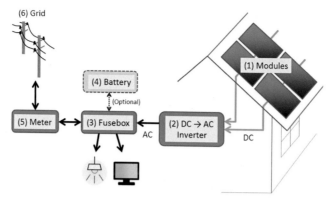

**FIG. 3.2** Schematics of a typical residential PV system. *Source: S-kei – Own work, CC0, https://commons.wikimedia.org/w/index.php?curid=17181966.*

# Balance of System

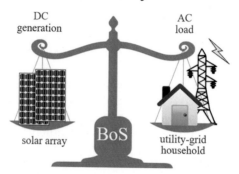

**FIG. 3.3** Balance of a solar PV system. *Source: Rfassbind – Own work based on the Solar Energy Technologies Program. U.S. Department of Energy archived website. Public Domain, https://commons.wikimedia.org/w/index.php?curid=35283841.*

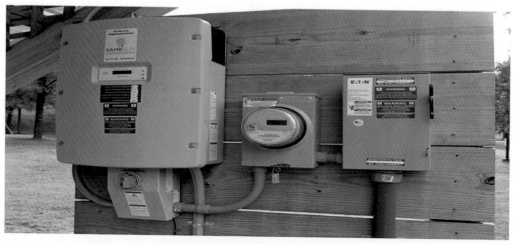

**FIG. 3.4** Solar string inverter and other BOS components in Vermont, USA. *Source: Cheeeeeese – Own work, CC0, https://commons.wikimedia.org/w/index.php?curid=28631474.*

- Power-conditioning equipment and structures for mounting;
- One or more DC to AC power converters, also known as inverters;
- An energy storage device;
- A racking system that supports the solar array;
- Electrical wiring and interconnections;
- Mounting for other components (Wikipedia Photovoltaic, 2020).

Optionally, the balance of a solar PV system may include one or more of the following components:

- Renewable energy credit revenue-grade meter;
- Maximum power point tracker (MPPT);
- Battery system and charger;
- GPS solar tracker;
- Energy management software;
- Solar irradiance sensors;
- Anemometer, or task-specific accessories designed to meet specialized requirements for a system owner;
- Optical lenses or mirrors and sometimes a cooling system (Wikipedia Photovoltaic System, 2020).

The use of solar PV to power mini-grids is an excellent way to bring electricity access to communities located in remote areas of a country far from the national power transmission lines. That is particularly true in developing countries located in the region of the world with excellent solar energy resources. Solar PV parks may be mounted in the following manner:

- Ground-mounted;
- Rooftop mounted;
- Wall-mounted;
- Floating.

"The mount may be fixed or use a solar tracker to follow the Sun across the sky. Solar PV has specific advantages as an energy source: once installed, its operation generates no pollution and no greenhouse gas emissions, it shows simple scalability in respect of power needs, and silicon has large availability in the Earth's crust" (Wikipedia Photovoltaic, 2020).

Solar PV has long been used as stand-alone installations and grid-connected PV systems since the 1990s (Bazilian et al., 2013). Solar PV modules were first mass-produced in 2000 when German environmentalists and the Eurosolar organization gained government funding to implement a program to install ten thousand solar PV panels in private houses and buildings. Advances in solar technology and increased manufacturing scale have significantly reduced solar PV panels' cost and increased solar PV parks' reliability and efficiency (Bazilian et al., 2013).

"The cost of manufacturing solar panels has plummeted dramatically in the last decade, making them not only affordable but often the cheapest form of electricity. Solar panels have a lifespan of roughly 30 years and come in a variety of shades depending on the type of material used in manufacturing" (IRENA Solar Energy, 2020).

It is projected that the total installations of solar PV cost projects will continue to decline in the next three decades. That decrease would make solar PV projects highly competitive in many energy markets. It is expected that solar PV project average costs will decrease between US$340 and US$834 per kW by 2030 and between US$165 and US$481 per kW by 2050. In 2018, the cost of solar PV parks was US$1210 per kW (IRENA, 2019)[b].

## Concentrated solar power

Concentrated solar power (CSP) uses lens mirrors and tracking systems to focus a large area of sunlight on a small beam. The concentrated heat is then used as a heat source for a conventional power plant for electricity production. A wide range of CSP technologies exists. There are, according to the Solar Energy (2009) and the Solar Thermal Electric Generation (2007) reports, the following:

- **Parabolic Trough Collectors**. This type of CSP system "combines a curved parabola mirror to maximize the amount of sunlight gathered by the collectors, with an absorber tube embedded along the center of the mirror." The absorber tube is filled with a specific fluid that can easily be heated. The hot fluid is then used to boil water and produce steam in a connected device, and the steam is transferred to a generator that can produce electricity. "A large array of connected parabolic trough collectors is needed to provide enough power for a generator";
- **Dish/Engine System**: This type of CSP system uses "an array of mirrors, arranged in the shape of a dish, to concentrate sunlight onto a receiver placed at the focal point of the dish. The heat produced by these systems is transferred to a heat engine, which converts the heat into mechanical energy. This energy then drives a generator to produce electricity";
- **Power Towers**: This type of CSP system "uses a circular array of mirrors that track the sunlight and concentrate it on a receiver, placed at the top of a central tower at the focal point of the array. The heat produced by the receiver is used to create steam, which then powers a generator" (see Fig. 3.6).
- **Hybrid System** (see Fig. 3.5): This type of CSP system "combines power towers with natural gas generators, creating a hybrid system that can continuously generate electricity, even when the Sun is not shining or at night." This hybrid technology is still under development. Experimental systems have been connected to several USA's southwest utilities.

One important CSP characteristic is that this type of solar energy can generate electricity and heating in large-scale power plants. The first commercial CPS power plant was developed in the 1980s, named "the Ivanpah Solar Power Facility," located in the Mojave desert of California, USA, with a 392 MW capacity (see Fig. 3.6). The facility mentioned above is the largest CPS plant in the world built until 2019.

[b] IRENA (2019); Future of Solar Photovoltaic: Deployment, investment, technology, grid integration and socio-economic aspects (A Global Energy Transformation: paper); International Renewable Energy Agency, Abu Dhabi; 2019.

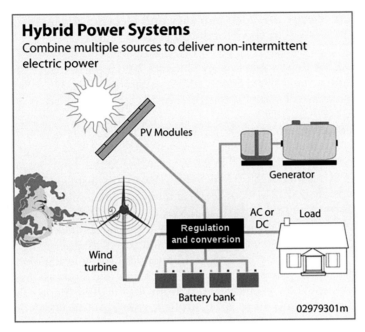

FIG. 3.5 Hybrid power system. *Source: Hybrid Wind and Solar Electric Systems From the US Department of Energy. https://www.energy.gov/eere/communicationstandards/copyright-laws-web.*

FIG. 3.6 Ivanpah Solar Electric Generating System Facility. *Source: US Geological Survey. https://www.usgs.gov/media/images/ivanpah-solar-electric-generating-system-facility.*

# Solar energy installed capacity and electricity generation in the North American region

According to IRENA Renewable Energy Statistics 2020 report, in 2019, the worldwide installed solar park capacity reached 584,842 MW, generating 562,033 GWh in 2018 (IRENA, 2020b). In the specific case of solar PV, the total world capacity reached, in 2019, 578,553 MW, generating 549,833 GWh in 2018.

In North America, the solar park capacity in 2019 reached 65,608 MW,[c] generating 154,594 GWh in 2018. Canada has a solar PV capacity of 3310 MW or 5.3% of the total, and the USA 62,298 MW or 94.7% of the total.[d] In 2019, according to the report mentioned above, the world capacity of CSP reached 6,28975 MW.[e] The CPS capacity installed in the North American region in that year reached 1772 MW[f] or 28.2% of the world total. All CPS capacity in the North American region is concentrated in the USA if Mexico is excluded (Mexico has only 14 MW of CSP capacity installed). Canada has no CSP capacity installed until December 2019.

Solar PV is the third renewable energy source in terms of global capacity installed in the North American region, after hydro and wind energies. The International Energy Agency (IEA) report entitled "Renewables 2019 Market Analysis and Forecast from 2019 to 2024" expects growth between 700 GW and 880 GW from 2019 to 2024 (Renewables 2019. Market Analysis and Forecast From 2019 to 2024, 2019). According to the IEA's forecast, solar PV could become the technology with the largest installed capacity in the North American region by the mid-2020s.

The evolution of the solar parks' capacity installed in the North American region during the period 2010–19 is shown in Fig. 3.7.

According to the data included in Fig. 3.7, the following can be stated: solar park capacity installed in the North American region increased by 12.2-fold during the period under consideration, rising from 3603 MW in 2010 to 65,608 MW in 2019. It is foreseen that North America's solar park capacity will continue to grow during the coming years, particularly in the USA, and preferable in solar PV.

The evolution of solar PV park installed capacity in the North American region during the period 2010–19 is shown in Fig. 3.8.

Based on the data included in Fig. 3.8, the following can be stated: solar PV park capacity in the North American region increased 20.4-fold during the period 2010–19, rising from 3130 MW in 2010 to 63,850 MW in 2019. It is expected that this trend will continue during the coming years, with more intensity in the USA, where more solar PV parks will be built in several states. The aim is to increase the role of solar PV in the state's energy mix.

[c] Excluding Mexico.

[d] Education and skills policies can help also to equip the work-force with adequate skills and would increase opportunities for local employment. Similarly, sound industrial policies that build upon domestic supply chain can enable income and employment growth by leveraging existing economic activities in support of solar PV industry development (IRENA, 2019).

[e] According to the IRENA Renewable Capacity Statistics 2020 report, in 2019, the world solar CSP capacity was 6275 MW.

[f] According to the IRENA Renewable Capacity Statistics 2020 report, in 2019, the solar CSP capacity in the North American region was 1758 MW.

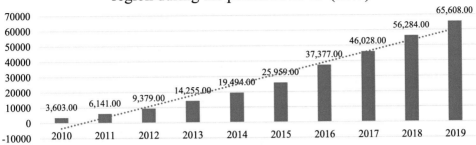

FIG. 3.7  Evolution of solar parks capacity installed in the North American region during the period 2010–19. *Source: IRENA, 2020. Renewable Energy Statistics 2020. The International Renewable Energy Agency, Abu Dhabi and Author own calculations.*

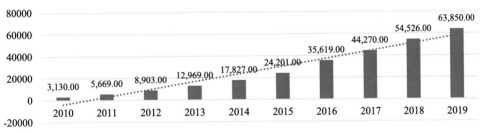

FIG. 3.8  Evolution of solar PV parks installed capacity in the North American region during the period 2010–19. *Source: IRENA, 2020. Renewable Energy Statistics 2020. The International Renewable Energy Agency, Abu Dhabi and Author own calculations.*

Besides, it is important to be aware that by 2050, solar PV for electricity generation would represent the second-largest power generation source at the world level and within renewable energy sources. Therefore, it is foreseen that solar PV will lead the way for transforming the global electricity sector at the world level. Solar PV would generate 25% of total electricity needs globally, become one of the prominent generation sources by that year. To achieve that goal, solar PV capacity should increase at least five-fold over the next ten years,[g] from

[g] In annual growth terms, this means an almost threefold rise in yearly solar PV capacity additions is needed by 2030 (270 GW per year) and almost 4-fold rise by 2050 (372 GW per year) compared to current levels (97.6 GW added in 2019), according to IRENA (2019) and IRENA Renewable Energy Statistics 2020.

580.2 GW in 2019 to 2840 GW by 2030 and 8519 GW by 2050; this represents an increase of almost 15-fold the level registered in 2018. An increase in solar PV investment is critical in accelerating solar PV installations' growth over the coming decades to support this effort. Globally this would imply a 68% increase in average annual solar PV investment of US$192 billion per year from now until 2050. In 2018, solar PV investment stood at US$114 billion, according to IRENA (2019).

The following is a list of operating solar PV parks with a 200 MW capacity or larger located in the North American region (see Table 3.1).

According to Table 3.1, the USA has 28 solar PV parks with a capacity of or above 200 MW in operation in 2019. Canada has no solar PV parks with this capacity. Canada's largest solar PV park is the Grand Renewable Solar Project, with a 102 MW capacity.

TABLE 3.1    List of solar PV parks larger than 200 MW capacity in operation in North America.

| Name | Country | Capacity $MW_P$ or $MW_{AC}$ | Land size $km^2$ | Year |
|---|---|---|---|---|
| Agua Caliente Solar Project | United States | 290 | 9.7 | 2014 |
| Antelope Valley Solar Ranch | United States | 230 | 8.5 | 2015 |
| Beacon Solar Project | United States | 250 | – | 2017 |
| Blythe Solar Energy Center | United States | 235 | 8.1 | 2016 |
| Buckthorn Solar 1 | United States | 202 | 5.1 | 2018 |
| California Flats Solar Project | United States | 280 | 11.7 | 2017 |
| California Valley Solar Ranch | United States | 250 | 7.96 | 2013 |
| Copper Mountain Solar Facility | United States | 552 | 16.2 | 2016 |
| Desert Sunlight Solar Farm | United States | 550 | 16.0 | 2015 |
| Escalante Solar Project | United States | 240 | 7.7 | 2016 |
| GA Solar 4 Project | United States | 200 | 8.1 | 2019 |
| Garland Solar Facility | United States | 200 | 8.1 | 2016 |
| Great Valley Solar | United States | 200 | 6.5 | 2018 |
| McCoy Solar Energy Project | United States | 250 | 9.3 | 2016 |
| Mesquite Solar project | United States | 400 | 9.3 | 2016 |
| Midway Solar | United States | 236 | 6.1 | 2019 |
| Moapa Southern Paiute | United States | 250 | 8.1 | 2016 |
| Mount Signal Solar | United States | 460 | 15.9 | 2018 |

**TABLE 3.1**   List of solar PV parks larger than 200 MW capacity in operation in North America—cont'd

| Name | Country | Capacity MW$_p$ or MW$_{AC}$ | Land size km$^2$ | Year |
|---|---|---|---|---|
| Roadrunner Solar Project | United States | 252 | – | 2019 |
| Roserock Solar | United States | 212 | 5.3 | 2016 |
| Silver State South Solar Project | United States | 250 | 11.7 | 2016 |
| Solar Star (I and II) | United States | 579 | 13.0 | 2015 |
| Springbok Solar Farm | United States | 350 | 5.7 | 2019 |
| Stateline Solar | United States | 250 | 6.82 | 2016 |
| Techren Solar Project | United States | 300 | 9.3 | 2019 |
| Topaz Solar Farm | United States | 550 | 19.0 | 2014 |
| Tranquillity Solar project | United States | 200 | 7.7 | 2016 |
| Upton Solar 2 | United States | 235 | 7.7 | 2017 |

*Source: 2020. Wikipedia List of Photovoltaic Power Stations 2020.*

In 2019, according to the IRENA Renewable Capacity Statistics 2020 report, the world CSP park capacity reached 6275 MW. The CPS capacity installed in 2019 in the North American region reached 1758 MW, or 28% of the world total. All CPS capacity installed in the North American region is located in the USA.

The evolution of CSP park installed capacity in the North American region during the period 2010–19 is shown in Fig. 3.9.

According to the data included in Fig. 3.9, the following can be stated: solar CSP installed capacity in the North American region grew by 3.7-fold during the period 2010–19, rising from 473 MW in 2010 to 1758 MW in 2019. However, it is important to single out that during the period 2015–19, no increased capacity installed in the region in CSP parks has been reported. It is expected that this situation will change during the coming years, particularly in the USA, because Canada has no CSP parks in operation or plans to build this type of facility shortly.

The CSP list of at least 50 MW capacity in operation in the North American region is shown in Table 3.2.

Based on the data included in Table 3.2, the following can be stated: there are only eight CSP facilities with at least 50 MW of capacity in the North American region, all located in the USA.

Finally, it is important to single out that solar PV parks can also be constructed to supply electricity outside the electrical grid in remote areas. However, about 90% of all US solar

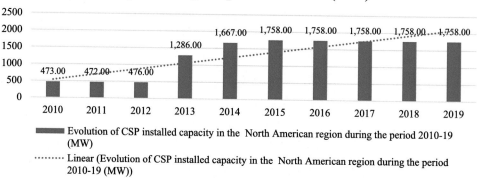

FIG. 3.9    Evolution of solar CSP park installed capacity in the North American region during the period 2010–19. *Source: 2020. IRENA Renewable Capacity Statistics 2020. International Renewable Energy Agency, Abu Dhabi and Author own calculations.*

**TABLE 3.2**    List of CSP facilities with at least 50 MW capacity in operation in the North American region.

| Name | Country | Location | Electrical capacity (MW) | Technology type | Storage hours |
|---|---|---|---|---|---|
| Ivanpah Solar Power Facility | United States | San Bernardino County, California | 392 | Solar power tower | – |
| Crescent Dunes Solar Energy Project | United States | Nye County, Nevada | 125 | Solar power tower | 10 |
| Solar Energy Generating Systems (SEGS) | United States | Mojave Desert, California | 310 | Parabolic trough | – |
| Solana Generating Station | United States | Gila Bend, Arizona | 280 | Parabolic trough | 6 |
| Nevada Solar One | United States | Boulder City, Nevada | 75 | Parabolic trough | – |
| Mojave Solar Project | United States | Barstow, California | 280 | Parabolic trough | – |
| Genesis Solar Energy Project | United States | Blythe, California | 280 | Parabolic trough | – |
| Martin Next Generation Solar Energy Center | United States | Indiantown, Florida | 75 | ISCC with parabolic trough | – |

*Source: Wikipedia Solar thermal Power Stations (CSP) 2020.*

power systems are connected to the electrical grid in the USA. In contrast, off-grid systems are somewhat more common in Australia and South Korea, according to the Global Market Outlook for Photovoltaics 2014–2018 (2014) report.

## Solar energy investment costs in the North American region

The world energy system is under a deep transformation from the use of conventional energy sources for electricity generation, heating, and water desalination to increase the use of renewable energy sources for the same purpose. This process is going extremely fast. New renewable energy capacity is now exceeding fossil fuel generation capacity by a widening margin. The global investment in renewable energy sources capacity carried out in 2018 to support the construction of new renewables capacity facilities reached US272.9 billion, "the fifth successive year in which it has exceeded US$250 billion" (Global Trends in Renewable Energy Investment 2019, 2019). However, it is important to single out that the figure reported in 2018 is 12% lower than the one that was reported in 2017, "due in part to a policy change that hit the financing of Chinese solar in the second half of the year" (Global Trends in Renewable Energy Investment 2019, 2019).

Solar power kept its position as the technology attracting the most capacity investment at US$141.1 billion in 2019. Even though this is a huge investment in solar power capacity, this figure is 21.6% lower than the total solar power investment reported in 2017 (US$180 billion). This decrease is due in part to a significant reduction in solar technology costs registered in recent years.

Without a doubt, renewable energy sources, particularly solar power, have become a compelling investment energy alternative in many countries. For this reason, investment into new solar power park projects has grown 39.6% during the period 2010–18, rising from around US$101 billion per year in 2010 to about US$141 billion per year in 2018 (see Fig. 3.10).

The evolution of the investment in solar energy projects at the world level during the period 2010–18 is shown in Fig. 3.10.

The IRENA's report entitled "Global Landscape of Renewable Energy Finance 2018" (IRENA and CPI, 2018) outlines key global investment trends between 2013 and 16, offering a comprehensive overview of capital flows by region and technology. The report also examines the differing roles and approaches of private and public finance, highlights the emergence of viable risk mitigation instruments, and provides an outlook for renewable energy finance in 2018 and beyond.

An updated key finding on renewable energy investment can also be found in the report entitled "Global Trends in Renewable Energy Investment 2019". This report has reported a relatively low capital flow from the world's major institutional investors to the renewable energy sector in recent years. Institutional and private equity investors have contributed less than 1% each to global renewable energy investment in the last three years. Their investment peaked in 2015 at around US$3 billion for institutional investors and US$2 billion for private investors. One possible reason for the lack of institutional and private equity investors' lack of commitment to low-carbon investment projects could be that the possible

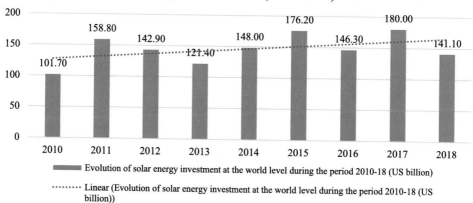

FIG. 3.10 Evolution of the investment in solar energy at the world level during the period 2010–18. *Source: 2019. Global Trends in Renewable Energy Investment 2019. UN Environment Program.*

investment's asset size is not significant enough to attract them. Once renewable energy project developers can present multi-billion dollar portfolios of operational projects, then perhaps institutional and private equity investors may be more inclined to finance this type of project.

These are some of the key findings included in the report mentioned above related to the period 2010–19:

- During the period under consideration, a total of US$2.6 trillion has been invested in renewable energy capacity (excluding large hydropower plants), more than three times the amount invested during the period 2000–09. Solar energy alone has attracted, during the period 2010–18, a total of US$1.3 trillion; this means 50% of the total investment in renewable energy projects registered in that period;
- The USA invested in renewable energy capacity during the period 2010–19, a total of US$356 billion, second to China with US$758 billion;
- The period 2010–19 has seen a spectacular improvement in the cost-competitiveness of renewables, with the LCOE for solar PV decreased by 81%, making this technology one of the cheapest options for electricity generation and heating in many countries around the world;
- Cost reductions in solar PV energy registered in last years are the outcome of a group of factors such as a combination of economies of scale in manufacturing, fierce competition along the supply chain, record-low costs of finance, and improvements in the efficiency of generating equipment;
- More solar energy capacity was installed during the period 2010–19 than any other generating technology. Solar new power capacity of some 638 GW during the period mentioned above is a significant figure given that there were only 25 GW of solar power capacity installed worldwide at the end of the previous decade (Global Trends in Renewable Energy Investment 2019, 2019);

- During the period 2010–19, a total of 2.4 TW of the power capacity of all types were installed, with solar energy capacity at the top, followed by coal and wind energy narrowly beating natural gas for third place (Global Trends in Renewable Energy Investment 2019, 2019).

Specifically, regarding the outcome of the energy sector in 2018, the following are the key findings, according to the World Bank Group (2019), IRENA Renewable Capacity Statistics 2020, and Global Trends in Renewable Energy Investment 2019:

- Global investment in renewable energy capacity in 2018 reached the amount of US$272.9 billion. That is the fifth successive year in which it has exceeded US$250 billion. Still, it is important to note that this amount is 12% lower than the one reported in 2017. The USA was the largest investor in renewable energy source capacity in 2018, which amounted to US$42.8 billion. However, this amount was 6% lower than the amount invested in the previous year, but still, the third-highest figure reported until 2019, and some way above the average for the last five years of just below US$40 billion (World Bank Group, 2019);
- The global investment amount for 2018 was achieved despite continuing falls in solar and wind power projects' capital costs. Solar power keeps its position as the technology attracting the most capacity investment, at US$141.1 billion, although this was down 21.6% concerning the level reached in 2017. The decline in solar investment had two reasons. One of these reasons was the continued reduction in capital costs for solar PV installations, with the global benchmark for solar PV systems without tracking decreasing from US$1.03 million per MW in 2017 to an average of US$0.82 million per MW in the second half of 2018; this means a reduction of almost 30%. It is important to single out that, in the most competitive markets, large solar PV projects are being built for much less than US$1 million per MW;
- Solar energy capacity investment declined 11% in the USA in 2018, reaching the amount of US$20.3 billion, the lowest amount since 2013. The USA solar energy sector trend was somewhat weaker in dollar terms, even if solar energy capacity added grew modestly by 10.1 GW in 2018 with respect to 2017, and by 9.1 GW in 2019 with respect to 2018 (IRENA Renewable Capacity Statistics 2020, 2020). Commitments to utility-scale solar PV fell by 9% to US$11.4 billion, but there was a 15% drop to US$8.9 billion in the financing of small-scale solar PV systems of less than 1 MW of capacity. Both figures reflect the continuing falls in global solar PV capital costs per MW;
  It is important to single out that solar energy secured US$96.7 billion of asset finance in 2018, but this amount is 25% lower than the amount registered in 2017. Within the above figure, solar PV got US$95.4 billion, some 24% lower than the amount reported in 2017, while CSP projects received just US$1.3 billion of investment, a decreased of 48% with respect to 2017. That amount represents the lowest figure since 2005;
- Canada invested US$579 million in renewable energy projects in 2018. That amount represented a decrease of 63% with respect to the level reached in 2017. It is the lowest level during the whole period since 2004. The peak investment figure in the renewable energy sector of US$7.4 billion was reached in 2011. The country's main renewable energy project development activity is in wind energy, with the province of Alberta holding renewable energy auctions late in 2018 for a total of 763 MW of capacity. Ontario province

saw much of the investment in wind and solar energy projects earlier in the decade, "but policy changes have dulled project activity there more recently." The solar energy capacity installed in Canada in 2019 was only 3310 MW and only in solar PV capacity. The new solar energy capacity installed in the country during the period 2018–19 was 210 MW, while in the USA, it was 9114 MW or 43.4 fold (Global Trends in Renewable Energy Investment 2019, 2019);

- Renewable energy capacity investment was more spread out across the globe than ever, with 29 countries, each investing in the renewable energy sector more than US$1 billion in 2018 (World Bank Group, 2019). China invested the most at US$88.5 billion, followed by Europe with US$59.9 billion, and the USA at US$42.8 billion. Investment in renewables energy capacity in 2018 was about three times global investment registered in coal and natural gas-fired generation capacity combined. That increase came despite further reductions registered in 2018 in the average capital cost per MW of solar and wind projects (Global Trends in Renewable Energy Investment 2019, 2019).

## Solar energy construction and generation costs in the North American region

The decade 2010–19 opened with non-hydro renewables widely seen as expensive energy sources for electricity generation compared to conventional energy sources unless a high carbon price reflects the pollution and greenhouse gas effects of fossil fuel power.

Taking into account different studies that several experts on the LCOE model carried out, it can be stated that the global benchmark for solar PV without tracking has decreased by 81%, falling from US$304 per MWh in the second half of 2009 to US$57 per MWh in the first half of 2019. The reduction in the cost of 1 MWh generated by solar energy facilities has resulted from several factors. These factors are, among others, the following:

- Fierce competition among solar manufacturers of components and equipment and solar technology developers to cut overheads as much as possible;
- The influence of renewable energy sales as a result of the fierce competition mentioned above;
- Developing solar technology with a higher level of efficiency;
- A strong downward trend in finance costs associated with the construction and operation of solar energy facilities;
- The massive production of solar energy components and equipment, particularly solar PV panels.

All renewable energy technologies involve significant upfront capital expenditure. Besides, in the specific case of solar energy, ongoing running costs are now extremely low compared to other energy sources. That situation makes the cost of the equity and debt used to finance the construction of new solar power capacity highly influential in calculating the total LCOE of new solar power projects. Without a doubt, record-low official interest rates in many countries during the period under consideration, and an increasing competition among investors and banks to participate in the development and implementation of renewable energy projects in many countries all over the world, particularly solar energy projects, has driven down the cost of both equity and debt as well as upfront capital expenditure. As a result of this situation, the use of solar power for electricity generation is a cheaper option for new generating capacity than fossil fuel sources in an increasing number of countries.

## The efficiency of the solar energy plants in the North American region

Efficiency is the most commonly used parameter to compare the performance of solar PV cells. The efficiency concept is defined as the solar PV cell's energy output ratio to input the Sun's energy. Based on the above concept, efficiency explicitly measures how much of a given resource's energy potential gets turned into electricity. In other words, the efficiency of a solar PV cell "refers to the portion of energy in the form of sunlight that can be converted, via photovoltaics, into electricity by the solar cell" (Wikipedia Solar Cell Efficiency, 2020).

For the majority of the solar PV cells available in the market, this is around 20%. However, the remaining 80% cannot be considered as expensive waste. Besides reflecting the solar PV cell's performance, it is important to single out that this type of cell's efficiency depends on the spectrum and intensity of the incident sunlight and the solar cell temperature. For this reason, conditions under which the efficiency of a solar cell is measured must be carefully controlled to compare the performance of one solar cell device to another in the most effective manner.

Without a doubt, efficiency is an important factor for non-renewable energy sources because they use fuel to produce electricity and heating, and, for this reason, it is easier to calculate how much fuel is converted into power. Solar energy does not use fuel for that purpose. The Sun shines regardless of whether there is a solar power park to capture the sunlight and generate power or not, and it is a free energy resource.

The solar cells' efficiency in a solar PV system, combined with latitude and climate, determines the system's annual energy output. For example, a solar panel with 20% efficiency and an area of one $m^2$ will produce 200 kWh per year under normal conditions, but it can produce more when the Sun is high in the sky and will produce less in cloudy conditions or when the Sun is low in the sky. In central Colorado, which receives annual insolation of 2000 kWh/$m^2$/year (Roberts, 2008), such a panel can be expected to produce 400 kWh of energy per year. However, in Michigan, which receives only 1400 kWh/$m^2$/year (Roberts, 2008), the annual energy yield will drop to 280 kWh for the same panel. At more northerly European latitudes, yields are significantly lower: 175 kWh annual energy yield in southern England under the same conditions (MacKay, 2017).

The solar cells' efficiency in a solar PV system of 20% could be considered a very low-efficiency percentage by many people compared to other energy source's efficiency. Still, there is an important difference that needs to be considered. Fossil fuels must be purchased in the market, and when they are burned, they produce other greenhouse emissions. But when the sunlight is used, which allows a solar panel to function, it is free and clean. It may seem that the remaining 80% is being wasted, and solar panels often get criticism for their low-efficiency percentage. However, it must be considered that without solar PV parks, 100% of sunlight is being wasted, and nothing can be done to avoid that this happens. It would be correct to say that solar PV panels obtain electricity from an abundant energy resource that is otherwise wasted because the Sun cannot be shut down. In the case of the use of fossil fuels for electricity generation and heating, the efficiency rate is important to be as high as possible because otherwise resulting in wasted fuel, high operating expenses, and emission of greenhouse gases. That is not the case for solar energy.

Solar power has another factor, which is important for economics. While the sunlight is free, it is not shining all the time hard enough to allow solar PV panels to generate electricity at maximum potential. Over a year, the ratio of actually generated electricity to the maximum potential electricity is called the "capacity factor." The capacity factor ranges from 15% to around 20%–30% for solar energy, depending on where the solar PV cells are located and whether the solar PV panels follow the Sun on (more expensive) trackers or not.[h]

The conversion efficiency of a solar PV cell is the percentage of solar energy shining on a solar PV device that is converted into usable electricity and heating. That is done by capturing the electrical current generated when sunshine interacts with the solar PV cells, and it is converted into AC (alternating current) energy. Improving this conversion efficiency is a key goal to make solar PV technologies cost-competitive with conventional energy sources. However, it is important to single out that not all sunlight that reaches a solar PV cell is converted into electricity or heating.

Several factors affect a solar cell's conversion efficiency value, including reflectance, thermodynamic efficiency, charge carrier separation efficiency, charge carrier collection efficiency, and conduction efficiency values (US DoE Photovoltaic Cell Conversion Efficiency Basics, 2014; Kumar, 2017). Because these parameters can be difficult to measure directly, other parameters are measured instead, including quantum efficiency, open-circuit voltage ($V_{OC}$) ratio, and Fill factor values.[i] Reflectance losses are accounted for by the quantum efficiency value, as they affect "external quantum efficiency." In 2019, the world record for solar PV cell efficiency at 47.1% was achieved using multi-junction concentrator solar PV cells, developed at National Renewable Energy Laboratory in Golden, Colorado, USA (Geisz et al., 2018). Solar PV cell efficiency of 40% was reached in the field in a practical test. Researchers had said that new solar PV technology could be fine-tuned shortly and reach up to 50% efficiency (Delbert, 2020). That is above the standard rating of 37% for polycrystalline photovoltaic or thin-film solar cells already registered (Shieber, 2018). However, it is important to stress that these

---

[h] Solar PV panels have been consistently increasing in efficiency at about 0.5% annually since 2010. Numerous scientific factors play into the solar PV cell efficiency equation. Three major components are panel material components, reflectance efficiency, and thermodynamic efficiency.

[i] The efficiency of a solar PV cell is determined as the fraction of incident power which is converted to electricity and is defined as:

$$P_{max} = V_{OC} I_{SC} FF$$

$$\eta = \frac{V_{OC} I_{SC} FF}{P_{in}} \quad FF = \frac{P_m}{V_{OC} \times I_{SC}}$$

Where:

VOC is the open-circuit voltage;

ISC is the short-circuit current;

FF is the fill factor

η is the efficiency

efficiency figures were achieved under laboratory conditions. Laboratory prototypes used for these tests are still very expensive and not viable for commercial use.

Multiple solar PV cell design factors also play relevant roles in limiting a cell's ability to convert sunlight into electricity and heating. Designing with these factors in mind is how higher efficiencies can be achieved. Among these factors are, according to the Solar Performance and Efficiency (2013) paper prepared by the Office of Energy Efficiency and Renewable Energy of the US DoE, the following:

- **Wavelength**: The sunlight that reaches the Earth's surface has wavelengths from ultraviolet to infrared. Some photons are reflected when light strikes a solar PV cell's surface, while others pass right through. Some of the absorbed photons have their energy turned into heat. The remainder has the right amount of energy to separate electrons from their atomic bonds to produce charge carriers and electric current;
- **Recombination**: One way for electric current to flow in a semiconductor is for a "charge carrier," such as a negatively charged electron, to flow across the material. Another such charge carrier is known as a "hole."[j] "When an electron encounters a hole, they may recombine and, therefore, cancel out their contributions to the electrical current. Direct recombination, in which light-generated electrons and holes encounter each other, recombine, and emit a photon, reverses the process from which electricity is generated in a solar cell", according to Solar Performance and Efficiency (2013);
- **Temperature**: "Solar PV cells generally work best at low temperatures. Higher temperatures cause the semiconductor properties to shift, resulting in a slight increase in current, but a much larger decrease in voltage", according to Solar Performance and Efficiency (2013);
- **Reflection**: A solar PV cell's efficiency can be increased by minimizing the amount of light reflected away from the solar PV cell's surface.

Another important element associated with the solar PV cell efficiency factor is the Sun's solar PV panel angle. That is a very important factor during the design process. The orientation of solar PV panels affects efficiency. Electricity production is increased when the solar PV panels are oriented correctly toward the Sun. Solar PV panels that receive sunlight head-on produce more electricity than those getting sunshine at a certain angle. For this reason, to increase efficiency, the solar PV panels should have an orientation that maximizes efficiency in the presence of direct sunlight.

## Solar energy and the impact on the environment

Solar energy systems do not produce air pollution, water pollution, or greenhouse gases during electricity and heating production. Without a doubt, the use of solar energy for electricity generation and heating can have a positive, indirect effect on the environment. The reason is simple: its use replaces or reduces the utilization of other energy sources for the same purposes but with larger negative effects on the environment. However, as with any other

---

[j] A hole represents the absence of an electron within the material and acts as a positive charge carrier.

type of power plant, "solar energy comes with its own environmental challenges regarding land use, water consumption, emissions, and the use of hazardous materials" (Leblanc, 2019). For this reason, large solar power plants can have a negative effect on the environment near the site where the mentioned plant will be built. These negative effects are, among others, the following:

- The land-use implications for solar energy projects depend on their scale. Small rooftop arrays in a house or building are not a significant concern. The problem is with larger-scale solar power projects that need a large area. The area of a solar power plant site cannot be used for any other purpose. This issue can be mitigated by utilizing low-value locations such as brownfields, abandoned mining sites, transportation, and transmission corridors;
- Clearing the site where the solar power plant will be built may have long-term effects on the habitats of native plants and animals;
- Solar PV parks do not use water for electricity generation, but CSP parks require water for cleaning solar collectors and concentrators or for cooling turbine generators.[k] The use of "large volumes of groundwater or surface water in some arid locations may affect the ecosystems that depend on these water resources" (Solar Explained. Solar Energy and the Environment, 2020). For this reason, one potential point of concern during the consideration of the sites to be selected to build a solar power park is that some of the best places for the construction of this type of power plant happen to have the driest climates and poorest water availability. Therefore, water supply is an important element that needs to be considered when it comes to the construction of solar power plant projects (Leblanc, 2019);
- The beam of concentrated sunlight in a solar power tower can kill birds and insects that fly into the beam with a negative impact on animal wildlife;
- Several hazardous materials are used during the solar PV cell manufacturing process. Chemicals are used for the most part to clean and purify the semiconductor surface of the panel (Leblanc, 2019). The final disposal of all toxic materials and chemicals used to make the solar PV panels, and for cleaning purposes, can have a negative impact on the environment if they are not treated and disposed of correctly;
- Some solar thermal systems use potentially hazardous fluids to transfer heat. Leaks of these materials could be harmful to the environment.

There is a general belief that the use of solar energy for electricity generation and heating does not produce greenhouse gases during operations. That is true. However, global warming emissions are created at other stages of the solar power plant lifecycle. These stages include, among others, the following:

- Resource extraction;
- The manufacturing process of the solar PV cells;
- Cleaning the site;

---

[k] According to Leblanc (2019), CSP plants using wet-recirculating technology with cooling towers withdraw between 600 and 650 gal of water per MWh of electrical production. Dry-cooling technology can cut water usage by 90% but can result in higher costs and lower efficiencies.

- Transportation of materials to the construction of the site;
- Installation of the solar PC cells and other equipment and facilities in the site;
- Maintenance of the different components of the solar power plant;
- Decommissioning and dismantling of the solar power plant.

According to Leblanc (2019), "most estimates show that solar energy, over its complete lifecycle, produces a lot less carbon dioxide equivalent than natural gas, and dramatically less than coal. Solar PV systems fall between 0.07 and 0.18 pounds of carbon dioxide equivalent per kWh, while CSP solar systems generate $CO_2$ equivalent in a range of 0.08 to 0.2 pounds. These figures are dramatically less than the lifecycle emissions for natural gas (0.6-2 lbs. of $CO_2E$/kWh) and coal (1.4-3.6 labs of $CO_2E$/kWh)."

## Advantages and disadvantages in the use of solar energy plants in the North American region

Some of the main benefits that solar energy offers for electricity generation in the North American region are, among others, the following:

- Reduces countries' dependence on the use of coal and oil to generate electricity and heating, the two most polluted energy sources available and used for this specific purpose;
- Since the Sun's energy is unlimited, there will be no shortage of electricity production; the only limitation is in the capacity of the solar parks and their level of efficiency for electricity generation;
- Solar energy can also reduce the energy bill in many countries, adding a new type of energy source to the country's energy mix if the current trend of a significant decrease in costs associated with this type of energy source continues;
- Since the Sun's energy is clean and pure, it will reduce health problems such as respiratory infections due to the pollution of cities and regions where other types of energy sources are used for electricity generation and heating;
- The use of solar energy for electricity generation and heating is one of the most effective forms of conservation and protection of the environment;
- The maintenance of the solar panel after been installed is easy and economical. There is no need to use any kind of fuel to operate a solar power installation, and there are, generally, no moving components. For this reason, it has no problems with mechanical parts breakage if the solar panels are fixed. Solar energy is more convenient for power generation in remote or isolated sites than other available energy sources (Schuessler, 2008).

The main advantages and disadvantages associated with the use of solar power for electricity generation and heating, according to Morales Pedraza (2012, 2015), Bolinger et al. (2019), and other experts' opinions, are briefly detailed in the following paragraphs:

A-**Advantages**
- Solar energy is an energy resource that is sustainable for energy consumption, and it is renewable indefinitely;
- Solar energy can be used to generate electricity and to heat water (solar water heaters);

- Solar panels require little maintenance. After installation and start operation, solar panels are exceptionally reliable because they actively generate electricity in just a few millimeters and do not require any kind of mechanical parts that could fail, in the case of fixed solar parks;
- Solar panels are also a silent energy producer of energy. That is an important characteristic if a town or city is located near the solar park;
- Many governments have introduced generous tax credits for individuals and companies to promote investment in solar systems for electricity generation and heating. For this reason, more companies are investing in the construction of new solar parks or installing solar panels in residential houses and buildings for the same purposes;
- Solar energy is becoming less expensive every year. With the decrease in the cost of solar energy in recent years (around 0.5% each year), it has become a long-term source of economic energy for many countries;
- Solar energy does not emit any greenhouse gases that affect the environment. Solar panels do not release any greenhouse gases into the atmosphere while generating electricity. That is something that cannot be expected from electricity generated from thermal and hydroelectric energy sources. Furthermore, there are efforts to find cheaper and safer technology to recycle solar panels, which is one of the main limitations in the use of this type of energy source for electricity generation;
- Solar energy is flexible in its configuration for electricity generation. Solar parks for electricity generation and solar panels used to power the entire house or building can be mounted with different configurations and capacities at the selected site. They can also be used for heating, to power garden lights, or something that requires less electricity;
- Solar energy can also be used in industry, to power satellites, and even to move cars and airplanes, to make these means of transportation more economical and less contaminating sources;
- In remote places, solar energy can be the ideal source of electricity. It is cheaper to install solar panels for electricity supply in such areas than to connect them to the electrical grid in many cases.

B-**Disadvantages**

- The main disadvantage of solar energy for electricity generation is that it cannot generate electricity at night or on very cloudy days. The energy generated can also be reduced during cloudy periods (although it is possible to produce energy on a cloudy day in certain regions). A regular supply of sunlight is essential to continue generating electricity by a solar park. Once the Sun goes down, the level of solar energy falls extremely fast. For this reason, it is important to have an energy storage device in the form of batteries to ensure an uninterrupted supply of electricity at night or on very cloudy days. Adding battery storage to a solar park is a key way to increase the value of solar energy and its use for electricity. It is also one important mechanism to remove one of the barriers that impede a wider use of this type of energy source for electricity generation and heating (Bolinger et al., 2019);
- Energy production through the use of solar panels is maximized when the panel faces the Sun directly. That means that fixed solar panels will see reduced power output

when the Sun is not at an optimal angle. Many solar parks reduce this limitation to the minimum by mounting the solar panels on mobile devices that can follow the Sun to keep the solar panel at optimal angles with the Sun throughout the whole day;

- Today, even the most efficient solar cells only convert around 20% or a little more sunlight into electricity. As advances in solar cell technology improve, this percentage is likely to increase;
- In addition to their low conversion efficiency, solar panels can represent a substantial initial capital investment. However, after its installation, solar panels produce unbelievably cheap energy;

It is important to single out that most of the disadvantages of solar energy are economic, even after much technological development registered in recent years. Solar panels used to produce electricity are still expensive for many countries, especially for the least advanced developing countries, even with the price reduction currently registered per 1 MW. On the other hand, a single solar panel can generate only a small amount of energy. That means that many solar panels must be installed to generate enough electricity for the electrical grid. For this reason, a large area of land that cannot be used to produce food can be occupied by the solar park. Although some critics argue that solar energy's future rests mainly on reducing its electricity generation cost, other experts consider solar power a sustainable and less contaminating energy source.

According to several experts' opinions, solar energy will be a competitive energy source for electricity generation and heating compared to other types of energy sources in the next decade in some regions. In any case, one thing is for sure, solar energy will be an indispensable energy source for electricity generation and heating in many countries, including the USA and Canada, during the coming years.

## Main barriers to the massive use of solar energy for electricity generation in the North American region

Today, several barriers exist to the use of solar energy for electricity generation and heating in the North American region. The multiple main barriers are, according to the Electrical power generation (2016) report, the following:

- Solar radiation has a low energy density relative to other conventional energy sources used for electricity generation and heating, such as oil, natural gas, or coal. Therefore, it requires a relatively large area to collect and produce a considerable amount of energy. Typical solar power plant designs require about five acres per MW of generating capacity. Likewise, a 30 MW thin-film solar PV array would require about 168 acres. For this reason, the construction of large solar power plants requires that land use issues be considered in detail;
- The need for water depends on the solar technology used for electricity generation and heating. Solar thermal electric technologies, such as central receiver and parabolic trough designs, require a considerable amount of water for cooling. Solar power plants based on solar PV and dish-Stirling engine designs and small-scale solar PV and CSP installations do not require water. These systems reduce water consumption by offsetting energy production from conventional generators, which do consume water;

- Adequate transmission lines are required to transport the power generated by solar power plants to urban load centers. Intermittent resources such as solar energy can pose unique problems in transmission planning and efficient utilization of transmission infrastructure, resulting in higher transmission costs, increased congestion, and even generation curtailments when adequate transmission capacity is not available;
- Lack of government policy supporting the use of solar power for electricity generation and heating;
- Lack of information dissemination and consumer awareness about the use of solar energy for electricity generation and heating;
- The high cost of solar technologies compared with conventional energy sources;
- Difficulty overcoming established energy systems for electricity generation, particularly fossil fuel power plants. That includes difficulty introducing innovative energy systems, particularly for distributed generation such as solar PV, because of technological lock-in, electricity markets designed for centralized power plants, and market control by established generators;
- Inadequate financing options for the implementation of solar power projects;
- Inadequate workforce skills and training available;
- Lack of stakeholder and community participation in solar energy choices;
- The strong dependency of the weather situation for electricity generation and heating;
- Instability of the supply of energy at the level requested in a given moment.

Besides the barriers mentioned above, it is important to know the following: hailstorms can be dangerous for solar systems. A hailstorm can have serious consequences for a solar PV system. On the one hand, there may be visible damage, in the worst case, glass breakage. On the other hand, certain damages cannot be detected by a simple visual inspection. These damages, in the long run, would reduce the performance of the solar PV system. For this reason, when choosing a solar PV system, special attention must be paid to the hail resistance class used in the production of solar panels (Aleo solar, 2018).

Although the cost of producing solar PV devices and electricity using solar panels continues to fall, the use of solar energy for electricity generation and heating is not increasing particularly rapidly in some countries all over the world, especially in developing countries. One important barrier that should be in mind is the so-called "sociotechnical barriers." That barrier is a complex issue and covers product quality, quality standards, and consumer concerns about the complexity, durability, efficiency, and safety of solar technology currently available. Without a doubt, a lack of knowledge about solar power technologies leads to some planners not recommending the use of solar PV for electricity generation in new buildings or to improper use and poor maintenance by adopters. In some countries and regions, climatic conditions and architectural constraints make the use of solar energy for electricity generation and heating less suitable than other energy sources (Karakaya and Sriwannawit, 2015).

Within the energy policy category, the lack of stability of incentives for the use of solar PV for electricity generation and heating is an important element to be considered. That includes inconsistencies between policy measures and socio-economic factors or the sudden removal of existing subsidies that make using this type of energy source less competitive than other energy sources, particularly conventional energy sources. While most countries have policy measures to support the use of renewable energies for electricity generation and heating, the market loses trust when policy decisions are reversed, such as the recent retrospective reduction of feed-in tariffs in some countries. "Failure to involve all the relevant stakeholders in energy policy planning and regulatory issues, such as difficulties acquiring building permits and lengthy decision processes, constituted further barriers" (Karakaya and Sriwannawit, 2015) to the increasing use of solar energy for electricity generation and heating in some countries.

Management barriers included an incorrect distinction between rural and urban situations or low-income and high-income business strategies. For example, fee-for-service and microcredit financial schemes could be used for the low-income access-oriented markets in rural areas but would not be suitable for the high-income adopters in cities, where solar is an alternative power supply. Other management barriers are, among others, the following:

- Poor after-sales service;
- Ineffective marketing and education campaigns among the population to promote the use of solar energy for electricity generation and heating;
- Lack of collaboration between the construction and solar PV industries;
- Lack of national infrastructure to support the massive use of solar energy for electricity generation and heating;
- Lack of policy backing the extensive use of solar energy for electricity generation and heating (Karakaya and Sriwannawit, 2015.

Economic barriers include high initial capital costs of the solar PV modules for large installation, maintenance, repair costs, and the low costs of competing for electricity generation and heating energy sources. Perception of the cost can be an additional economic barrier, as well as "uncertainties in the funding process, inadequate government subsidies compared to competing for energy sources, including fossil fuels, and the unwillingness of banks to fund medium- or long-term investments in shrinking economies" (Karakaya and Sriwannawit, 2015.

While the use of solar power for electricity generation and heating has been encouraging in the USA with subsidies and tax incentives, there are still many barriers left that solar power must overcome before this source of power becomes more widely used in the country. One of these barriers is the adoption of limits on net-metered solar power. The reason for adopting this measure is that the cost of solar energy is nearing the point where it has become cheaper than the cost of electricity from the grid. Thus, the utility companies see solar power as a threat and are encouraging to limit the use of solar energy for electricity generation in the energy industry.

Without a doubt, solar PV is a mature energy source for electricity generation and heating that can compete with conventional energy sources in many countries. However, there are still barriers to using solar energy for these specific purposes in high-income economies, including the USA and Canada, and low-income economies.

# The future of solar energy in the North American region

The future of solar energy in the North American region depends on the USA energy policy related to using this type of energy source for electricity generation and heating in the coming years because Canada uses solar energy, mainly solar PV, on an exceedingly small scale. In 2019, Canada had only 3310 MW of solar PV capacity installed, representing 5% of the total solar PV capacity installed in the region (65,608 MW). Besides, Canada has no CSP facilities for electricity generation and heating and has no plans for its future use. On the other hand, it is important to know that the current Trump administration's energy policy denies climate change and is responsible for the lack of government support for using eco-friendly energy sources for electricity generation and heating. This situation increases the solar industry representatives' concern about the solar power industry's future in the country and the region.

Besides, the future of solar energy in the region depends heavily on uncertain future market conditions and public policies to be adopted by the US and Canadian governments, including but not limited to policies aimed at mitigating global climate change, something that the US government is reluctant to recognize.

Undoubtedly, despite the US energy policy, the US solar industry is currently an expanding industry that employs about 242,000 professionals and workers and generates tens of billions of dollars. By the end of September 2019, the USA had deployed over two million solar PV systems, totaling 62,298 MW of solar capacity (95% of the total solar PV capacity installed in the region in that year), and generated over 90 TWh of electricity. In 2018, solar power generated about 1.5% of US electricity and is expected to grow the fastest renewable energy source from now to 2050 (see Fig. 3.11) (Rhodes, 2020).

According to the Annual Energy Outlook 2019 (2019) report, the USA is expected to become a net energy exporter by 2020. Thus, the country will become a net exporter of petroleum liquids as US crude oil production increases and domestic consumption of petroleum products decreases and continue to be a net exporter of natural gas and coal

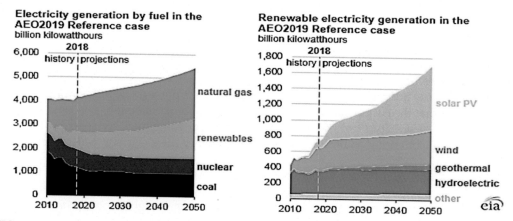

FIG. 3.11    Renewable electricity generation in the USA and projections until 2050. *Source: 2019. Annual Energy Outlook 2019. US Energy Information Administration (EIA). https://www.eia.gov/about/copyrights_reuse.php.*

(including coal coke) through 2050. It is important to stress that renewable energy shares are expected to grow in US electricity generation during the coming years. The renewables share, including hydroelectric generation, is expected to increase from 18% in 2018 to 31% in 2050, driven largely by wind and solar power generation growth. Without a doubt, renewables are projected to become a larger share of US electricity generation than nuclear energy and coal in less than a decade (Yen, 2019), despite the current US administration energy policy supporting the use of conventional energy sources for electricity generation and heating.

In view of the US Bureau of Labor Statistics, the USA's total installed solar PV capacity is expected to more than double during the next five years. It estimates that a solar PV installer will be the fastest-growing job between 2018 and 2028 (Rhodes, 2020). In 2019, the USA ranked second in solar PV deployment behind China (see Fig. 3.12).

Looking at Fig. 3.12, the following can be stated: China was, in 2019, by far the country with the highest cumulative solar PV deployment at the world level. Its totals are higher than the USA, Japan, and Germany's cumulative solar PV deployment together.

Estimates for global solar PV growth are expected to exceed 100 GW per year from 2019 and more than triple by 2050. The North American region is responsible for about 20% of the total solar PV deployment. Besides, the US solar energy generated, in 2018, a total of 1.5% of the total electricity produced by the country and can manufacture about 6 GW of solar panels per year, with plans to expand to about 9 GW shortly to respond to projected high demand for solar panels within the country. The top three US states for solar PV manufacturing are Ohio, Georgia, and New York. These three states account for almost 60% of US solar PV manufacturing capability (Rhodes, 2020).

One important challenge facing the US solar industry is the import tariffs placed on solar panels by the Trump administration. According to some estimates, the new import tariffs have provoked a loss of almost US$20 billion in private investment in the solar sector. The new interest tariffs "also have a disproportionate impact on projects in less-sunny areas and raise the Levelized cost of electricity for projects by non-negatable amounts" (Rhodes, 2020).

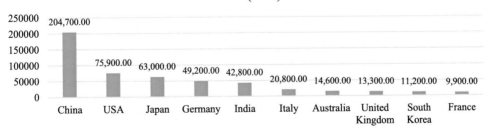

FIG. 3.12   Solar PV deployment by a selected group of countries in 2019. *Source: 2020. IEA PV Snapshot 2019. International Energy Agency. Retrieved 2 May 2020.*

Finally, it is important to single out the following: electricity generation and heating by solar power parks is one of the very few low-carbon energy technologies with the potential to grow to an exceptionally large scale in the future. That is true not only for the North American region but also in other regions as well, according to an interdisciplinary MIT study entitled "The Future of Solar Energy" (The Future of Solar Energy: An Interdisciplinary MIT Study, 2015).

Without a doubt, new technological breakthroughs happen every day. For this reason, it can be expected that solar PV panels will continue to be even more efficient, cheaper, and simpler to produce electricity in the future than today with the current solar technology. Several countries and the energy industry worldwide are working hard on improving the solar technologies available today.

## Solar energy capacity in Canada

It is a fact that the solar energy industry is having speedy growth in Canada. However, the solar energy capacity already installed in Canada and the use of this type of energy source for electricity generation and heating is still extremely low in comparison with the US solar energy capacity already installed, the US solar industry, and the use of this type of energy source for electricity generation and heating in that country. In 2019, the solar energy capacity installed in Canada totaled 3310 MW, which represents 3.3% of the total renewable energy source capacity installed in the country in that year, and 5% of the solar energy capacity installed in the North American region. Canada ranked nine according to its solar power capacity installed at the world level, representing 1% of the world's total (Chakrabarti, 2019). According to the National Energy Board Database 2020, the use of solar energy for electricity generation will account for 1.2% of Canada's total energy by 2040, a very small percentage compared to the projected percentage of other energy sources available in the country.

Canada's largest solar facility is the Loyalist Solar Project, with 54 MW capacity in Ontario, the Canadian province with 95.6% of the national solar power generation capacity installed (2964.8 MW). However, Canada, despite having substantial solar energy resources unexploited due to the vast extension of the country, its latitude causes a relatively low level of solar irradiance. Besides, due to cloudy days, its capacity factor is only 6%, which is extremely low compared to the USA's capacity factor (15%). Until today, the country's most valuable solar energy resources have been found in the provinces of Alberta,[1] Manitoba, Ontario, and Saskatchewan[m] (Chakrabarti, 2019). Other areas with plentiful solar energy sources are southern Quebec, New Brunswick, southern Nova Scotia, and western Prince Edward Island (Wikipedia Solar Power in Canada, 2020). It is important to single out that Canada's total solar energy capacity is in solar PV, and almost all of them are connected to the electrical grid.

---

[1] "Alberta, known for being Canada's oil and gas province, is one of the best province in the country for solar power – scoring high in almost every category. Alberta has high sunlight levels and energy needs, incoming financing options, and solar rebate programs in some jurisdictions" (Rahimi, 2019).

[m] "Canada's sunniest province, Saskatchewan, is the lowest-ranked province in the country, mostly due to a recently-altered poor net metering program and an almost complete lack of financing and incentive options" (Rahimi, 2019) for the promotion of the use of solar energy for electricity generation and heating.

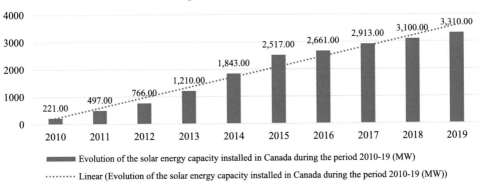

FIG. 3.13  Evolution of the solar power installed capacity in Canada during the period 2010–19. *Source: 2020. IRENA Renewable Capacity Statistics 2020. International Renewable Energy Agency, Abu Dhabi.*

The evolution of the solar power installed capacity in Canada during the period 2010–19 is shown in Fig. 3.13.

The following can be stated, according to the data included in Fig. 3.13: solar installed capacity in Canada increased almost 15-fold during the period 2010–19, rising from 221 MW in 2010 to 3310 MW in 2019. Regrettably, compared with the leading solar energy-producing countries globally, Canada does not have any solar electricity policy at the federal level that can provide excellent opportunities to the country's solar energy industry. It is a reality that solar energy technology for electricity generation and heating has not been exploited as expected. As a result, the capacity installed and the electricity generation using this renewable energy source are relatively small. It is expected that solar energy participation in the Canadian energy mix in the future will be almost the same as today.

The total solar power capacity installed in Canada by provinces in 2018 is shown in Table 3.3.

At Present, in Canada, remote off-grid solar systems are reporting a very moderate growth. However, it can be a viable market by bringing about incentives. As a result, the country's solar power market looks more like a small market segment than the market segment of other renewable energy sources for electricity generation and heating. However, micro-grid options offer positive growth, particularly in the case of micro-grids that use multiple energy sources located in remote regions. Despite this possibility that needs to be explored in the coming years, it is expected that solar energy participation in the country's energy mix will continue without significant changes during the coming years.

## Solar energy electricity generation in Canada

Historically, the main uses of solar energy technologies in Canada "have been non-electric active solar system applications for space heating, water heating and drying crops and lumber. As a result, these systems presently comprise a small fraction of Canada's energy use."

**TABLE 3.3**   Solar power capacity installed in Canada by provinces in 2018 (MW).

| | |
|---|---|
| Ontario | 2964.8 |
| Alberta | 61.7 |
| Saskatchewan | 22.2 |
| British Columbia | 14.4 |
| Manitoba | 14.2 |
| Quebec | 6.4 |
| Nova Scotia | 4.9 |
| Prince Edward Island | 2.5 |
| Yukon | 2.2 |
| Other provinces and territories | 4.82 |

*Source: 2020. IRENA Renewable Capacity Statistics 2020. International Renewable Energy Agency, Abu Dhabi and 2019. Statista 2019.*

However, "some government studies suggest they could make up as much as 5% of the country's energy needs by the year 2025" (Wikipedia Solar Power in Canada, 2020).

Solar PV cells are increasingly used as stand-alone units, mostly as off-grid distributed electricity generation to powered isolated towns, villages, homes, telecommunications equipment, oil and pipeline monitoring stations, and navigational devices. The Canadian solar PV market has grown quickly in recent years, and Canadian companies make solar modules, equipment controls, specialized water pumps, high-efficiency refrigerators, and solar lighting systems. In recent years, grid-connected solar PV systems have grown significantly and reached over 3 GW of capacity installed in 2019.

According to a forecast by the International Energy Agency (IEA) database 2020, by 2050, solar electricity could account for 27% of the global electricity mix. By 2020, and according to the database mentioned above, solar electricity in Canada is expected to:

- Produce nearly 1% of the total electricity generation in the country[n];
- Create approximately 65,000 jobs every year, employing a labor force of nearly 10,000 people per year. The primary sectors of employment will be construction, manufacturing, operations, and maintenance;
- Reduce nearly 1.5 million tons of greenhouse gas emissions annually. That figure is the equivalent of removing 250,000 cars and trucks off the road every year.

Canada's solar electricity industry could be sustainable and commercially viable without direct subsidies and operating in a supportive environment and with a favorable regulatory policy.

[n] In 2017, Canada generated 3% of the total solar electricity generated at the world level (Natural Resources Canada Electricity Facts, 2020).

In Canada, in 2019, the use of renewable energy sources for electricity generation and heating provides about 17.3% of Canada's total primary energy supply (Energy Facts, 2019; Renewable Energy Facts, 2019) and about 67% of its electricity production (Natural Resources Canada Electricity Facts, 2019). According to the world electricity production, the country's renewable energy ranks seven on an international scale, preceded by China with 14%, India with 11%, the USA with 8%, Brazil and Nigeria with 6% each, and Indonesia with 4% (Renewable Energy Facts, 2019).

The main renewable energy sources used for electricity generation and heating in Canada are hydro, wind, tidal, geothermal, solar, and biomass. The primary renewable energy source used in Canada for electricity generation and heating is hydropower. It supplied 67.1% of total electricity generated in the country in 2017, making Canada the second-largest producer of hydroelectric at the world level (Renewable Energy Facts, 2019).

Solar energy alone generated 1% of the total electricity produced in the country in 2019 (Energy Facts Book 2020–2021, 2020). The evolution of solar electricity generation in Canada during the period 2010–18 is shown in Fig. 3.14.

According to Fig. 3.14, the following can be stated: the solar electricity generation in Canada during the period 2010–18 increased almost 15-fold, rising from 255 GWh in 2010 to 3802 GWh in 2018. It is expected that this trend will continue in the future, increasing the role of solar energy in the country's energy mix, but moderately. Undoubtedly, hydropower will continue to be the main energy source used in Canada for electricity generation and heating during the coming years.

## Solar energy capacity in the Univted States

Without a doubt, the USA has the technology, talent, and environmental conditions required for the successful deployment of large-scale solar energy to meet the ever-increasing demand for energy. According to the USA Energy Information Administration, renewable

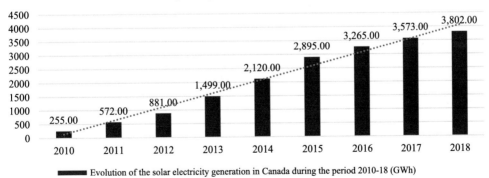

**Evolution of the solar electricity generation in Canada during the period 2010-18 (GWh)**

■ Evolution of the solar electricity generation in Canada during the period 2010-18 (GWh)

········· Linear (Evolution of the solar electricity generation in Canada during the period 2010-18 (GWh))

**FIG. 3.14** Solar electricity generation in Canada during the period 2010–18 (GWh). *Source: IRENA, 2020. Renewable Energy Statistics 2020. The International Renewable Energy Agency, Abu Dhabi.*

energies provided 17.61% of electricity generation in the USA in 2018. Electricity generation from different renewable energy sources provided a record of 6,586,124 GWh of electricity in 2018, almost 1.6-fold the 4,202,026 GWh produced in 2010. It is important to stress that almost 90% of the increase in electricity generated in the USA between 2008 and 2018 within renewables came from wind and solar energies (Reve, 2019). In the case of solar energy, this type of energy source provided about 2% of the total USA electricity in 2019 (Electricity explained, 2020).

Among the USA's renewable energy sources, solar energy has had the biggest rise during the last years, being 2016 the first best year of solar energy in the country. A total of 14,626 MW of new solar PV capacity was installed throughout the year, almost doubled its annual record previously reached. It grew 95% on the results achieved in 2015 when it also broke a record with the installation of 7493 MW of new solar PV capacities. In 2018, solar capacity increased by 51 GW, from less than 1 GW in 2008; this increase represents more than 50-fold the capacity installed in 2008. The existing capacity includes 1.8 GW of CSP, 30 GW of utility-scale solar energy, and 20 GW of small-scale solar energy. In the USA, solar power generation rose to 85,184 GWh in 2018, from 3942 GWh in 2010. Utility-scale solar generation accounted for 69%, or 67 million MWh, of total solar generation in 2018 (Reve, 2019).

According to the Solar Energy Industries Association (Solar Energy Industries Associations, 2019), the USA is the home to more than two million solar PV installations (2,073,346 solar PV installations) in 50 states across the country. The mark comes three years after the industry completed its first one million solar PV installations and two million in 2019. It is expected that solar PV installations will reach three million in 2021 and four million in 2023. The solar industry has completely reshaped the energy mix in the USA, and the US$17 billion solar industry is on track to double again in five years. According to the president of SEIA opinion, 2020–2030 will be the decade that solar energy is expected to become the dominant new form of energy generation.

The number of solar PV installations increased significantly in the USA during the period 2010–2016. However, since 2016, it has decreased until 2019, with some increase registered in 2017.

Finally, it is important to single out that no new utility-scale CSP projects have come online in the USA since 2015. There are also no CSP plants currently under construction or at any stage of development. "The only new CSP data reported in 2019 relates to the capacity factors of existing CSP plants. On that front, two recent solar trough projects without storage have largely matched ex-ante capacity factor expectations, while two solar power tower projects and a third solar trough project with storage continue to underperform relative to projected long-term, steady-state levels" (Bolinger et al., 2019).

The evolution of solar energy, solar PV, and CSP installed capacities in the USA during 2010–19 is shown in Figs. 3.15–3.17.

According to Fig. 3.15, the USA's solar energy installed capacity increased by 18.4-fold, rising from 3382 MW in 2010 to 62,298 MW in 2019. Therefore, it is projected that solar energy will continue to grow in the USA during the coming years, being the energy source with the fasted growth in the country among all other energy sources.

Looking at Fig. 3.16, the following can be stated: the USA's solar PV installed capacity increased by 20.8-fold during the period 2010–19, rising from 2909 MW in 2010 to

## Evolution of solar energy installed capacity in the USA during the period 2010-19 (MW)

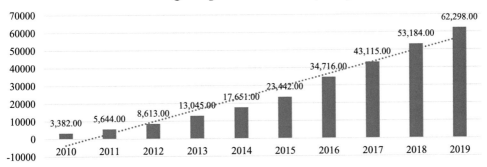

Evolution of solar energy installed capacity in the USA during the period 2010-19 (MW)

········ Linear (Evolution of solar energy installed capacity in the USA during the period 2010-19 (MW))

**FIG. 3.15** Evolution of the solar energy installed capacity in the USA during the period 2010–19 (MW). *Source: IRENA, 2020. Renewable Energy Statistics 2020. The International Renewable Energy Agency, Abu Dhabi.*

## Evolution of solar PV installed capacity in the USA during the period 2010-19 (MW)

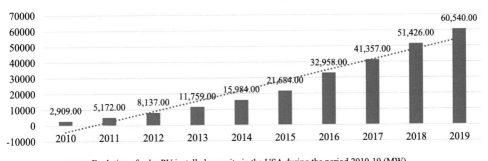

Evolution of solar PV installed capacity in the USA during the period 2010-19 (MW)

········ Linear (Evolution of solar PV installed capacity in the USA during the period 2010-19 (MW))

**FIG. 3.16** Evolution of the solar PV installed capacity in the USA during the period 2010–19 (MW). *Source: IRENA, 2020. Renewable Energy Statistics 2020. The International Renewable Energy Agency, Abu Dhabi.*

60,540 MW in 2019. Furthermore, it is expected that this trend will continue without change during the coming years.

Finally, the USA's solar CSP installed capacity evolution during the period 2010–19 is shown in Fig. 3.17.

According to Fig. 3.17, the following can be stated: CSP installed capacity in the USA increased by 3.7-fold during the period 2010–19, rising from 473 MW in 2010 to 1758 MW in 2019. It is important to single out that the major increase in CSP installed capacity was reported during the period 2010–15. Since 2015, no new CSP capacity was reported that had

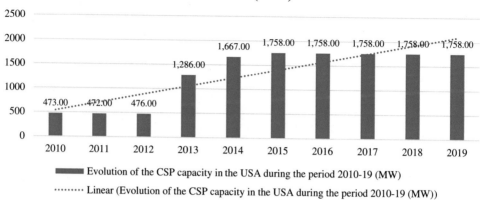

FIG. 3.17  Evolution of the solar CSP installed capacity in the USA during the period 2010–19 (MW). *Source: IRENA, 2020. Renewable Energy Statistics 2020. The International Renewable Energy Agency, Abu Dhabi.*

been built in the country. However, it is expected that this trend could change during the coming decades, and new CSP capacities will be added to the existing ones. Despite that possibility, it is important to stress that the major increase in solar energy use for electricity generation and heating in the USA will be registered in solar PV parks and not in CSP facilities.

## Solar energy electricity generation in the United States

In the USA, the use of solar energy for electricity generation and heating is the following: the USA has, in 2019, more than two million solar PV installations (SEIA, 2019), with a generation capacity of 60.5 GW. Furthermore, it is expected that by 2023, solar PV installations could be double that in 2019.[°] The evolution in the growth of solar PV installations in the USA is shown in Fig. 3.16.

According to the Office of Energy Efficiency and Renewable Energy in the US DoE, "solar power is now more affordable, accessible, and prevalent than in the past. Since 2008, US solar energy installations have grown 35-fold. The solar power capacity installed is enough to power the equivalent of 12 million average American homes" (SEIA, 2019). In addition, since the beginning of 2014, the average cost of solar PV panels has dropped by nearly 50% (Feldman and Margolis, 2019). According to the SunShot Vision Study prepared in 2012, it is projected that the price of solar technologies will decline by about 75% in 2020.

---

[°] Wood Mackenzie forecasts that there will be three million of solar PV installations in 2021 and four million in 2023, continuing the swift rise of solar energy for electricity generation and heating. By 2024, "there will be on average, one solar installation per minute," said Michelle Davis, Senior Solar Analyst with Wood Mackenzie, that's up from one installation every 10 min in 2010." Around 2.5% of all USA homes will have a solar installation for electricity generation in 2024. "The rapid growth in the solar industry has completely reshaped the energy conversation in this country," said Abigail Ross Hopper, SEIA president and CEO.

Based on the report mentioned above, solar technology installed system prices are assumed to reach:

- US$1 per watt for utility-scale solar PV systems;
- US$1.25 per watt for commercial rooftop solar PV systems;
- US$1.50 per watt for residential rooftop solar PV systems;
- US$3.60 per watt for CSP systems with up to 14 h of thermal energy storage capacity.[P]

As a result of the price reduction forecast mentioned above, solar technologies are projected to play an increasingly important role in meeting USA electricity demand over the next 20–40 years, satisfying around 14% of the country's electricity demand by 2030 and 27% by 2050. In 2018, solar power parks generated 1.5% of the total electricity produced by the country. In terms of solar technology, the focus will be on solar PV and less on CSP. "Markets for solar energy are maturing rapidly around the country, and solar electricity is now economically competitive with conventional energy sources in several states, including California, Hawaii, and Minnesota," according to the Office of Energy Efficiency & Renewable Energy in a material entitled "Solar Energy in the United States (2019)".

Moreover, the solar industry is a proven incubator for job growth throughout the nation. Solar jobs have increased by nearly 160% since 2010, which is nine times the national average job growth rate in the last five years. According to the Solar Foundation, "there are more than 242,000 solar workers in the United States, with manufacturing being the second largest sector in the solar industry" (National Solar Jobs Census, 2018).

Increased solar energy utilization for electricity generation and heating offers a variety of benefits for the USA. Solar's abundance and potential throughout the country are large solar PV panels on just 0.6% of the nation's total land area that could supply enough electricity to power the entire country (SunShot Vision Study, 2012). Undoubtedly, CSP is another solar system that can capture energy from the Sun to produce electricity and heating. According to the report entitled "Solar Energy in the United States (2019)" already mentioned, "seven southwestern states have the technical potential and land area to site enough CSP to supply more than four times the current US annual electricity demand."

Despite this impressive progress in the use of solar PV for electricity generation and heating in the USA, market barriers and grid integration challenges continue to limit a greater use of this type of facility in the country. Without a doubt, solar power technology is advancing. However, innovative solutions are still needed to:

- Increase the efficiency of solar PV systems;
- Decrease further the costs associated with the electricity generation and heating using solar power;
- Enable utilities to rely on solar energy for baseload power in combination with other energy sources.

---

[P] Note that throughout the SunShot Vision Study report all "$ per W" units refer to 2010 US dollars per peak watt-direct current (DC) for solar PV and 2010 US dollars per watt-alternating current (AC) for CSP, unless otherwise specified.

Finally, the following conclusion can be stated: in the USA, solar energy in general but solar PV, in particular, will be the fastest-growing energy source to be used for electricity generation and heating at least until 2050.

During the period 2010–18, solar PV and CSP electricity generation evolution are shown in Figs. 3.18 and 3.19.

Based on the information included in Fig. 3.18, the following can be stated: the solar PV electricity generation in the USA during 2010–18 increased by 26.5-fold, rising from 3063 GWh

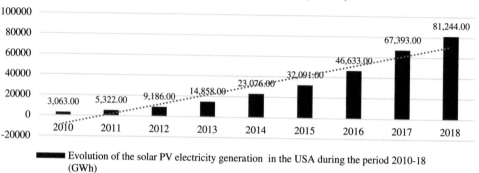

FIG. 3.18   Evolution of solar PV electricity generation in the USA during the period 2010–18 (GWh). *Source: IRENA, 2020. Renewable Energy Statistics 2020. The International Renewable Energy Agency, Abu Dhabi.*

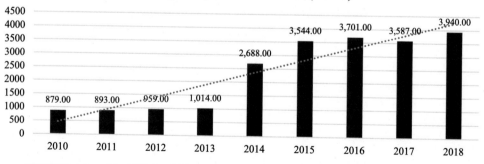

FIG. 3.19   Evolution of the CSP electricity generation in the USA during the period 2010–18 (GWh). *Source: IRENA, 2020. Renewable Energy Statistics 2020. The International Renewable Energy Agency, Abu Dhabi.*

in 2010 to 81,244 GWh in 2018. Therefore, it is expected that the trend shown in Fig. 3.19 will continue without change during the coming years.

According to the information included in Fig. 3.19, the following can be stated: CSP electricity generation in the USA during the period 2010–18 increased by 4.5-fold, rising from 879 GWh in 2010 to 3940 GWh in 2018. The period under consideration can be divided into two parts. In the first part covering the period 2010–13, the electricity generation using CSP facilities increased only 15.4%. In the second part covering the period 2014–18, the electricity generation using CSP facilities increased by 3.9-fold. It is expected that the trend shown in Fig. 3.19 after 2014 will continue without change during the coming years, and the role of CSP facilities in the energy mix of the country will increase.

# References

Aleo solar, 2018. Hail Storms Can Be Dangerous for Solar Systems.

Anon., 2019. Annual Energy Outlook 2019. US Energy Information Administration (EIA). https://www.eia.gov/about/copyrights_reuse.php.

Bazilian, M., Onyeji, I., Liebreich, M., 2013. Re-considering the economics of photovoltaic power. Renew. Energy 53. Archived (PDF) from the original on 31 August 2014; Retrieved 31 August 2014.

Bolinger, M., Seel, J., Robson, D., 2019. Utility-Scale Solar: Empirical Trends in Project Technology, Cost, Performance, and PPA Pricing in the United States. Lawrence Berkeley National Laboratory. 2019 Edition.

Chakrabarti, S., 2019. Solar Power Statistics in Canada 2019. Solar Feeds.

Delbert, C., 2020. This solar panel just set a world record for efficiency. Popular Mechanics.

Delucchi, M., Jacobson, M., 2013. Meeting the world's energy needs entirely with wind, water, and solar power. Bull. At. Sci. 69 (4), 30–40.

Electrical power generation, 2016. The Current Situation and Perspectives on the Use of Solar Energy For Electricity Generation: Main Barriers to the Massive Use of Solar Power Technology for the Generation of Electricity. machineryequipmentonline.com/electrical-power-generation/the-current-situation-and-perspectives-on-the-use-of-solar-energy-for-electricity-generationmain-barriers-to-the-massive-use-of-solar-power-technology-for-the-generation-of-electricity.

Electricity explained, 2020. Electricity in the United States. USA EIA.

Anon., 2019. Energy Facts. Natural Resources Canada.

Anon., 2020. Energy Facts Book 2020–2021. Natural Resources Canada.

Feldman, D., Margolis, R., 2019. Solar Industry Update. National Renewable Energy Laboratory.

Geisz, J.F., Steiner, M.A., Jain, N., Schulte, K.L., France, R.M., McMahon, W.E., Perl, E.E., Friedman, D.J., 2018. Building a six-junction inverted metamorphic concentrator solar cell. IEEE J. Photovoltaics 2156-3403. 8 (2), 626–632. https://doi.org/10.1109/JPHOTOV.2017.2778567. OSTI 1417798.

Anon., 2014. Global Market Outlook for Photovoltaics 2014–2018. EPIA—European Photovoltaic Industry Association. www.epia.org. Archived from the original on 12 June 2014; Retrieved 12 June 2014.

Anon., 2019. Global Trends in Renewable Energy Investment 2019. UN Environment Program.

IRENA, 2019. Future of Solar Photovoltaic: Deployment, Investment, Technology, Grid Integration and Socio-Economic Aspects (A Global Energy Transformation: Paper). International Renewable Energy Agency, Abu Dhabi.

IRENA, 2020a. Insight Renewables. IRENA homepage,.

IRENA, 2020b. Renewable Energy Statistics 2020. The International Renewable Energy Agency, Abu Dhabi.

IRENA and CPI, 2018. Global Landscape of Renewable Energy Finance, 2018. International Renewable Energy Agency, Abu Dhabi.

Anon., 2020. IRENA Renewable Capacity Statistics 2020. International Renewable Energy Agency, Abu Dhabi.

Anon., 2020. IRENA Solar Energy. International Renewable Energy Agency, Abu Dhabi.

Karakaya, E., Sriwannawit, P., 2015. Barriers to the adoption of photovoltaic systems: the state of the art. Renew. Sust. Energ. Rev. 49, 60–66. https://doi.org/10.1016/j.rser.2015.04.058.

Kumar, A., 2017. Predicting efficiency of solar cells based on transparent conducting electrodes. J. Appl. Phys. 0021-8979. 121 (1), 014502. Bibcode: 2017JAP...121a4502K https://doi.org/10.1063/1.4973117.

Leblanc, R., 2019. What Is the Environmental Impact of Solar Power Generation? The Balance Small Business.

MacKay, D.J.C., 2017. Sustainable Energy—Without the Hot Air. inference.org.uk. Retrieved 20 November 2017; Solar photovoltaics: data from a 25-m$^2$ array in Cambridgeshire in 2006.

Morales Pedraza, J., 2012. Nuclear Power: Current and Future Role in the World Electricity Generation. Nova Science Publishers, ISBN: 978-1-61728-504-2.

Morales Pedraza, J., 2015. Electrical Energy Generation in Europe: The Current Situation and Perspectives in the Use of Renewable Energy Sources and Nuclear Power for Regional Electricity Generation. Springer.

Anon., 2018. National Solar Jobs Census. The Solar Foundation.

Anon., 2019. Natural Resources Canada Electricity Facts. The Government of Canada.

Anon., 2020. Natural Resources Canada Electricity Facts. The Government of Canada.

Rahimi, M., 2019. Solar Power Guides and Ranking.

Anon., 2019. Renewable Energy Facts. Natural Resources Canada.

Anon., 2019. Renewables 2019. Market Analysis and Forecast From 2019 to 2024. International Energy Agency (IEA) report 2019.

Reve, 2019. Renewable Energy Provided 17.6% of Electricity Generation in the United States in 2018.

Rhodes, J., 2020. The Future of US Solar Is Bright. Forbes.

Roberts, B., 2008. Photovoltaic Solar Resource of the United States. National Renewable Energy Laboratory. Retrieved 17 April 2017.

Schuessler, F., 2008. Energy partnership Africa-Europe: concentrated solar power between technical realization and ethic responsibility. Erdkunde 62 (3), 221–230.

SEIA, 2019. United States Surpasses 2 Million Solar Installations. Solar Energy Industries Association (SEIA).

Shieber, J., 2018. A New Solar Technology Could Be the Next Big Boost for Renewable Energy. TechCrunch.

Solar Energy, 2009. Solar Energy. Status, Trends & Opportunities. Clixoo, India.

Anon., 2019. Solar Energy in the United States. Office of Energy Efficiency and Renewable Energy in the US Department of Energy (DoE).

Anon., 2019. Solar Energy Industries Associations. SEIA.

Anon., 2020. Solar Explained. Solar Energy and the Environment. US Energy Information Administration.

Anon., 2013. Solar Performance and Efficiency. Office of Energy Efficiency and Renewable Energy in the US Department of Energy (DoE).

Anon., 2007. Solar Thermal Electric Generation. http://uspowerpartners.org/Topics/SECTION2TopicSolarThermalElec.htm. (Accessed 22 June 2007).

Anon., 2012. SunShot Vision Study. US Department of Energy (DoE).

Teske, S., 2008. La Energía Solar Puede Dar Electricidad Limpia a Más de 4.000 millones de Personas para 2030. Greenpeace España.

Anon., 2015. The Future of Solar Energy: An Interdisciplinary MIT Study. Massachusetts Institute of Technology (MIT).

Anon., 2014. US DoE Photovoltaic Cell Conversion Efficiency Basics. US Department of Energy (DoE). Retrieved 6 September 2014.

Anon., 2020. Wikipedia Photovoltaic.

Anon., 2020. Wikipedia Photovoltaic System.

Anon., 2020. Wikipedia Solar Cell Efficiency.

Anon., 2019. Wikipedia Solar Energy.

Anon., 2020. Wikipedia Solar Power in Canada.

World Bank Group, 2019. Capacity Investment—Developed Countries. (Chapter 5) https://olc.worldbank.org/system/files/Chapter%205_1.pdf.

Yen, T., 2019. EIA's Annual Energy Outlook 2019. Projects Growing Oil, Natural Gas, Renewables Production. Today in Energy, US EIA.

# 4

# The use of wind energy for electricity generation

## Introduction

It is undeniable that energy production, especially electricity generation and its sustained growth, is crucial to ensure any country's progress independent of the region where this country is located. The well-being of people, industries, and the economy depends on safe, secure, sustainable, clean, and reasonably energy prices. In the modern world, energy is a daily need, and many people in the North American region have taken it for granted.

The energy system and its organization evolved over centuries, using different types of energy sources and distribution systems to cover basic energy needs. These basic energy needs are, among others, food preparation, protection against winter temperatures, and the production of tools and equipment. Over the last century, basic energy needs included delivering heat and warm water as well as industrial and transportation fuels and electricity to consumers. Without a doubt, there has been a significant increase in energy production and consumption at the world level over the last 100 years. Energy consumption provides more comfort and individual freedom to customers, but at the same time, some conventional energy sources are polluting the environment and reducing existing fossil fuel reserves (Impact Assessment, 2010).

Undoubtedly, the use of all energy sources for electricity generation will continue to increase gradually at the world level at least until 2050. If this is true, then the main question that needs to be answered is: How high will the foreseen increase in energy demand, and how this energy demand will be met? According to different studies made by several experts on energy demand, it is expected that by 2030, the world energy demand will be double, and it is most likely that it could be triple by 2050.

Until 2030, the primary energy demand at the world level is projected to increase annually at 1.7%, which is 0.4% smaller than the world growth of 2.1% registered during the last three decades. It is also expected that, during the coming decades, around 90% of the increase in the world energy demand will be satisfied with oil, coal, and gas, particularly the last one. However, the fastest-growing energy sources at the world level are, and will continue to be in the future, renewable energy sources, including in the North American region and, on a lower scale and in a limited number of countries, nuclear power. Wind energy is likely to be the fastest-growing energy source among renewable energy sources during the coming decades.

"The renewable share of total energy use is expected to rise from 11% in 2010 to 15% in 2040, an increase of 4%, and the nuclear share is expected to grow in the same period from 5% to 7%, an increase of 2%. Undoubtedly, renewable energy sources are projected to be the fastest-growing sources of electricity generation during the next two decades, with expected annual increases averaging 2.8% per year from 2010 to 2040" (Electrical Power Generation, 2016).

Different types of renewable energy sources can be used for electricity generation and heating at the world level and in the North American region. According to IRENA Renewable Energy Statistics 2020 report,[a] the participation of wind energy connected to the electrical grid at the world and the North American region levels was the following: wind power continues to be number two within all renewable energy sources. It shares around 19.2% (1,262,914 GWh) of the total electricity produced in the world in 2018 by all renewable energy sources. In 2019, the total wind farm capacity reached 116,997 MW in the North American region, generating 307,682 GWh in 2018. In 2019, the USA had a wind farm capacity of 103,584 MW. It generated, in 2018, 275,834 GWh, representing 89.6% of the total regional electricity generation using this type of renewable energy source (307,683 GWh). In 2019, Canada had a wind farm capacity of 13,413 MW and generated, in 2018, 31,848 GWh, representing 10.4% of the regional total. The onshore wind farm installed capacity increased approximately 3.3-fold, rising from 177,790 MW in 2010 to 594,253 MW in 2019, generating 1,194,718 GWh in 2018. In North America, the onshore wind farm capacity increased 23.7-fold during the period 2010–19, rising from 43,102 MW in 2010 to 116,968 MW in 2019, generating 307,581 GWh in 2018. In the specific case of offshore wind farms, the capacity installed during the period 2010–19 increased 9.2-fold, rising from 3056 MW in 2010 to 28,155 MW in 2019, generating 68,196 GWh in 2018. Only the USA has a very small offshore wind farm capacity installed since 2016 (29 MW) within the region, generating 102 GWh in 2019;

Wind is a particular form of solar energy. The unequal heating of the atmosphere by the Sun, the anomalies of the Earth's surface, and its rotation are, among others, the primary producers of winds. Wind flow patterns are modified by the Earth's topography, areas of water, and vegetation. The wind flow, when collected by modern wind turbines, can be used to generate electricity.

According to Morales Pedraza (2016), wind turbines, like windmills, are mounted on a tower to contact the wind flow. They can take advantage of the faster and less turbulent wind at 30 m or more above ground (up to 110 m). According to the Office of Energy Efficiency & Renewable Energy, USA Department of Energy (DoE), "wind turbines work on a simple principle: instead of using electricity to make wind—like a fan—wind turbines use the wind to make electricity. The wind turns the propeller-like blades of a turbine around a rotor, which spins a generator, which creates electricity" (How Do Wind Turbine Works?, 2014).

Usually, two or three blades are mounted on a shaft to form a rotor. A blade acts much like an airplane wing. When the wind blows, a pocket of low-pressure air is built on the blade's downwind side. The low-pressure air pocket then pulls the blade toward it, causing the rotor to turn. That movement is called "lift" The force of the lift is much stronger than the wind's force against the front side of the blade, which is called "drag" (see Fig. 4.1). The combination of lift and drag causes the rotor to spin like a propeller, and the turning shaft rotates a generator to generate electricity.

[a] IRENA (2020), Renewable Energy Statistics 2020. The International Renewable Energy Agency, Abu Dhabi.

**FIG. 4.1**  San Gorgonio Pass Onshore wind farm in California, USA. *Source: Courtesy of Erik Wilde from Berkeley, CA, USA – harvesting wind, CC BY-SA 2.0, https://commons. wikimedia.org/w/index.php?curid=51105579.*

According to Renewable Energy World, wind turbines can be used in three different forms:

- As a stand-alone installation;
- As a wind farm and connected to an electric grid;
- Combined with a solar PV system for electricity production.

Many wind turbines are usually built close together for utility-scale wind energy to form what is known as a "wind farm" (see Figs. 4.1 and 4.2). It is important to know that "many electricity providers today use wind farms to supply their customers' power. Stand-alone wind turbines are typically used for water pumping or communications. However, home-owners, farmers, and ranchers in windy areas can also use wind turbines as a way to produce electricity and cut their electric bills. Small wind systems also have potential as distributed energy resources"[b] (Wind Power Technology, 2020).

The tower of the wind turbine carries the gondola and the rotor. According to the Danish Wind Industry Association (n.d.), there are several types of wind towers for large wind turbines. These types are the following:

- Steel conical tubular towers;
- Lattice towers;
- Ferroconcrete towers;

---

[b] Distributed energy resources refer to a variety of small, modular power generating technologies that can be combined to improve the operation of the electricity delivery system, such as a wind farm or solar parks, among others.

FIG. 4.2    Offshore wind farm. *Source: Courtesy of David Dixon/Walney Offshore Windfarm/CC BY-SA 2.0 From geo-graph.org.uk*

- Guyed towers;
- Hybrid towers.

According to the wind industry association mentioned above, most large wind turbines are produced with steel conical tubular towers because of their elegant look and strength. It can be manufactured in sections between 20 and 30 m with flanges at either end or bolted together on the site. The wind towers conical structure (i.e., with their diameter increasing toward the base) increases their strength and saves materials simultaneously (see Fig. 4.3). The thickness of this type of wind tower depends on the wind farm's size and tower height.

Lattice wind towers are manufactured using welded steel skeletons (see Fig. 4.4). It is like a kind of transmission tower used in the high voltage power transmission industry. This type of wind tower is constructed so that it consists of a winder foot compared to another type of wind tower structure. According to the Danish Wind Industry Association (n.d.), the primary advantage of lattice wind towers is the cost "since a lattice tower requires only half as much material as a free-standing tubular tower with similar stiffness." Lattice wind towers' construction is pretty much straightforward, and transportation is easy due to its configuration. According to some experts, the primary disadvantage of lattice wind towers is their visual appearance, although that issue is debatable. However, it is important to single out that lattice wind towers' strength is not quite good compared to other wind towers. Besides, safety is another key concern to have in mind when the structure of the wind power tower is considered. For the above reasons, lattice wind towers are no longer used for large, modern wind turbines for optical and safety reasons.

Ferroconcrete tubular towers are made using the material called "Ferroconcrete," which also consists of steel bars for additional strength and protection. This type of tubular tower

**FIG. 4.3** Wind tubular steel towers. *Source: Courtesy CarstenE; Concrete Tower Segments of a Enercon E-101 Turbine in the Freiensteinau Wind Farm (Hesse, Germany) Ready for Mounting. In the Right, a Segment Is Pulled up by the Crane. https:// commons.wikimedia.org/wiki/File%3AWP_Freiensteinau24.JPG.*

**FIG. 4.4** Wind lattice tower. *Source: Courtesy HardyS Pixabay.*

structure is higher than steel towers, and it is even cheaper. One of the disadvantages of such types of wind towers is that the materials used in their construction cannot be used later for other purposes like steel towers.

"Many small wind turbines are built with narrow pole towers supported by guy wires." The guyed towers are anchored by using the ground to three or four sides, erected by tilting up by deflection rods. However, the strength of such towers is not significant. "The advantage is weight savings and thus cost. The disadvantages are challenging to access around the towers, which makes them less suitable in farm areas. According to the Danish Wind Industry Association (n.d.), this type of tower is more prone to vandalism, thus compromising overall safety."

Finally, it is possible to have a hybrid tower. This type of wind tower uses a combined technology that utilizes concrete and steel tubular to build wind towers. The lower part of the wind turbine is made using concrete, and the upper part uses a steel tower arrangement. Such towers' strength is higher, and the cost is also cheap because of this hybrid arrangement. The foundation of such hybrid towers is constructed by using concrete.

The global wind energy market was valued, according to Doshi (2017), at US$81,147 million in 2016, and it is projected to reach US$134,600 million by 2023, representing an increase of 65.9%. It is expected to grow by 7.2% on average during the period 2017–23.

The new wind energy capacity added at the world level during 2010–19 was 441,562 MW, representing a growth of approximately 3.4-fold. In the North American region, the new wind energy capacity added during the same period was 73,895 MW. The wind energy capacity growth in the North American region during the period 2010–19 was 2.7-fold, which is 0.7 lower than the growth reported at the world level. The world's electricity production by wind farms during the period 2010–18 increased by 920,083 GWh, representing a growth of 3.7-fold. In the specific case of the North American region, this growth was 201,329 GWh. During the same period mentioned above, the North American region's increase in electricity generation was approximately 3-fold.

Strong growth in onshore wind capacity was registered at the world level during the period 2010–19, with the addition of 416,463 MW new capacities. That increase represents 94.3% of the total increase recorded in that period and at that level. The new wind onshore capacity added during the same period in the North American region was 72,826 MW. That increase represents 17.5% of the new wind onshore capacity added during the period 2010–19.

The electricity production by onshore wind farms at the world level during the period 2010–18 increased by 859,281 GWh, representing 93.4% of the overall growth recorded in that period.[c] In the North American region, the rise in electricity generation during the same period was 201,227 GWh. That growth in electricity generation in the North American region represents 23.4% of the total electricity generation increase reported in the region during the period 2010–18.

In offshore wind farms, the capacity added during the period 2010–19 was 25,099 MW, representing a growth of 9.2-fold. This growth represents 5.7% of the overall growth reported

---

[c] The world's electricity generation using renewable energy sources as a fuel increased by 2,384,098 GWh during the period 2010–18 for a total growth of 56.7%. The major contributor to this increase is wind energy with 920,083 GWh (38.6% of the total), followed by hydropower with 733,896 GWh (30.8% of the total), and solar energy with 528,220 GWh (22.2% of the total).

at the regional level in offshore wind farms in the period 2010–19. In the North American region, the new capacity of offshore wind farms added during the above period was only 29 MW, all in the USA.

The electricity production at the world level by offshore wind farms during the period 2010–18 registered an increase of 60,802 GWh, representing a growth of 9.2-fold. This growth represents 6.6% of the overall electricity generation growth reported in the period 2010–18. In the North American region, the increase in electricity generation by offshore wind farms during the period 2010–18 was 102 GWh, all in the USA.

Without a doubt, 2019 was a solid year for the wind energy sector at the world level, with 59,222 MW new capacity installed,[d] representing an increase of 10.5% compared to 2018 (1% higher than the growth reported in 2018, which was 9.5%). In the North American region, the new wind energy capacity installed in 2019 reached 9764 MW. The increase in electricity production at the world level in 2018 compared with 2017 was 129,291 GWh, representing 11.4% growth. However, this growth in electricity production is 7.2% lower than the increase reported in 2018 (18.6%). In the North American region, the increase in electricity generation was 21,658 GWh in 2018 with respect to 2017.

New capacity in the onshore wind market reached 54,696 MW in 2019 concerning 2018, representing 10.1% growth. In the North American region, the new onshore wind capacity added in 2018 was 9764 MW. In Canada's case, its onshore wind power capacity has grown steadily in the last ten years, reaching 13,413 MW in 2019. In the USA, the onshore wind capacity also grows each year of the period 2010–2019 steadily, reaching 103,555 MW in 2019. In the USA, the consumption of nonhydroelectric renewable sources more than doubled from 2000 to 2018, following state and federal government requirements and incentives (World Energy Council Issues Monitor, 2020).

The electricity generation of onshore wind farms at the world level increased by 117,747 GWh in 2018 with respect to 2017; this represents a 10.9% growth. In the North American region, onshore wind farms' electricity generation increased by 21,658 GWh in 2018, with respect to 2017.

Offshore wind energy began in the 1990s and had been slowly growing in capacity ever since. However, in recent years, offshore wind capacity growth has accelerated and in 2019 grew 19.2% with respect to 2018. From being 1% of global wind installations by capacity in 2009, the offshore wind capacity grew to over 10% in 2019 with respect to 2018. Undoubtedly, offshore wind energy is now a mature industry, but it is only beginning its worldwide expansion. Given that the sea covers more than 70% of the planet and that wind speeds are considerably stronger offshore than onshore, the future of offshore wind farms for electricity generation is promising at the world level and in the North American region in particular. However, it is important to single out that offshore wind farms for electricity generation, in 2019, in the USA are minimal and inexistent in Canada.

The global offshore wind market installed 4526 MW new capacities in 2019 with respect to 2018, representing 19.2% growth. In the North American region, there was no increase in offshore wind capacity reported in 2018. The electricity generation by offshore

---

[d] In 2018, the new wind energy capacity installed at the world level was 48,810 MW, which represents an increase of 9.5% compared to 2017. This increase is lower than the growth reported in 2019.

wind farms increased, in 2018 with respect to 2017, by 11,544 GWh, representing 20.4% growth. There was no increase in electricity production by offshore wind farms in the North American region in 2018.

Finally, it is important to single out that the use of renewable energy sources for electricity generation, particularly the use of wind energy sources, has been growing at the world level significantly over the past years due to several benefits. Some of these benefits are the following:

- From the perspective of energy security, the use of wind energy for electricity generation can provide opportunities for the energy mix diversification, which is very important for economies heavily dependent on the imports of fossil fuels;
- With the increased use of wind energy for electricity generation, the environmental impact (decrease $CO_2$ emissions per unit of GDP and reduce air pollution) can be reduced significantly;
- There could be economic considerations behind the use of wind energy sources for electricity generation;
- The use of wind energy for electricity generation can be one of the most effective tools in solving energy access in several countries, particularly for energy supply in remote rural areas.

The key findings in the study of the importance of wind energy sources for electricity generation worldwide are the following:

- The use of wind energy sources for electricity generation is becoming more competitive in comparison with other renewable energy sources and, therefore, it is likely to be used for this specific purpose on an increasingly broad scale in the future;
- The wind energy capacity installed at the world level has grown swiftly over the past decade, rising from 180,846 MW of installed capacity at the end of 2010 to 622,408 MW of installed capacity at the end of 2019; that represents an increase of around 3.4-fold. The onshore wind energy installed capacity increased approximately 3.4-fold, rising from 177,790 MW installed capacity in 2010 to 594,253 MW installed capacity in 2019. In the specific case of offshore wind energy, the capacity installed during the period 2010–19 increased 9.2-fold, rising from 3056 MW in 2010 to 28,155 MW in 2019.
  Without a doubt, 2019 was one of the best years for the global offshore wind energy industry, with an increase of 4526 MW new capacities with respect to 2018. In the North American region, in 2019, no new offshore wind energy capacity was added.
  The global offshore wind energy market outlook to 2030 has grown more promising over the past years as governments raise their ambition levels and new countries join the market. With a projected average annual growth rate of 18.6% until 2024 and 8.2% up to the end of 2030, new yearly offshore wind energy installations are expected to reach 20 GW in 2025 and 30 GW in 2030. For the period 2020–24, the majority of growth in offshore wind farms outside of Europe will primarily come from China and Taiwan, with the contribution from the USA becoming sizeable from 2024 (Lee et al., 2020);
- Wind power electricity generation during the period 2010–18 increased approximately 3.7-fold, rising from 342,831 GWh in 2010 to 1,262,914 GWh in 2018. In the specific case of onshore wind energy, the electricity generation during the

period 2010–18 increased approximately 3.6-fold, rising from 335,437 GWh in 2010 to 1,194,718 GWh in 2018.

In the specific case of offshore wind energy, electricity generation during the same period at the world level increased approximately 9.2-fold, rising from 7394 GWh in 2010 to 68,196 GWh in 2018. In the North American region, offshore wind energy electricity generation reached 102 GWh in 2019.

- The growth in the use of onshore and offshore wind farms for electricity generation has been driven by larger wind turbines and improving efficiency in the wind turbines already installed;
- High construction costs can make the total cost to build and operate wind farms higher than building and operating conventional power plants. This situation is affecting the increased use of wind energy for electricity generation in several countries;
- Wind energy's specific characteristics can further hinder this type of energy source's economic competitiveness. They are not operator-controlled and are not necessarily available when they would be of the most significant value to the electrical grid. However, improving battery storage technology and dispersing wind generating facilities over wide geographic areas could mitigate many of the problems associated with intermittent energy supply over a period.

## Types of wind energy

Wind turbines can be grouped into two categories. These groups are, according to the configuration of the rotating axis of rotor blades, the following:

- Horizontal axis wind turbine (HAWT);
- Vertical axis wind turbine (VAWT).

The HWAT is the most used design configuration in wind turbines with rotors similar to aircraft rotors. HAWT captures kinetic wind energy with a propeller-type rotor. Their rotational axis "is parallel to the direction of the wind stream. HAWTs are available in many sizes ranging from a few hundred watts up to a hundred kilowatts. These types of wind turbines are typically used under streamlined wind conditions where a constant stream and direction of the wind is available in order to capture the maximum wind energy" (Difference Between Horizontal and Vertical Axis Wind Turbine, 2020). It is important to single out that "HAWTs are not effective where the wind is turbulent. For this reason, they are generally located in areas where there is constant directional airflow" (Difference Between Horizontal and Vertical Axis Wind Turbine, 2020).

There are two types of HAWTs. These are:

- The upwind turbine;
- The downwind turbine.

In an upwind turbine, the rotor faces the wind first. Today most of the HAWTs are manufactured using this design. This type of wind turbine must be inflexible and placed at some distance from the tower. The main advantage of this type of turbine is that it can avoid wind shade behind the tower. It requires a yaw mechanism so that its rotor always faces the wind. In the downwind turbine, the rotor is present at the downside of the tower. In this type of wind turbine, the wind first faces the tower, and after that, it faces the rotor blades. The yaw mechanism is absent in this type of wind turbine. The rotors and gondolas in a downwind turbine are designed so that the gondola allows the wind to flow in a controlled manner. It receives some fluctuation in wind power because the rotor passes through the tower's wind shade. The rotor is present after the tower's gondola, which creates a wind power fluctuation (Mishra, 2017).

In the other type of wind turbine, VAWT, the rotating axis is perpendicular to the wind stream. "VAWT is probably the oldest type of windmills in which the drive shaft axis is perpendicular to the ground. The blades of the VAWTs rotate with respect to their vertical axes that are perpendicular to the ground. VAWTs are typically used in areas with turbulent wind flow, such as coastlines, rooftops, cityscapes, etc." (Difference Between Horizontal and Vertical Axis Wind Turbine, 2020).

There are essential differences between the two types of wind turbines mentioned above. These differences are, according to the paper called "Difference Between Horizontal and Vertical Axis Wind Turbine (2020)", the following:

- **Design**: In HAWTs, the blades' rotating axis is parallel to the direction of the wind. In VAWTs, the axis of the drive shaft is perpendicular to the ground;
- **Machinery**: The HAWTs "have the entire rotor, gearbox, and generator mounted at the top of the tower, which must be turned to face the wind direction. In VAWTs, the wind generator, gearbox, and other main wind turbine components can be set up on the ground, which simplifies the wind tower design and construction, and consequently reduces the turbine cost" (Difference Between Horizontal and Vertical Axis Wind Turbine, 2020);
- **Wind Conditions**: "HAWTs are generally used under streamlined wind conditions where a constant stream and direction of the wind is available to capture the maximum wind energy. VAWTs, on the other hand, are mainly beneficial in areas with turbulent wind flow such as rooftops, coastlines, etc. VAWTs can operate even in low wind speeds, and they may be built at locations where tall structures are prohibited" (Difference Between Horizontal and Vertical Axis Wind Turbine, 2020)

## Wind energy installed capacity and electricity generation in the North American region

Approximately 2% of the solar energy striking the Earth's surface is converted to wind's kinetic energy. Wind turbines convert the wind's kinetic energy into electricity without greenhouse gas emissions (Gustavson, 1979). Undoubtedly, 2019 was a big year for the global wind industry, with new installations reaching 60.4 GW, surpassing the 60 GW milestone for only the second time in history, and showing year-over-year growth of 19%. The 60.4 GW of new wind power installations brings global cumulative wind power capacity up to 651 GW. In the onshore market, 54.2 GW was installed, almost 90% of the total wind capacity installed globally, representing an increase of 17% compared to 2018. China and the USA remained the world's largest onshore market, together accounting for more than 60% of new onshore wind farm additions (GWEC Global Wind Report 2019, 2020).

Wind energy capacity installed at the world level rose by approximately 3.4-fold during the period 2010–19. In the North American region, the evolution of the wind energy installed capacity during the period 2010–19 is shown in Fig. 4.5.

According to Fig. 4.5, the wind energy capacity installed in the North American region increased 2.7 fold during the period 2010–19, rising from 43,102 MW in 2010 to 116,997 MW in 2019. It is important to single out that the increase in wind energy capacity installed in the North American region during the period under consideration is lower than the increased wind energy capacity reported at the world level in that period.

In 2019, new wind farm capacity additions grew 8.4% in North America compared to the 2018 increase reported last year (9.8 GW). Last year, the USA saw an installation rush within the region, with nearly 9.2 GW new wind farm capacity installed. The USA installed its third-largest volume of the onshore wind farm in 2019 at around 9 GW (9167 MW), just behind its previous records of 10 GW in 2009 and 13.4 GW in 2012 (13,399 MW), reaching a total of over 103.6 GW wind installed capacity (IRENA, 2020). In Canada, new wind farm installed capacity increased in 2019 by 4.7% compared to 2018 (597 MW). All new wind-installed capacities were in onshore wind farms.

It is easy to see, in Fig. 4.5, that wind energy capacity in the region increased every year of the period under consideration. It is projected that this trend will continue without change during the coming years. The electricity generated by wind farms in the North American region during the period 2010–18 also grew every year of the mentioned period. The evolution of the electricity generated by wind farms in the North American region during the period mentioned above is shown in Fig. 4.6.

Based on the data included in Fig. 4.6, the following can be stated: electricity generation in the North American region by wind farms increased almost 3-fold during the period 2010–18, rising from 103,872 GWh in 2010 to 307,683 GWh in 2018. According to Fig. 4.6, electricity production in the region grew every year of the period under consideration. It is expected that this trend will continue without change during the coming years.

Evolution of wind energy capacity in North America during the period
2010-19 (MW)

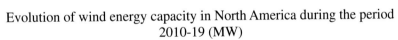

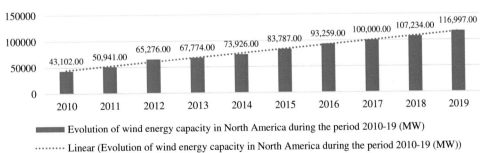

**FIG. 4.5**   Evolution of the wind energy installed capacity in the North American region during the period 2010–19. *Source: IRENA, 2020. Renewable Energy Statistics 2020. The International Renewable Energy Agency, Abu Dhabi and Author own calculations.*

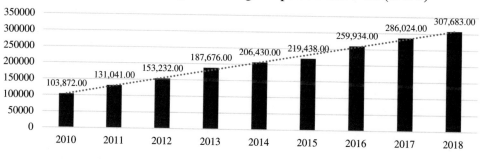

FIG. 4.6   Evolution of the wind energy electricity generation in the North American region during the period 2010–18. *Source: IRENA, 2020. Renewable Energy Statistics 2020. The International Renewable Energy Agency, Abu Dhabi and Author own calculations.*

The onshore wind energy installed capacity in the North American region increased every year of the period 2010–19 (see Fig. 4.7).

According to Fig. 4.7, the following can be stated: onshore wind energy installed capacity in North America increased by approximately 2.7-fold during the period 2010–19, rising from 43,102 MW in 2010 to 116,698 MW in 2019. It is projected that this trend will continue, without change, during the coming years, due to the increasing number of onshore wind farms projected to be constructed in the region.

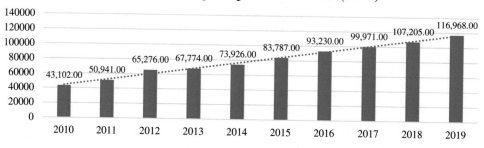

FIG. 4.7   Evolution of the onshore wind energy installed capacity in the North American region during the period 2010–19. *Source: IRENA, 2020. Renewable Energy Statistics 2020. The International Renewable Energy Agency, Abu Dhabi and Author own calculations.*

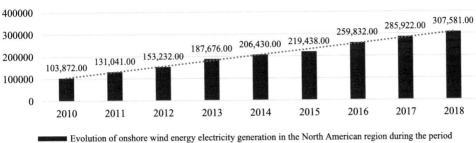

FIG. 4.8   Evolution of the onshore wind energy electricity generation in the North American region during the period 2010–18. *Source: IRENA, 2020. Renewable Energy Statistics 2020. The International Renewable Energy Agency, Abu Dhabi and Author own calculations.*

The evolution of the onshore wind energy electricity generation in the North American region during the period 2010–18 is shown in Fig. 4.8.

Based on the data included in Fig. 4.8, the following can be stated: onshore wind energy electricity generation in the North American region increased approximately 3-fold during the period 2010–18, rising from 103,872 GWh in 2010 to 307,581 GWh in 2018. It is foreseen that this trend will continue, without change, during the coming years, due to the entry into operation of new onshore wind farms in the region.

It is important to know that North America installed its first test offshore wind turbine, a 1/8th geometric scale of a 6 MW turbine, off the coast of Maine, USA, in 2013, and connected its first commercial offshore wind farm to the grid in Rhode Island in December 2016. As of the end of 2019, a total of 29 MW of offshore wind energy capacity was installed in North America, making it the only region with commercial offshore wind farms outside of Europe and Asia. This offshore wind energy capacity is installed in the USA. Without a doubt, the offshore market in the USA is progressing, with the first large-scale installations expected after 2023 and more than 10 GW (10.61 GW) expected to be built by 2026 (see Table 4.1). Based on GWEC Market Intelligence's global offshore wind project pipeline, no utility-scale offshore wind farm will come online in the USA before 2024. In total, 23 GW of offshore wind farm capacity is projected to be built in the North American region in this decade, of which less than 1 GW is expected to come from Canada, despite its high technical resource potential[e] (Lee et al., 2020).

The evolution of the offshore wind energy installed capacity in the North American region during the period 2010–19 is shown in Fig. 4.9.

According to Fig. 4.9, the offshore wind farm installed capacity began in 2016, but only in the USA. This offshore wind farm is the Deepwater Wind's Block Island Wind Farm project.

---

[e] In 2024, according to the Global Offshore Wind Report 2020 (2020), the North American region will require 14,300 workers to build new offshore wind farms projected.

**TABLE 4.1** Projected increase in offshore wind farms in the North American region during the period 2020–30 (GW).

| Year | USA | Canada | Total |
|------|-----|--------|-------|
| 2020 | 0.01 | 0 | 0.01 |
| 2021 | 0 | 0 | 0 |
| 2022 | 0 | 0 | 0 |
| 2023 | 0.3 | 0 | 0.3 |
| 2024 | 2.9 | 0 | 2.9 |
| 2025 | 4.2 | 0 | 4.2 |
| 2026 | 3.2 | 0 | 3.2 |
| 2027 | 3 | 0 | 3 |
| 2028 | 3 | 0 | 3 |
| 2029 | 3 | 0 | 3 |
| 2030 | 3 | 0.4 | 3.4 |

*Source: 2020. Global Offshore Wind Report 2020. Global Wind Energy Council*

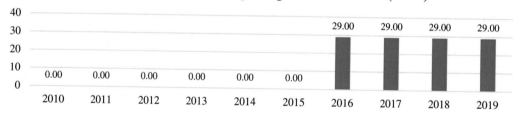

## Evolution of the offshore wind energy installed capacity in North America during the period 2010-19 (MW)

■ Evolution of the offshore wind energy installed capacity in North America during the period 2010-19 (MW)

**FIG. 4.9** Evolution of the offshore wind energy installed capacity in the North American region during the period 2010–19. *Source: IRENA, 2020. Renewable Energy Statistics 2020. The International Renewable Energy Agency, Abu Dhabi.*

Located off the coast of Rhode Island, the offshore wind farm has five wind turbines with a total capacity of 29 MW. The offshore wind energy capacity installed in 2016 has remained stable until 2019. However, it is important to single out that North America has only 29 MW offshore wind energy capacity in operation, all installed in the USA, but deployment will accelerate in the coming years in the USA and in Canada. The offshore wind energy industry is now moving to a project construction planning and execution phase in the USA. More than 15 offshore wind energy farms are expected to be built by 2026.

The projected increase in offshore wind farms in the North American region during the period 2020–30 is shown in Table 4.1.

Evolution of the offshore wind energy electricity generation in North
America during the period 2010-18 (GWh)

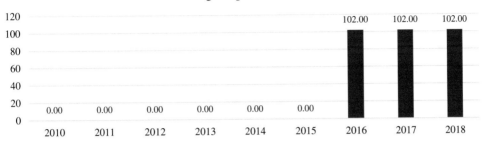

■ Evolution of the offshore wind energy electricity generation in North America during the period 2010-18 (GWh)

FIG. 4.10   Evolution of the offshore wind energy electricity generation in the North American region during the period 2010–18. *Source: IRENA, 2020. Renewable Energy Statistics 2020. The International Renewable Energy Agency, Abu Dhabi.*

Based on the data included in Table 4.1, the following can be stated: by 2030, it is expected to have 23.01 GW of new offshore wind farm capacity installed in the region. The new offshore wind farm capacity to be installed by 2030 will represent 98.3% of the total in the USA. During the period mentioned above, only 0.4 GW of the new offshore wind farm capacity will be installed in Canada (1.7% of the total).

The evolution of the offshore wind farm electricity generation in the North American region during the period 2010–18 is shown in Fig. 4.10.

Based on the data included in Fig. 4.10, the following can be stated: offshore wind farm electricity generation began in 2016 only in the USA. The offshore wind electricity generation in the USA has remained stable until 2018. However, it is expected that with new offshore wind farms to be installed during the current decade in the region, particularly in the USA, the electricity generated by offshore wind farms in the North American region will increase significantly in comparison with the electricity produced by these types of wind farms during the period 2010–19.

## Wind energy investment costs in the North American region

Global investments in wind energy technologies reached roughly US$131.5 billion in 2018. Investment in wind energy has risen over the past two decades, where global funding was recorded at just under US$19 billion in 2004; this represents an increase of almost seven-fold. In the case of the USA wind projects, private investment in 2019 represents US$14 billion. On the other hand, wind projects pay over US$1.6 billion to state and local governments and private landowners every year.

The evolution of global investment in wind energy during the period 2010–18 is shown in Fig. 4.11.

According to Fig. 4.11, the following can be stated: during the period 2010–13, global investment in wind energy decreased by 15.4%, falling from US$98.3 billion in 2010 to US$83.2

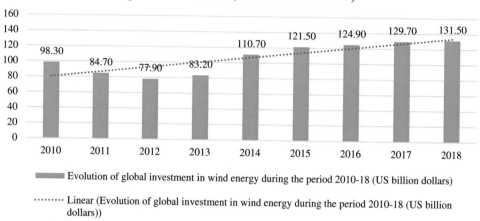

FIG. 4.11   Evolution of global investment in wind energy during the period 2010–18. *Source: 2020. Statista 2020.*

billion in 2013. However, since 2014 the global investment in wind energy has increased by 58.1%, rising from US$83.2 billion in 2013 to US$131.5 billion in 2018. It is expected that this trend will continue without change during the coming decades in order to implement the Paris agreement on climate change,[f] particularly in Canada, due to the opposition of the Trump administration to this agreement.

Global offshore wind financings increased by 19% to a record of US$29.9 billion in 2019. Investment in onshore wind advanced 2% to US$108.3 billion, the highest ever, with the USA one of the busiest markets as developers sought to qualify projects for the Production Tax Credit (Global Trends in Renewable Energy Investment 2020, 2020).

Without a doubt, wind energy is booming in the USA and the North American region as a whole. In 2018, this growth continued in the whole region, "signaling to manufacturers and suppliers that wind energy production is ramping up further. Onshore and offshore, the length of blades on wind turbines is increasing to capture more energy. The longer and larger the blades become, the more composite materials are required for production" (Expancel Newsletter, 2018), and this will increase the initial investment cost associated with the construction of a wind farm.

Achieving the Paris agreement climate change goals would require a substantial transformation in renewable energy source technologies and the electrical sector. Wind power, along with solar energy, would lead the way to transform the global electricity sector at the world level. Onshore and offshore wind farms are expected to generate 35% of total electricity needs, becoming the prominent generation source by 2050, among other energy sources. That requires increasing the global installed capacity of onshore wind farms to 1787 GW by 2030 and 5044 GW by 2050. In offshore wind farms, the global installed capacity would need

---

[f] It is important to recall that the Trum administration refused to ratify the Paris agreement on climate change, despite the fact that the USA is one of three most polluted country in the world.

to increase to 228 GW by 2030 and 1000 GW by 2050. To achieve the goal mentioned above, annual capacity additions for onshore wind farms would need to increase to 200 GW per year during the next two decades. In the specific case of offshore wind farms, the annual growth required is even higher, 45 GW per year by 2050, from 4.8 GW added in 2018 (IRENA, 2019).

Scaling up wind energy investments is key to accelerating the growth of global wind power installations over the coming decades. That would imply increasing global average annual onshore wind power investments by more than 2-fold from now until 2030 (US$146 billion per year) and more than three-fold over the remaining period to 2050 (US$211 billion per year) compared to 2018 investments (US$67 billion per year). For offshore wind farms, global average annual investments would need to increase 3-fold from now until 2030 (US$61 billion per year) and more than 5-fold over the remaining period to 2050 (US$100 billion per year) compared to 2018 investments (US$19 billion per year) (IRENA, 2019).

The price of a tower for a wind turbine could cost around 20% of the wind turbine's total price. For a wind tower around 50 m height, the additional cost of another 10 m of the tower is about US$15,000. Therefore, a wind farm must build towers with the most effective high. In other words, the high of a wind tower should ensure the production of the highest level of electricity with the lower possible cost.

To reduce the initial investment cost associated with the construction of a wind farm, it is important to know that lattice towers are the cheapest to build than other wind towers because it requires about half the amount of steel used, for example, by a tubular steel tower.

## Wind farms construction and generation costs in the North American region

IHS Markit[g] estimates that, by 2021, wind farms' operational and maintenance costs will cost the industry about US$7.5 billion annually. The wind industry is expected to increasingly focus its activities on providing services and minimizing costs at existing wind farms than planning and constructing new ones (McNulty, 2018).

According to the report mentioned above, one of its key findings is that newer wind farms in the North American region have operational and maintenance costs averaging 25% less per MWh than older wind farms using smaller wind turbines installed before 2010. Overall, the report showed that there is a steady increase in annual costs per MW installed after about three years of operation of a wind farm. However, there is a great range of expenses from one wind farm to another. Age, location, and operational and maintenance costs and strategy are important factors to consider when constructing a new wind farm in the future. Large wind farms are not immune to this trend, though the effect appears more modest for them in comparison to smaller wind farms.

Analysis and studies carried out on wind farms indicate that for the first year, operational and maintenance of wind turbine costs are on the decline, falling from an average of US$46,000 per MW during the period 2008–13, to US$38,000 per MW during the period 2014–17; this means a decrease of 17.4% for the whole period under consideration. Over the

---

[g] 2018 IHS Markit Wind O&M Benchmarking in North America: Aging Turbines, Rising Cost.

years, wind farms' operational and maintenance costs increase, "on average, to as much as US$58,000 per MW annually, with direct costs (the costs of actually maintaining the wind turbines) rising sharply by the end of the first decade of operation" (McNulty, 2018). Rising total operational and maintenance costs of a wind farm "are due to increasing direct costs, which are partially but not fully offset by declining indirect costs (e.g., general site administration and other business services, taxes, and royalties)" (McNulty, 2018).

Onshore wind farms' total installation cost is expected to decline during the next three decades at the world level. The average cost is likely to fall between US$800 and US$1350 per kW by 2030 and between US$650 and US$1000 per kW by 2050. It is important to recall that, in 2018, the global-weighted average reported was US$1497 per kW. The average total installation cost for offshore wind farms is likely to drop further in the coming decades between US$1700 and US$3200 per kW by 2030 and between US$1400 and US$2800 per kW by 2050 (IRENA, 2019).

It is important to stress that the electricity costs for onshore wind farms are already competitive compared to the electricity cost generated by oil, coal, and gas-fired power plants. It is foreseen that the electricity cost will decline further as installed costs and onshore wind farms' performance continue to improve. At the world level, it is expected that the cost of electricity for onshore wind farms will continue to fall from the current average of US$0.06 per kWh registered in 2018 to between US$0.03 and US$0.05 per kWh by 2030, and between US$0.02 and US$0.03 per kWh by 2050. Without a doubt, the cost of electricity of offshore wind farms is already competitive in specific European markets, such as Germany and the Netherlands, with zero-subsidy projects and lower auction prices. The cost of electricity for offshore wind farms is projected to be competitive in other markets worldwide by 2030, falling in the low range of coal and natural gas costs. The cost of electricity of offshore wind farms is likely to drop further from an average of US$0.13 per kWh registered in 2018 to an average between US$0.05 and US$0.09 per kWh by 2030 and between US$0.03 and US$0.07 per kWh by 2050 (IRENA, 2019).

The wind energy industry's experience has found that around 25% of all wind turbines' gearboxes and generators each need replacement during the first ten years of operation. These component breakdowns result in many wind turbine failures rising from almost 20% for the third year of operation to 65% by the tenth year of service.

## The efficiency of the wind farms in the North American region

Wind turbines seem to work like magic: they take energy from the wind and convert it into mechanical or electrical energy. It is like taking power out of thin air. However, that energy comes at a price. "German physicist Albert Betz established some important wind energy facts and proved, in 1919, that the most efficient wind turbine could only be 59% efficient at best" (Wind Energy Facts, 2016). However, any wind farm which operates with an efficiency of 30% is considered efficient today.

The fact is that, ideally, even modern wind turbines can only produce between 20% and 45% of electricity, depending on several factors. Wind farm layouts can reduce wind turbines' efficiency if not properly designed. Builders do not always put wind turbines in the places with the highest wind speeds, where they will generate the most power. Wind turbine spacing is also important because wind turbines create drag that lowers wind speed. For this reason, the first wind turbines to catch the wind will generate more power than those that come after.

To build more efficient wind farms, designers must take several factors into account. These factors are, among others, the following:

- Wind speed and wind turbine spacing;
- Land size;
- The geography of the site;
- Number of wind turbines installed;
- Amount of vegetation that exists in the site;
- Meteorological conditions;
- Building costs (Penn State, 2018).

The fact that a power source with an efficiency of 30% can be considered as a power source of the future could surprise many people. The reason is very simple: the best commercial or utility-scale wind turbines available today in the market are approximately 45% efficient. There is room for improvement, though not by considering the current wind energy technology available in the market.

## Types of incidents in wind farms in the North American region

While the wind energy industry and the installation of wind turbines are growing, the drawbacks of wind energy are not always considered and evaluated properly. One of the main accidents in the use of wind energy for electricity generation is wind turbine accidents. Wind turbine accidents include a multitude of ways in which wind turbines fail due to mechanical problems, nature, or human involvement (Asian et al., 2017).[h]

Wind farm incidents can be classified, according to Michigan Wind Farm Accidents (2020) and Windpower Engineering and Development (2020), as follows:

A-Human failure

- **Human health and injury**: this type of incident include ill-health and effects due to turbine noise, shadow flicker, incidents involved wind industry or construction and maintenance workers, and members of the public or workers not directly dependent on the wind industry (e.g., firefighters and transport workers), among others. Such incidents are expected to increase significantly as wind turbines are increasingly approved and built-in unsuitable locations, close to people's homes;
- **Fatality**: this type of incident involved the wind industry and direct support workers such as divers, construction workers, maintenance workers, engineers, technicians, and small turbine workers/operators, among others. This group also includes public fatalities, including workers not directly dependent on the wind industry (e.g., transport workers);

B-System/equipment failure

- **Blade failure**: this type of incident is by far the highest number reported until 2020. Blade failure can occur from several possible sources and results in either whole blades or pieces of a blade being thrown from the turbine;

[h] Asian, S., Ertek, G., Haksoz, C., Pakter, S., Ulun, S., 2017. Wind turbine accidents: a datamining study. IEEE Syst. J. 11(3), 1567–1578. doi: 10.1109/JSYST.2016.2565818.

- **Fire**: this is the second most common wind turbine incident reported until 2020. Fire can happen from many sources. Besides, some wind turbine types seem more susceptible to fire than others;
- **Structural failure**: this type of incident is assumed to be a major component failure under conditions that components should be designed to withstand. That mainly concerns storm damage to wind turbines and tower collapse. However, poor quality control, lack of maintenance, and component failure can also be responsible for this type of incident;
- **Transport**: most accidents involve wind turbine sections falling from transporters, through wind turbine sections have also been lost at sea during the construction of offshore wind farms or transportation of wind tower section;
- **Drops incidents**: according to Windpower (2020), this type of incident involved, mainly, offshore wind farms. "Vibrations from the operation, corrosion from seawater, and strong winds can cause components to come loose from their fittings and fall, potentially striking assets, vessels, or personnel below. Technicians routinely work at height, carrying tools to raised areas" (Windpower, 2020);

C-Nature

- **Environmental**: this type of incident includes oil and solvent spills, crop or habitat damage, and wildlife death due to wind turbine positioning;
- **Ice throw**: this type of incident resulting in property damage or evacuation of the population. Incidents are listed here unless they have caused the human injury;

D-Others

- This group of incidents includes components or mechanical failure if there has been no consequential structural damage. It also includes a lack of maintenance and electrical failure (not led to fire or electrocution). Construction and construction support accidents and lightning strikes are included when a strike has not resulted in blade damage or fire.

It is important to single out that wind turbine manufacturers, owners, and contractors collect data about their operations, including data on accidents and incidents in wind farms during their operation and construction. Still, they do not publicly share most of this data, especially the accident-incident data. The reason for keeping these data private might be not only due to confidentiality but also for preserving a positive public perception of wind energy (Malnick and Mendick, 2011). Industry organizations, such as the American Wind Energy Association (AWEA), also have not made a significant collection of data on wind turbine accidents-incidents publicly available.

## Wind energy and the impact on the environment

Releasing the massive potential of wind energy for electricity generation and reducing the use of fossil fuels for this specific purpose is crucial to achieving the Paris agreement climate change targets. That is only possible by overcoming several barriers that exist today, such as technology, economic, socio-political, and environmental barriers that could impede the deployment of new wind energy capacities in the next three decades. The following are the key barriers that should be taken into consideration:

- **Grid access.** In the last decades, a steadily increasing number of wind farms have been connected to the electrical grid. However, a distinctive feature of electricity produced by wind farms is unstable and cannot be stored today. This situation makes it somehow difficult to ensure a balance between electricity production and demand. Storage technologies such as batteries and pump storage have one common characteristic, i.e., the electric energy to be stored is converted to other forms, such as chemical (batteries) and potential energy in the form of water in high storage.

  Wind farms produce electricity when the wind blows. This characteristic is of little if any importance when the amount of wind power is modest compared to the total installed (and spinning) capacity of controllable power plants, but it changes into a major technical obstacle as the wind energy part (termed penetration) grows to cover a large fraction of the total demand for electricity in the system;

- **Public acceptance.** Public acceptance is an important barrier that should be overcome in order to increase the use of wind farms for electricity generation at the world level and in the North American region during the coming years as well;

- **Planning procedures and uncertainties.** Wind resource estimates are useful only if their associated uncertainty is well defined. It is impossible to construct a sound financial model for a wind project investment unless the resource analyst can offer a degree of confidence that the wind resource falls within a specified range. The uncertainty present in all wind energy estimates are mainly related to wind speed measurements, the historical climate adjustment, potential future climate deviations, wind cut, and the spatial wind resource distribution;

- **Economies of scale.** The economic consequences of using larger wind turbines in a wind farm and improved cost-effectiveness are clear. For an offshore wind farm site, the average cost has decreased from around €0.09 per kWh for the 95 kW turbine installed in the mid-1980s to around €0.05 per kWh for a fairly new 2000 kW wind turbine; that means an improvement of more than 40%. Using the costs per kWh produced as the basis, "the estimated progress ratios range from 0.83 to 0.91, corresponding to learning rates of 0.17 to 0.09. That means that when the total installed capacity of wind power doubles, the costs per kWh produced for new wind turbines go down by between 9% and 17%" (Krohn et al., 2009).

  Offshore wind farms currently account for a small amount of the world's total installed wind power capacity. It is important to single out that offshore wind farm capacity is still around 50% more expensive than onshore wind capacity (Krohn et al., 2009);

- **Access to finance.** The initial capital investment in the construction of wind farms is higher in comparison with other energy sources. For this reason, sufficient financial resources should be available if the USA and Canadian governments decide to implement a massive construction of wind farms in the North American region during the coming years, in parallel to the closure of old and inefficient fossil fuel power plants, particularly old and ineffective coal-fired power plants.

  Although the investment costs are considerably higher for offshore than for onshore wind farms, they are partially offset by higher total electricity production from the wind turbines due to higher offshore wind speeds. The energy production indicator is normally around 2000–2500 full load hours per year for an onshore wind installation utilization. In comparison, for a typical offshore wind installation, this figure could reach up to 4000 full load hours per year, depending on the site (Krohn et al., 2009);

- **Subsidies for traditional energy.** In the USA, the Trump administration is cutting all subsidies associated with developing renewable energy sources, particularly wind farms, while stimulating the use of fossil fuels, particularly coal, for electricity generation.

Overcome the current barriers immediately, through a range of supportive energy policies and implementation measures, including innovative business models and financial instruments, "are vital to boost the future deployment of wind capacities to enable the transition to a low-carbon, sustainable energy future" (IRENA, 2019).

In the USA, the use of wind energy for electricity generation avoided, only in 2019, a total of 198 million metric tons of $CO_2$ emissions, and saves 103 billion gallons of water consumption. However, recent studies have shown that wind turbines can disturb bats' echolocating abilities, causing the bats to fly into the wind turbines. "The sad result is that wind turbines kill an estimated 600,000 bats each year. The actual numbers could be much higher, and environmental advocates are now advocating against the installation of wind turbines in areas where there is a large bat population" (IRENA, 2019).

Undoubtedly, the use of wind turbines for electricity generation has quickly become popular because they are considered a clean source of energy since they do not require the burning of any fossil fuel that produces polluting waste or greenhouse gases. However, its use is not without environmental impact. Wind farms' location, often in remote areas of high ecological value, can negatively affect the environment. These harmful effects are, among others, the following:

- Negative visual impact on the horizon line caused by the wind towers;
- The large area occupied by a wind farm due to the necessary separation between the wind turbines;
- Noise generated by the blades;
- The necessary infrastructures that need to be built to construct the wind farm and to transport the energy produced to the consumers;
- Bird and bat deaths due to the collision with the blades.

Some experts assure that the mortality rates are high and that there is a risk of significantly altering the birds' habitat in the region where the wind farms will be installed. According to the outcome of some studies, it can be stated that about one dozen birds and four dozens of bats die each year for each MW installed on a wind farm.

The barrier effect produced by wind farms makes it difficult for birds to move by fragmenting the connection between sites, which leads to greater energy expenditure for them as they have to make detours to avoid encounters with wind turbines. Some environmental studies recommend developing programs that collect biological and behavioral information on migratory and resident birds to mitigate the fauna's impact and implement preventive measures to protect them. These programs generally foresee establishing observation points and radars, telemetry studies, ultrasonic detectors with recordings, acoustic monitoring, ultraviolet lights, and digital video cameras to determine nesting areas and trajectories heights, seasons accurately, and peak hours bird flights. Exist also the possibility of stopping the wind turbines when the birds fly within the rotor sweep area.

Although the adoption of additional measures is necessary and may become effective to reduce the killing of birds and bats, the central issue remains unanswered: how wind turbines have affected the species' behavior and their seasonal and interannual variability.

On the other hand, wind farms' construction can result in the fragmentation of contiguous habitat extensions, mainly affecting fauna and other slow-moving organisms. That can lead to species movement and an increase in soil temperature, which would affect the flow of surface water and basic trophic processes such as the relationships in the food chain between plants, insects, and predators.

The wind farm's effects on the flora will be greater during the wind farm's construction phase. That is due, mainly, to the following:

- Earth movement during the preparation of the site;
- The construction of the wind turbines' foundations;
- The construction of control buildings and other infrastructures necessary for the smooth running of the wind farm.

Depending on the climatic conditions on the site, erosion problems may appear. For this reason, some assumptions must be adopted in the early stages of project development to carry out the relevant hydrology and rainfall, road tracing, analysis of troughs and water-courses in order to minimize, as much as possible, the negative impact of the wind farm on them in a later phase.

Wind turbines produce noise derived from their operation. Four factors determine the degree of discomfort that this can cause on the population:

- The noise produced by the wind turbine;
- The position of the wind turbines within the wind farm;
- The distance between the residents in the area and the wind turbines;
- The existing background sound.

There are two sources of noise in a running wind turbine. These sources are the following:

- **Mechanical noise**. This type of noise comes from the generator, the gearbox, and the connections and can be easily reduced using conventional techniques;
- **Aerodynamic noise**. This type of noise, produced by the blades' movement, is more difficult to reduce by conventional methods but can be reduced by lowering the rotor's speed.

It is also estimated that wind farms can affect telecommunications through the refraction or rotational curvature of the electromagnetic waves produced by wind turbines' movement. That can interrupt, obstruct, degrade, or limit the performance, transmission, and reception of electronic equipment signals such as televisions, radios, microwaves, cell phones, and radars.

On the other hand, wind turbine blades' movement in sunny conditions produces moving shadows on the ground, resulting in light intensity changes. This phenomenon, which lasts for about half an hour, is called "shadow flicker" and can become a distraction for drivers and cause car accidents. That is the main risk since it has been ruled out that the frequency of the flickering of the shadow of a wind turbine, of the order of 0.6–1.0 Hz, represents a health problem since only frequencies above 10 Hz can cause attacks epileptics. Even so, the analysis of this impact must be incorporated into environmental impact studies because its intensity depends on several factors, such as:

- The location of the population with respect to the wind turbine;
- The speed and direction of the wind;
- The diurnal variation of sunlight;
- The geographic latitude of the site;
- The topography.

Other impacts that cannot be ruled out, even if they seem remote, are:

- The risk of fire in wind turbines due to mechanical and electrical failures that could hardly be controlled due to the great heights at which the wind turbines are located;
- The risk of blades colliding with low-flying aircraft due to detachment as a result of mechanical failure.

Visual intrusion into the landscape is the most frequently raised objection against the use of wind turbines for electricity generation. That is the main factor determining public attitudes against the utilization of wind energy for this specific purpose. That is why it is a problem that cannot be ignored in its development, being, in turn, the least quantifiable environmental impact of wind turbines and the least investigated compared to other types of environmental disturbances. This is because the environmental impact is often subjective and, in any case, difficult to estimate and quantify.

It is important to single out that wind turbines create an intrusion in the landscape since they are vertical structures in a landscape of horizontal components. That makes their visual impact, even if it exists, less than if they were linear structures that occupy large areas. It must be borne in mind that moving objects attract the observer's attention; therefore, the wind turbines' rotating blades are dominant points in the landscape.

A wind farm's visual effects depend on its characteristics: its size, height, material, and color. Vegetation can be used to reduce visual interference, both at great and close range. That, together with the correct design of the wind farm, reduces the visual impact.

## Wind energy and the public opinion in the North American region

Over the last three decades, the use of wind energy for electricity generation in the North American region has evolved from a fringe, isolated, experimental concept into a mainstream and viable source of electricity. Wind power supplied about 7.1% of USA electricity generation (4401.3 TWh, according to BP Statistical Review of World Energy 2020) in 2019, and about 32.9 TWh or 6% of the total electricity generated in Canada (660.4 TWh according to the same report) in that year. The year 2019 saw the completion of five projects that added 597 MW of new wind installed capacity, representing over US$1 billion of investment. Canada is home to the world's ninth-largest wind generating fleet (Wikipedia Wind power in Canada, 2020).

The rapid growth in wind energy deployment will likely continue in the whole region, particularly in the USA. For example, in that country, recent market analysis suggests that annual wind power capacity additions are expected to continue rapidly in the coming five years

(Wiser and Bolinger, 2016), driven by expected lower prices (Wiser et al., 2016). Meanwhile, the USA Department of Energy's recent Wind Vision Report, which outlines wind energy pathways, indicated that this type of energy source could provide up to 35% of the nation's electrical demand by 2050.

Multiple acceptance factors can impact the use of wind energy for electricity generation in the North American region. According to Wüstenhagen et al. (2007), there are three acceptance levels in wind energy use for electricity generation in the region. These levels are the following:

- Sociopolitical acceptance. That means acceptance of policymakers and key stakeholders;
- Market acceptance. That means acceptance of investors and consumers;
- Community acceptance. That means public acceptance.

However, as Sovacool (2009) points out, these social, technical, economic, and political dimensions of acceptance all influence each other in an integrated manner. For example, community acceptance of wind energy use for electricity generation can affect market acceptance and vice versa. Indeed, this has been the case when local opposition has delayed or derailed proposed wind projects in the past (Corscadden et al., 2012; Fast, 2015; Shaw et al., 2015). It is important to mention that community acceptance of wind power for electricity generation is affected by perceptions of the need to use wind power for this specific purpose, the potential for wildlife impacts, perceptions of sound, and visual impact.

For years, debates around the use of wind energy for electricity generation in the North American region focused on sociopolitical and market acceptance, pertaining largely to technological innovation, economic incentives, and impacts on the operations and resiliency of the electric grid, with less attention paid to community acceptance (Lantz and Flowers, 2011; Phadke, 2010). However, the rapid growth in wind energy use for electricity generation in the North American region has increased the footprint of wind developments, increasing local conflicts, and bringing the issue of community acceptance to the forefront (Lantz and Flowers, 2011).

Despite broad public support for the use of wind energy for electricity generation in the North American region in general, local wind developments have been challenged by vocal opposition within host communities (Bidwell, 2013; Bohn and Lant, 2009; Lantz and Flowers, 2011). In some parts of the USA, the opposition and negative attitudes toward the use of wind for electricity generation dismayed the wind industry, which had anticipated that local acceptance would be consistent with the generally favorable opinions toward the use of wind power for electricity generation in the whole country (Pasqualetti, 2001).

As a result of certain communities' opposition and negative attitudes around some of the earliest experimental wind farms in the USA, community acceptance is now widely perceived by wind energy practitioners as a significant barrier to wind energy deployment in the region (Lantz and Flowers, 2011).

A new Pew Research Center survey carried out in 2016 found that 65% of Americans prioritize developing renewable energy sources, particularly wind and solar energies, compared with 27%, who would emphasize the expanded use of fossil fuels for electricity generation. Support for concentrating on renewable energy increased 5% since the last survey carried out in December 2014. At that time, 60% of Americans expressed their support for developing renewable energy sources, giving this type of energy source for electricity generation the highest priority (Kennedy, 2017).

According to the last survey mentioned above, there are also differences in public priorities about using different electricity generation sources by age. Americans under the age of 50 are more likely to support the use of renewable energy sources for electricity generation over expanding the use of fossil fuels for this specific purpose. About 73% of those ages 18 to 49 say developing renewable energy sources for electricity generation should be the most important priority for the government and the energy industry. In comparison, 22% say expanding the use of fossil fuels for electricity generation should have the highest priority. Older adults are more divided in their views. They also give the highest priority to using renewable energy sources for electricity generation than the use of fossil fuels for this specific purpose. Among those 50 and older, 55% say renewable energy development is more important than the development of fossil fuels, while 34% say it is more important to expand the use of fossil fuels for electricity generation (Kennedy, 2017).

However, consider the whole USA population, the percentage of those that support the use of renewable energy sources for electricity generation and those that support the use of fossil fuels change with time significantly.

According to Holm (2017), the evaluation of selected surveys from major industrialized countries shows that in Canada, 65% of Canadians support the use of renewable energy sources for electricity generation,[i] 19% the use of fossil fuels, and 16% the use of nuclear energy. In the USA, around 67% support the use of fossil fuels for electricity generation, 20% for the use of nuclear energy, and 13% for the use of renewable energy sources (see Fig. 4.12).

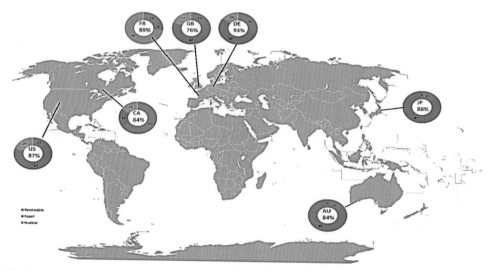

**FIG. 4.12**    Infographics and visual thinking focused on renewable energy and climate change. *Source: Strom-Report Blog, 2019. Infographics and visual thinking focused on renewable energy and climate change. What is the Public's Opinion on Renewables? Survey Results on Acceptance Worldwide.*

[i] In Canada, the main renewable energy source used for electricity generation is hydro, followed by wind energy (onshore wind energy).

According to Fig. 4.12, the following can be stated: In Canada, the majority of the Canadians (65%) supports the use of renewable energy sources for electricity generation, while in the USA, the majority of Americans (67%) supports the use of fossil fuels for this specific purpose. Within the industrialized countries included in Fig. 4.12, Canada has the highest percentage of renewable energy sources for electricity generation, and the USA has the lowest percentage (similar to Japan). Likely, this situation will not change significantly during the coming years, despite the energy industry's measures to increase the role of renewable energy sources in the USA energy mix. The Trump administration's energy policy promoted the use of conventional energy sources for electricity generation, particularly coal, and eliminated all main measures approved by the previous USA administration supporting the increased use of renewables within the country's energy mix.

## Advantages and disadvantages in the use of wind energy for electricity generation in the North American region

Wind energy for electricity generation has become the fastest-growing renewable energy source lately in the world (exceeding solar energy). The reason is straightforward. Wind energy offers many more advantages than disadvantages.

These are some of the advantages of the use of wind energy for electricity generation:
A-Advantages

- Wind energy is the fastest-growing renewable energy source globally (Savin, 2018), mostly because it is more efficient than some other renewable energy sources. For this reason, wind energy is one of the most promising renewable energy sources for electricity generation at the world level and in the North American region in the future. Without a doubt, wind energy will play a significant role in the transition from fossil to renewable energy for electricity generation at the world level and the North American region as well. However, the degree to which wind energy is usable in an efficient manner depends on the geographic conditions of the area where the wind farm is going to be built (Piramal, 2020);
- Wind energy technology is more efficient than some other renewable energy sources. Since technology will improve even further, wind energy might be even more efficient in the future. For this reason, the amounts of energy that could be produced from wind turbines may be even higher in the future than it is today (Piramal, 2020);
- Wind energy is considered a renewable energy source because the Sun is the source that produces the wind, and the Sun is regarded as a continuous source of clean energy. The wind is an energy source produced by the Sun by heating the atmosphere's different layers unevenly. Without a doubt, the wind is an excellent source of green energy for humankind (Savin, 2018);
- Wind energy is a renewable, non-polluting energy source. Wind energy is a free, renewable energy source, so no matter how much is used today for electricity generation, there will be ready to be used for the same purpose in the future. "Wind energy is also a source of clean, non-polluting electricity" (Daraniya, 2012). Unlike conventional power plants, wind farms "emit no air pollutants or a large amount of greenhouse gases" (Daraniya, 2012).

On this subject, it is important to single out that the construction of wind farms and the related infrastructure associated with them will certainly produce a certain amount of greenhouse gas emissions. But these emissions will be reduced to the minimum during the operation of the wind farms. A wind farm also produces not much waste during the electricity generation;

- Electricity price. Onshore wind farms produce one of the cheapest, if not the cheapest, energy on the market today (with a cost between US$0.02 and US$0.06 per kWh). It is predicted that the price of the electricity generated by wind farms in the North American region can fall even more in the future if the wind generation capacity significantly increases at the world level (Savin, 2018);
- The wind energy sector creates hundreds of thousands of new jobs every year. Only in the USA, more than 100,000 new workers have been employed in 2017 in the wind industry;
- The increased use of wind energy for electricity generation at the world level will reduce the use of coal and oil for this specific purpose at that level and in the North American region as well;
- Wind energy requires low maintenance and has low operating costs. A wind turbine requires a relatively low maintenance service. However, older wind turbines could have some reliability issues, requiring more attention from the maintenance service team. Because wind energy is a free energy source, the wind turbine will have low operating costs (maintenance costs are the only costs associated with the turbine) (Savin, 2018);
- Wind turbines could be placed among farmland. Apart from a negative impact on the environment and population living in the area where the wind farm will be built, wind power could significantly improve the financial conditions of farmers by making an additional income for the growing of crops and the raising of cattle (Piramal, 2020);
- Wind farms could be a source of tax income for local authorities and municipalities located in a rural area far from the electric grid;
- The use of wind energy for electricity generation could save water compared to the use of other energy sources for the same purpose. For example, the use of nuclear energy for electricity production consumes around 500 times more water in comparison with the use of wind energy for the same purpose and for the production of the same amount of energy;

There is no fuel involved in a wind farm's operation, only the wind, a free energy source. For this reason, the transportation cost of fuel and mining of resources to ensure the operation of the wind farm is non-existent. For this reason, there are no greenhouse gas emissions associated with the transportation of fuel for the operation of a wind farm;

B-Disadvantages

The use of wind energy for electricity generation also has disadvantages that need to be considered. Some of the main disadvantages associated with the use of wind energy for electricity generation are the following:

- Wind energy is not a continuous source of power. It depends somehow on weather and geographical conditions. However, it is important to single out that wind energy can work day and night if there is some wind to blow. There are wind turbines that can generate electricity with low-speed winds. Other types of turbines require stronger winds to generate electricity;

- Cost issues. Even though the cost of wind power has decreased dramatically in the past ten years, constructing a wind farm requires a higher initial investment than some fossil-fueled power plants. Roughly 80% of the wind farm cost is the machinery, with the balance being site preparation and installation.[j] If wind generating systems are compared with fossil-fueled systems on a life-cycle cost basis, "then wind costs are much more competitive with other generating technologies because there is no fuel to purchase and minimal operating expenses" (Wind Energy, 2020). Without a doubt, "wind energy is the cheapest form of new electricity generation available today" (Wind Energy, 2020);

- Construction time. Apart from the high construction costs of wind farms, wind turbines' construction will also take plenty of time. It might even take several years until a single wind turbine could be used for energy production. Thus, investors who have to invest large amounts of money in constructing a wind farm will have to wait many years until they see any return from their investment. This situation may scare off institutional investors since they may seek alternative investment opportunities instead (Piramal, 2020).

- Environmental concerns. Although wind power plants have relatively little impact on the environment than fossil fuel and hydropower plants, there is some concern over the noise produced by the rotor blades,[k] visual impacts, and birds and bats killed by flying into the rotors. Most of these problems have been resolved or significantly reduced through technological development or by properly siting wind farms;

- Supply and transport issues. The significant challenge for using wind energy as a power source is that it is intermittent and does not always blow when electricity is needed. In principle, wind energy cannot be stored (although wind-generated electricity can be stored if batteries are used but they still are not very efficient), and not all winds can be used to meet the timing of electricity demands;

- Good wind farm sites are often located in remote locations far from areas of electric power demand (such as cities or major towns), and this could increase transmission losses;

[j] Wind turbines have significant economies of scale. A large-scale wind turbine (i.e., greater than 600 kW) costs approximately US$1000 per kilowatt of nameplate capacity. That means a hypothetical 1000 kW (1 MW) turbine will cost approximately US$1 million full installed. Smaller wind farm or residential scale wind turbines cost less overall, but are more expensive per kilowatt of energy producing capacity. Wind turbines under 100 kW cost roughly US$3000–US$5000 per kW of capacity. That means a 10 kW machine (the size needed to power an average home) might cost US$35,000–US$40,000 (Wind Energy, 2020).

[k] According to Savin (2018), "the propeller of a large turbine rotates and produces noise, which is quite strong, 105 dB (A). Even at a distance of 500 yards (0.3 miles) from the wind turbine, the noise produced is still at about 38 dB (A), which is the equivalent to the noise produced by a refrigerator. Wind turbines produce noise pollution, maybe that is the reason why the large wind turbines are usually installed in remote locations (preferably without natural obstacles that can block the wind)".

- Wind resource development may compete with other land uses, which might be more highly valued than electricity generation. However, wind turbines can be located on a property that is also used for grazing or even farming;
- Maintenance costs can be significant for wind farms located in remote areas. In quite remote areas, the maintenance costs of wind turbines and wind farms, in general, might be quite high. The maintenance costs of a wind turbine and a wind farm also vary significantly depending on geography and climatic conditions (Piramal, 2020);
- Wind farms could be affected by severe natural disasters events such as hurricanes and earthquakes.

According to Bates (2020), even though the offshore wind market in the North American region, particularly in the USA, is full of opportunities, several challenges still need to be addressed and overcome. Federal regulation is a primary concern. For this reason, the USA administration has slowed the federal offshore wind permitting process in 2019. It also decided that its environmental analysis applied to Avangrid's Vineyard Wind I offshore energy project should be expanded into a broader review of all planned offshore wind projects' potential impacts.

One issue that forced the USA administration to apply environmental analysis to all offshore wind farms under revision concerns the USA commercial fishing industry's potential impact. The USA Coast Guard is also studying possible shipping issues, mainly how offshore wind farms are sited within shipping lanes and safe operating clearances for ships traveling over turbine towers. All of the studies mentioned above look beyond the impacts of the Vineyard Wind project "to address the broader effects of offshore wind on industries, the environment, and marine life across the Northeast region. The study found significant adverse impacts on commercial fishing and scientific research, moderate effects on marine life, and minor impact on air quality" (Bates, 2020).

In commercial fishing, it is important to single out that the National Marine Fisheries Service and the National Oceanic and Atmospheric Administration had refused to endorse the Bureau of Ocean Energy Management (BOEM) Draft Report for Vineyard Wind I (n.d.). Both offices complain that in the BOEM's draft report, fishing concerns were not addressed adequately. This situation helped trigger the government's ongoing analysis of offshore wind's cumulative impacts in the region (Davidson, 2020).

One issue under discussion is the transit lanes of four nautical miles (7.4 km) wide and up to 70 nautical miles (130 km) long for a wind-development area off Massachusetts and Rhode Island. The project developers that would be impacted—Vineyard Wind, Mayflower Wind, Ørsted/Eversource, and Equinor—had already proposed a compromise that wind turbines be spaced one nautical mile (1.9 km) apart without the transit lanes. However, according to Ørsted and Eversource, the proposed transit lanes would lose more than 50 wind turbines from their South Fork, Revolution, and Sunrise Wind projects. That equates to nearly a 25% loss in total wind-turbine locations. In addition, the transit lanes may render the billions of dollars of investment economically unfeasible. The other main sticking point is access to fishing grounds within proposed wind arrays, and this issue is also not yet solved (Davidson, 2020).

"Another concern focuses on the substantial increase in energy to the coastal transmission system" (Bates, 2020). Offshore wind farms could add an enormous amount of power to the existing electrical grid. Perhaps the current network capacity may be unable to handle this increase without affecting the energy supply to customers.

Another issue under consideration "is the current offshore solicitation model, which gives developers the right to generate power and transmit it onshore. This approach could result in numerous separate transmission lines that serve offshore wind farms instead of a more centralized system where lines deliver electricity from multiple wind projects. Although that model has become the norm, some states now are discussing whether to separate transmission from generation" (Bates, 2020).

Without a doubt, the increased use of wind energy for electricity at the world level, particularly in the North American region, could bring additional socioeconomic benefits. The wind industry is projected to employ 3.74 million people by 2030 and more than six million people by 2050, a figure nearly three times higher and five times higher, respectively, than the 1.16 million jobs reported to work in the wind industry in 2018. A holistic policy framework is needed to maximize the outcomes of the energy transition from fossil fuels to renewable energy sources for electricity generation.

"Deployment policies within the energy sector will need to coordinate and harmonize with integration and enabling policies. Under the enabling policy umbrella, a particular focus is required on industrial, labor, financial, education, and skills policies to maximize the transition benefits. For example, education and skills policies can retain and reallocate existing expertise in the oil and gas sectors to support offshore wind foundation structures. Similarly, sound industrial and labor policies that build upon domestic supply chains can enable income and employment growth by leveraging existing economic activities in support of wind industry development" (IRENA, 2019).

## The future of wind energy in the North American region

According to wind industry analysts, the global offshore wind energy industry is expected to grow between six to eight-folds over the next decade. By 2030, the forecasted cumulative offshore global wind capacity will range from 140 GW (McNulty, 2018) to 190 GW (Global Offshore Wind Report 2020, 2020). "Potentially, global onshore and offshore wind power at commercial turbine hub heights could provide 840,000 TWh of electricity each year. Similarly, the USA annual wind potential of 68,000 TWh greatly exceeds annual USA electricity consumption" (Wind Energy Factsheet, 2019).

Wind energy and solar energy would lead the way for transforming the global electricity sector during the coming years. Onshore and offshore wind would generate more than 35% of the world's electricity needs.[1] It is projected that wind energy will become the prominent electricity generation source by 2050. To achieve the goal mentioned above, "the global cumulative installed capacity of onshore wind farms should reach more than 3-fold by 2030 up to 1787 GW and 9-fold by 2050 up to 5044 GW compared to installed capacity in 2019 (542 GW). For offshore wind power, the global cumulative installed capacity would increase

[1] In 2010, the participation of wind energy in the global energy mix was 1.7% and increased to 6% in 2018. It is projected that the participation of wind energy in the global energy mix will raise to 21% in 2030 and to 35% in 2050.

almost 10-fold by 2030 up to 228 GW and substantially towards 2050, with total offshore wind installation nearing 1000 GW by 2050" (IRENA, 2019).[m]

By region, the onshore wind power industry domination is as follows: "Asia (mostly China) would continue to dominate the onshore wind power industry, with more than 50% of global installations by 2050, followed by North America (23%), and Europe (10%)". For the offshore wind energy industry, "Asia would take the lead in the coming decades, with more than 60% of global wind installations by 2050, followed by Europe (22%) and North America (16%)" (IRENA, 2019).

As a result of a 2015 study carried out by the US Department of Energy, it was found "that wind energy could provide 20% of the country's electricity by 2030 and 35% by 2050" (USA Wind Vision Report, 2015). According to this report, the following are some of the key findings included in it:

- Wind power can be a viable source of renewable electricity in all 50 states by 2050;
- Wind power has the potential to support over 600,000 jobs in manufacturing, installation, maintenance, and supporting services by 2050;
- Wind energy is affordable. By reducing national vulnerability to price spikes and supply disruptions with long-term pricing, wind power is anticipated to save consumers US$280 billion by 2050;
- Wind energy use for electricity generation avoided over 250,000 metric tons of air pollutants, including sulfur dioxide, nitric oxide, nitrogen dioxide, and particulate matter, in 2013. By 2050, wind energy could avoid the emission of 12.3 gigatonnes of greenhouse gases;
- Wind energy preserves water resources. By 2050, wind energy can save 260 billion gallons of water—the equivalent to roughly 400,000 Olympic-size swimming pools—that would have been used by the electric power sector;
- Wind energy deployment increases community revenues. Local communities will be able to collect additional tax revenue from land lease payments and property taxes, reaching US$3.2 billion annually by 2050 (USA Wind Vision Report, 2015).

North American offshore wind farms' market potential is promising, including the so-called "floating offshore wind farms." This new offshore wind technology will be ready for commercial expansion, possibly in the next couple of years, eliminating current depth limitations (i.e., requiring shallow ocean depths). "There are already plans to use this technology in the Gulf of Maine, the Gulf of Mexico, the California coast, and the Oregon coast, while additional studies are looking at using the technology in the Great Lakes" (Bates, 2020). The use of offshore wind floating technology "will be limited only by the availability of continuous wind sources and interconnection constraints, and as such, the opportunity may even dwarf the available lease areas currently under consideration in the Northeast and Mid-Atlantic USA" (Bates, 2020).

---

[m] To reach 5044 GW of cumulative onshore wind energy capacity by 2050, it is crucial to have an annual capacity increase of more than 200 GW per year in the next 20 years, compared to 45 GW added in 2018. In the case of offshore wind energy capacity, annual growth of 45 GW per year would be required by 2050. In 2018, the offshore wind energy capacity increased only 4.5 GW.

As regulations continue to evolve and wind technology continues to advance, offshore wind farms are likely to be the next major energy frontier in the USA. In 2018, the American Wind Energy Association (AWEA) announced new offshore wind projects with 2 GW capacity in the USA. By 2019, that amount had increased to 26 GW of planned projects; this means a ten-fold increase. These new targets indicate that the offshore wind industry is likely at a tipping point (Bates, 2020).

It is important to stress that until 2019, there is only one offshore wind farm operating in the USA and non in Canada. Located off Rhode Island's coast, the Deepwater Wind's Block Island Wind Farm project has five wind turbines with a 29 MW capacity.

Within the North American region, the USA has policies that promote the construction of new offshore wind farms. Different USA states are establishing offshore wind procurement targets between 21.4 GW and 23.7 GW by 2030. A total of 15 offshore wind farms is expected to be built, in the USA alone, by 2026. Considering the high number of wind farms under consideration, seek to sign contracts with the US Department of the interior's BOEM to implement such projects. The reason is straightforward. Most offshore wind energy projects will be implemented in federal waters.[n]

## Wind energy capacity in Canada

Canada's power generation was already met by 80% clean energy in 2015, mainly because of the country's extensive hydroelectric capacity already installed. However, Canada plans to further increase its hydroelectric capability by 2025, in addition to increasing wind and solar capacity, according to the paper entitled "Renewables share of North America electricity mix expected to rise" (2016).

EIA's International Energy Outlook (2016) projects reduced coal use in Canada between 2015 and 2025, consistent with the Canadian government's plans to gradually phase out old and ineffective coal-fired power plants for electricity generation in the country. "However, the combined share of renewables and nuclear in Canada's total generation is expected to fall to 75% by 2025 because of increases in natural gas use and projected retirements of existing nuclear capacity. Overall, Canada's electricity generation currently represents about 13% of the North America total generation" (EIA (USA Energy Information Administration), 2016).

Canada finished 2019 with 13,413 MW of wind energy capacity installed in the country, enough to power approximately 3.4 million homes.

The evolution of wind energy capacity installed in Canada during the period 2010–19 is shown in Fig. 4.13.

Based on the data included in Fig. 4.13, the following can be stated: the wind energy capacity installed in Canada during the period 2010–19 increased 3.4-fold, rising from 3967 MW in 2010 to 13,413 MW in 2019. According to the Canadian Wind Energy Association (CanWEA) (2020), the average annual growth of wind energy in Canada in the last decade was 16%. It is expected that this trend will continue during the coming years but at a different rate. At

---

[n] BOEM has already awarded 13 commercial wind energy leases off the Atlantic coast, and recently announced two more areas available off the coast of Massachusetts.

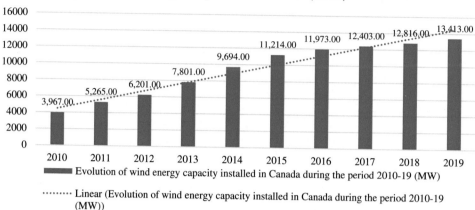

FIG. 4.13    Evolution of wind energy capacity installed in Canada during the period 2010–19. *Source: IRENA, 2020. Renewable Energy Statistics 2020. The International Renewable Energy Agency, Abu Dhabi.*

the end of 2019, 301 wind farms were operating in Canada, including wind farms located in two of the three northern territories. The total wind turbines installed in the country reached 6771 MW in 2019. All of them are onshore wind farms.[o]

However, despite this growth, Canada's wind energy installed capacity in 2019 represented only 11.5% of the total wind energy installed capacity in the North American region in that year (116,997 MW). The wind energy installed capacity in Canada, in 2019, by province is shown in Table 4.2.

The list of Canadian wind farms with a capacity of at least 100 MW is included in Table 4.3.

According to Table 4.3, there are 37 wind farms in operation in Canada with at least 100 MW capacity. All of these farms are onshore wind farms.[P] The larger wind farm is the Seigneurie de Beaupré wind farm located in Quebec with 363.5 MW capacity. The wind farm mentioned above was built in 2013. The total wind farm capacity installed in Canada at the end of 2019 was 13,413 MW.

It is important to single out that more wind farms have been constructed in Canada between 2009 and 2019 than any other energy source used for electricity generation. Canada ranks 9th in the world in terms of wind power capacity installed. In the country, the wind energy industry is continuing a trend of stable and robust growth. "It finished 2017, with ten new projects completed, representing US$800 million of investment" (Expancel Newsletter, 2018).

[o] Without a doubt, the Canadian wind industry's performance, in 2019, continued a trend of stable and robust growth. During the period 2015–19, the wind energy capacity installed in the country increased significantly reaching a growth of 19.6%. With Canada's unparalleled wind energy potential, there are still opportunities to do more to maximize the economic, industrial development, and environmental benefits associated with wind energy development, according to the Canadian Wind Energy Association.

[P] Canada has 31 offshore wind farm projects but none of them are today in the construction phase.

TABLE 4.2   Wind energy installed capacity in Canada, in 2019, by province.

| Province/territory | Wind energy installed capacity (MW) | Number of wind turbines installed | Number of projects |
|---|---|---|---|
| Ontario | 5436 | 2681 | 94 |
| Quebec | 3882 | 1990 | 47 |
| Alberta | 1685 | 957 | 38 |
| British Columbia | 713 | 292 | 9 |
| Nova Scotia | 616 | 309 | 78 |
| New Brunswick | 314 | 119 | 6 |
| Manitoba | 258 | 133 | 4 |
| Saskatchewan | 241 | 153 | 8 |
| Prince Edward Island | 204 | 104 | 10 |
| Newfoundland and Labrador | 55 | 27 | 4 |
| Northwest Territories | 9.2 | 4 | 1 |
| Yukon | 0.8 | 2 | 2 |
| Total | 13,413 | 6771 | 301 |

*Source: 2020. Wikipedia Wind Power in Canada and Canadian Wind Energy Association (CanWEA), 2020. Installed Capacity. CanWEA, Canada.*

## Wind energy electricity generation in Canada

At the end of 2019, Canada's wind power generating capacity was 13,413 MW, providing about 6% of Canada's electricity demand, according to the Canadian Wind Energy Association (CanWEA) (2020). The total electricity generation of wind farms in Canada at the end of 2018 was 31,848 GWh, enough to meet the needs of over three million Canadian homes (3.4 million homes). The future Canadian strategy for wind energy foresees to reach 55 GW capacity by 2025, an increase of 4.1-fold the capacity installed in 2019, and to meet 20% of the country's energy needs; this represents an increase of 14% with respect to the level reached in 2019.

The evolution of wind farm electricity generation in Canada during the period 2010–18 is shown in Fig. 4.14.

Based on the data included in Fig. 4.14, the following can be stated: electricity generation by wind farms in Canada, during the period 2010–18, increased by 3.7-fold, rising from 8724 GWh in 2010 to 13,848 GWh in 2018. The only year, within the period under consideration, that the electricity generation decreased was in 2017. The reduction registered was 5.6% with respect to 2016.

Without a doubt, wind energy has been the largest source of new electricity generation in Canada for more than a decade. "It has provided – and will continue to provide – substantial economic and social benefits to local and Indigenous communities across Canada" (A Wind Energy Vision for Canada, 2019).

**TABLE 4.3**    The list of Canadian wind farms with a capacity of at least 100 MW.

| | Name | Province | Capacity (MW) | Year |
|---|---|---|---|---|
| 1 | Seigneurie de Beaupré | Quebec | 363.5 | 2013 |
| 2 | Rivière-du-Moulin | Quebec | 350 | 2014 |
| 3 | Blackspring Ridge | Alberta | 300 | 2014 |
| 4 | Henvey Inlet | Ontario | 300 | 2019 |
| 5 | Lac Alfred | Quebec | 300 | 2013 |
| 6 | Niagara Region | Ontario | 230 | 2016 |
| 7 | Gros-Morne | Quebec | 211.5 | 2011 |
| 8 | Amaranth | Ontario | 199.5 | 2008 |
| 9 | Wolfe Island | Ontario | 197 | 2009 |
| 10 | Prince Township | Ontario | 189 | 2006 |
| 11 | Meikle | British Columbia | 184.6 | 2016 |
| 12 | Underwood | Ontario | 181.5 | 2009 |
| 13 | Armow | Ontario | 180 | 2015 |
| 14 | Kent Hills | New Brunswick | 167 | 2008 |
| 15 | Comber | Ontario | 165.6 | 2012 |
| 16 | Centennial | Saskatchewan | 150 | 2006 |
| 17 | Massif du Sud | Quebec | 150 | 2013 |
| 18 | Halkirk | Alberta | 149.4 | 2012 |
| 19 | Dokie Ridge | British Columbia | 144 | 2011 |
| 20 | Quality | British Columbia | 142.2 | 2012 |
| 21 | Le Plateau | Quebec | 138.6 | 2012 |
| 22 | St. Joseph | Manitoba | 138 | 2011 |

**TABLE 4.3** The list of Canadian wind farms with a capacity of at least 100 MW—cont'd

| | Name | Province | Capacity (MW) | Year |
|---|---|---|---|---|
| 23 | Jardin d'Eole | Quebec | 127 | 2009 |
| 24 | Summerhaven | Ontario | 124.4 | 2013 |
| 25 | St. Leon | Manitoba | 120 | 2006 |
| 26 | Baie-des-Sables | Quebec | 109.5 | 2006 |
| 27 | Carleton | Quebec | 109.5 | 2008 |
| 28 | Port Dover and Nanticoke | Ontario | 105 | 2013 |
| 29 | Bear Mountain | British Columbia | 102 | 2009 |
| 30 | South Canoe | Nova Scotia | 102 | 2015 |
| 31 | Chatham | Ontario | 101.2 | 2010 |
| 32 | Montérégie | Quebec | 101.2 | 2012 |
| 33 | Port Alma | Ontario | 101 | 2008 |
| 34 | Anse-à-Valleau | Quebec | 100.5 | 2007 |
| 35 | Mont-Louis | Quebec | 100.5 | 2011 |
| 36 | Belle River | Ontario | 100 | 2017 |
| 37 | North Kent | Ontario | 100 | 2018 |

*Source: 2020. Wikipedia List of Wind Farms in Canada.*

According to the paper mentioned above, Canada's wind energy industry has attracted more than US$23 billion in investment in the construction of wind farms. Besides, it has created over 58,000 person-years of employment in wind farms' construction and operations, directly benefited more than 299 communities in 12 provinces and territories within the country. In addition, several factory-made blades, towers, and other components within the wind turbine supply chain also benefited.

Furthermore, it is important to stress that wind energy costs are falling so sharply at the world level, particularly in Canada, where some new wind energy facilities have begun to generate electricity even more cost-effectively than many other existing generating facilities in the country.

Without a doubt, the wind is an energy source that is in a position to make a major contribution to ensuring that Canada's future electricity grid is affordable, reliable, and environmentally sustainable and will allow Canada to fulfill its commitments with Paris agreement on climate change.

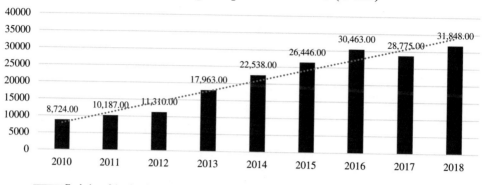

**Evolution of the electricity generation by wind farms in Canada during the period 2010-18 (GWh)**

FIG. 4.14   Evolution of the wind farm electricity generation in Canada during the period 2010–18. *Source: IRENA, 2020. Renewable Energy Statistics 2020. The International Renewable Energy Agency, Abu Dhabi.*

Summing up, the following can be stated:

- Wind energy is the lowest cost option for new electricity supply without subsidies in Canada. Reductions in total installed costs drive the fall in the LCOE for wind power technology to varying extents. That has been most notable for onshore wind farms, among others (IRENA, 2018);
- It is expected a reduction of 48% in world wind energy costs by 2050. According to the "IRENA 2018 Renewable Power Generation Costs in 2017" report,[q] wind turbine prices have fallen by around half over the period 2010–17, "depending on the market, leading to cheaper wind power globally. Onshore wind electricity costs have dropped by almost a quarter since 2010, with average costs of US$0.06 per kWh in 2017" (Wind Power in Ohio, 2020). Though still in their infancy in terms of deployment, offshore wind power costs fall between 2010 and 2017. For example, the global weighted average of the LCOE of offshore wind projects commissioned in 2017 was US$0.14 per kWh. However, in 2016 and 2017, public sale results for offshore wind projects to be commissioned in 2020 and beyond signal a step-change, with costs falling between US$0.06 and US$0.10 per kWh.
  According to the IRENA report mentioned above, "by 2020, all mainstream renewable power generation technologies can be expected to provide average costs at the lower end of the fossil-fuel cost range" (Quick Facts Ren Energy, 2019). Besides, "several wind power projects will provide some of the lowest-cost electricity from any source"[r] (IRENA, 2018);

[q] IRENA (2018), Renewable Power Generation Costs in 2017, International Renewable Energy Agency, Abu Dhabi.
[r] Onshore wind projects commissioned in 2017 largely fell within the range of generation costs for fossil fuel based electricity. Without a doubt, onshore wind is one of the most competitive sources of new generation capacity. Recent bidding carried out in Canada have resulted in onshore wind power LCOE as low as US$0.03 per kWh. In the case of offshore wind, recent bidding results suggest that this type of wind energy will provide electricity for between US$0.06 and US$0.10 per kWh by 2020, according to the IRENA 2018 Renewable Power Generation Costs in 2017 report.

- Innovation in wind turbines is pushing the price down. Larger wind turbines can generate more power from a given resource, and technology advances in monitoring and optimization allow more efficient wind farms' operations.
"Sharp cost reductions – both recent and anticipated – represent remarkable deflation rates for various wind options. Based on project and auction data, learning rates[s] for the 2010–20 period, are estimated at 14% for offshore wind, and 21% for the onshore wind" (IRENA, 2018).

Finally, it is important to stress the following. With 13,413 MW of installed wind power capacity, Canada has committed to reducing its greenhouse gas emissions by 30% from 2005 levels by 2030. It recognizes the importance of a clean electricity grid in support of that goal. Today, Canada's electricity grid is 80% emission-free, and the federal government has set a goal for that increase to reach 90% by 2030.

## Wind energy capacity in the United States

The use of wind energy in the USA has expanded quickly over the latest several years. In 2019, 275.8 TWh were generated by wind farms, representing 6.3% of all electricity produced in the USA in that year (4401.3 TWh) (IRENA Renewable Energy Statistics 2020 and BP Statistical Review of World Energy 2020). It is important to single out that in 2018, the US's hydropower plants generated 317,004 GWh and wind farms 275,834 GWh.

The evolution of the wind energy capacity installed in the USA during the period 2010–19 is shown in Fig. 4.15.

According to Fig. 4.15, the following can be stated: wind energy capacity installed in the USA during the period 2010–19 increased by approximately 2.7-fold, rising from 39,135 MW in 2010 to 103,584 MW in 2019. The major increase in wind energy capacity installed in one year within the period under consideration was registered in 2012.

In the case of onshore wind energy, the evolution in the use of this type of wind energy for electricity generation in the USA during the period 2010–19 is shown in Fig. 4.16.

Based on the data included in Fig. 4.16, the following can be stated: the onshore wind energy capacity installed in the USA grew approximately 2.7-fold during the period 2010–19, rising from 39,135 MW in 2010 to 103,555 MW in 2019. Furthermore, according to the Canadian government's decision to increase the role to be played by renewable energy sources within the country's energy mix, it is expected that this trend will continue without change during the coming years.

In 2019, the USA wind industry created around 120,000 jobs across all states. In that year, the USA wind industry added 9167 MW of new wind capacity, the third strongest year ever for new wind energy installations, and the second major growth during the period 2010–19. At the end of 2019, approximately 60,000 wind turbines with a combined capacity installed of 103,555 MW were operating across 41 states.

The wind power capacity installed in the country has more than doubled over the past ten years (2.7-fold) and, in 2019, was the largest source of renewable energy in the USA.

---

[s] The learning rate is the percentage cost reduction experienced for every doubling of cumulative installed capacity.

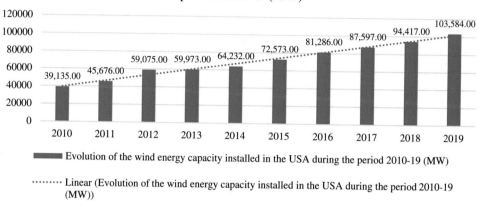

FIG. 4.15   Evolution of the wind energy capacity installed in the USA during the period 2010–19. *Source: IRENA, 2020. Renewable Energy Statistics 2020. The International Renewable Energy Agency, Abu Dhabi.*

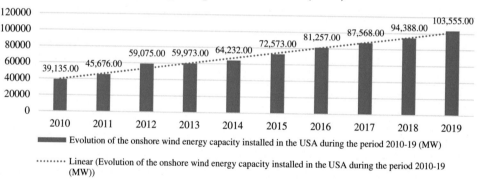

FIG. 4.16   Evolution of the onshore wind energy capacity installed in the USA during the period 2010–19. *Source: IRENA, 2020. Renewable Energy Statistics 2020. The International Renewable Energy Agency, Abu Dhabi.*

Wind energy generated 6.3% of the nation's electricity in 2019, enough to power 27.5 million homes. In six USA states, wind farms provide over 20% of the electricity produced by them. It is important to single out that the USA has enough installed wind capacity to power 32 million homes. Wind supports economic development in local areas, providing more than US$1.6 billion each year in state and local tax payments and rental payments to landowners. Wind supports domestic manufacturing with more than 530 factories in 43 states employing 26,000 people.

As of January 2020, the total installed wind power generating capacity in the USA was 105,583 MW, according to AWEA 3rd Quarter 2019 Public Market Report (2020). This capacity is exceeded only by China and the EU. It is also important to single out that wind power's largest growth in capacity during the period 2010–19 was registered in 2012 when 13,399 MW of new wind power capacity was installed in the country.

By September of 2019, a total of 19 states had over 1000 MW of wind installed capacity, with five states (Texas, Iowa, Oklahoma, Kansas, and California) generating over half of all wind energy production in the country, according to the AWEA 3rd Quarter 2019 Public Market Report (2020). Texas continues to dominate the USA's wind energy production, adding far more generating capacity than any other state in 2017 and having more installed wind power capacity than all but five countries in the world (Druzin, 2018). Texas also had more wind farms under construction than any other state within the USA. However, the state within the USA generating the highest percentage of energy from wind farms is Iowa, with 42% of total energy production, according to the AWEA 3rd Quarter 2019 Public Market Report (2020). At the same time, North Dakota is the USA state with the highest per-capita wind electricity generation. The Alta Wind Energy Center in California is the largest wind farm in the USA, with a capacity of 1548 MW (see Table 4.5).

In offshore wind farms, a new demonstration project in the USA, a 12 MW pilot project in Virginia, was completed in June 2020. This project is the first offshore wind farm approved by BOEM and installed in federal waters since 2016 when the first USA offshore wind farm was operational. In addition, the level of offshore wind development activity remains impressively high. As of the end of 2019, BOEM has auctioned 16 active commercial leases for offshore wind farm development to support more than 21 GW generating capacity.

The evolution of the offshore wind farm installed capacity in the USA during the period 2015–19 is shown in Fig. 4.17.

According to Fig. 4.17, the following can be stated: before 2015, no offshore wind farms were operating in the USA. Since 2016 and until 2019, only one offshore wind farm has been

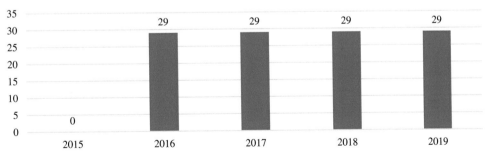

**FIG. 4.17** Evolution of the offshore wind farm installed capacity in the USA during the period 2015–19. *Source: IRENA, 2020. Renewable Energy Statistics 2020. The International Renewable Energy Agency, Abu Dhabi.*

operating in the USA with a 29 MW capacity. After June 2020, the total offshore wind capacity in the USA reached 41 MW.

On the state level, the East Coast cluster consists of Maine, Connecticut, Massachusetts, New York, New Jersey, Delaware, Maryland, Virginia, and North Carolina, driving strong demand for offshore wind energy. The total announced offshore wind procurement targets in these states reached 28.1 GW in the first quarter of 2019. GWEC Market Intelligence predicts a total of 22.6 GW of offshore wind could be built in the USA by the end of this decade.

To realize offshore wind potential, however, the following two key challenges need to be addressed:

- Slow project permitting processes. The current permitting processes have delayed the USA wind offshore industry's ramp-up, with more than 2000 GW of technical resource potential. Without a doubt, to ramp up the USA offshore wind industry, the current permitting processes must be streamlined;
- Establishing a local supply chain and fostering investment and long-term planning in the electrical grid and port infrastructure must be achieved across states through a collaborative approach.

In the USA, independent forecasts indicate that the offshore wind power industry capacity could grow between 11 GW and 16 GW by 2030. A recent white paper called "Special Initiative for Offshore Wind (SIOW)" projects that wind power could hit 18.6 GW in seven states on the Atlantic Seaboard by 2030 (Special Initiative on Offshore Wind (SIOW), n.d.). According to the report mentioned above, a nearly US$70 billion revenue opportunity for businesses in the offshore wind power supply chain is expected over the next decade.

Without a doubt, the future of the USA offshore wind power industry looks promising. According to the USA Department of Energy's National Offshore Wind Strategy (2016) report, the USA has an offshore wind power potential of more than 4000 GW – more than four times the wind energy generating capacity of the current USA electrical system (103.7 GW in 2019). In 2019, more than 10,000 MW of offshore wind farm projects was in various development stages along the country's Eastern seaboard. "Avangrid, in a joint venture with Copenhagen Infrastructure Partners, has the Vineyard Wind project, located 15 miles south of Martha's Vineyard off the coast of Cape Cod, Massachusetts. Once complete, the project will deliver 800 MW to power more than 400,000 homes" (Bates, 2020).

Other regions of the country have shown interest in offshore wind power construction for electricity generation in the future. BOEM has identified significant wind energy potential along the west coast, particularly in Oregon and California. Besides, attention is also spreading inland, with a 20 MW Lake Erie wind farm project currently going through federal approvals.

## Wind energy electricity generation in the United States

In 2019, according to BOEM sources, electric power generation from wind farms was 10% or more in 14 states. These states are Colorado, Idaho, Iowa, Kansas, Maine, Minnesota, North Dakota, Oklahoma, Oregon, South Dakota, Vermont, Nebraska, New Mexico, and Texas. According to Equinor's Empire Wind (n.d.) source, Iowa, South Dakota, North Dakota,

Oklahoma, and Kansas, more than 20% of their electric power generation comes from wind farms. Twenty states have more than 5% of their electricity generation coming from wind farms. Iowa becomes the first USA state to generate 40% of its electricity from wind power in late 2019. The five US states with the most wind capacity installed at the end of 2019 were the following:

- Texas, with a total wind installed capacity of 28,843 MW;
- Iowa, with a total wind installed capacity of 10,201 MW;
- Oklahoma, with a total wind installed capacity of 8172 MW;
- Kansas, with a total wind installed capacity of 6128 MW;
- California, with a total wind installed capacity of 5973 MW.

The top five US states corresponding to the percentage of generation by wind farms in 2019 were the following, according to BOEM sources:

- Iowa, with 41.7%;
- Kansas, with 36.4%;
- Oklahoma, with 31.7%;
- North Dakota, with 25.8%;
- South Dakota, with 24.4%.

The US offshore wind market has picked up strong momentum since the Block Island Wind Project came online in Rhode Island in December 2016. Despite a complex regulatory scene with differing rules across the offshore states, large-scale projects are advancing, and developer appetite has been at a fever pitch. And with technical resource potential for US offshore wind farms exceeding 4000 GW, there is vast room to increase the offshore wind farm capacity and its participation in the US energy mix in the coming years.

The evolution of offshore wind farms electricity generation in the USA during the period 2015–18 is shown in Fig. 4.18.

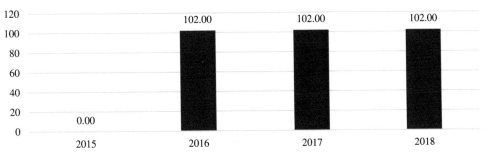

Evolution of the offshore wind farms electricity generation in the USA during the period 2015-18 (GWh)

■ Evolution of the offshore wind farms electricity generation in the USA during the period 2015-18 (GWh)

**FIG. 4.18**  Evolution of offshore wind farms electricity generation in the USA during the period 2015–18. *Source: IRENA, 2020. Renewable Energy Statistics 2020. The International Renewable Energy Agency, Abu Dhabi.*

Based on the data included in Fig. 4.18, the following can be stated: The first offshore wind farm was constructed in the USA in 2016. In that year, it began generating electricity. Until 2018, the offshore wind farm electricity generation was stable at 102 GWh. It is expected that with the installation of new offshore wind farms during the coming years, electricity production using this type of wind farm will grow in the country significantly.

In 2009, the Department of the Interior (DOI) announced final regulations for the Outer Continental Shelf (OCS) Renewable Energy Program. These regulations provide a framework for issuing contracts, easements, and rights-of-way for OCS activities that support energy production and transmission from sources other than oil and natural gas. Since the regulations were enacted, BOEM has issued 16 commercial offshore wind energy leases, including three commercial contracts in 2019. In total, the 16 offshore wind leases could support more than 21 GW of generating capacity.

BOEM is now in the planning stages for leasing areas off the coast of New York, South Carolina, California, and Hawaii and held two lease sales in 2020, one in the Atlantic in the New York Bight and one in California's Pacific coast.

Table 4.4 includes the USA wind farms with a generating capacity of at least 250 MW (97 wind farms).

Finally, Table 4.5 includes the offshore wind farms operating in the USA (9 offshore wind farms).

**TABLE 4.4**  List of USA wind farms with a generating capacity of at least 250 MW.

| Wind farm | State | Installed capacity (MW) |
| --- | --- | --- |
| Alta Wind Energy Center I-XI | California | 1548 |
| Altamont Pass Wind Farm | California | 576 |
| Amazon Wind Farm | Texas | 253 |
| Arbor Hill Wind Farm | Iowa | 250 |
| Balko Wind | Oklahoma | 300 |
| Beaver Creek Wind Farm I & II | Iowa | 340 |
| Bethel Wind Farm | Texas | 276 |
| Biglow Canyon Wind Farm | Oregon | 450 |
| Bison Wind Energy Center | North Dakota | 497 |
| Blue Canyon Wind Farm | Oklahoma | 423 |
| Blue Creek Wind Farm | Ohio | 302 |

**TABLE 4.4** List of USA wind farms with a generating capacity of at least 250 MW—cont'd

| Wind farm | State | Installed capacity (MW) |
|---|---|---|
| Brady Wind Energy Center (I & II) | North Dakota | 299 |
| Broadview Energy Wind | New Mexico | 324 |
| Buffalo Gap Wind Farm | Texas | 523 |
| Camp Springs Energy Center | Texas | 250 |
| Canadian Hills Wind Farm | Oklahoma | 299 |
| Capricorn Ridge Wind Farm | Texas | 662 |
| Cedar Creek Wind Farm | Colorado | 550 |
| Cedar Point Wind Farm | Colorado | 252 |
| Chisholm View Wind I & II | Oklahoma | 300 |
| Cimarron Bend Wind Farm | Kansas | 400 |
| Crystal Lake Wind Farm | Iowa | 416 |
| Desert Wind Farm | North Carolina | 208 |
| Diamond Vista Wind Farm | Kansas | 299 |
| El Cabo Wind Farm | New Mexico | 298 |
| Flat Ridge Wind Farm | Kansas | 570 |
| Fowler Ridge Wind Farm | Indiana | 600 |
| Grande Prairie Wind Farm | Nebraska | 400 |
| Grandview | Texas | 211 |
| Gratiot Farms | Michigan | 213 |
| Great Western Wind Farm | Oklahoma | 225 |
| Green Pastures Wind Farm | Texas | 300 |

*Continued*

**TABLE 4.4**    List of USA wind farms with a generating capacity of at least 250 MW—cont'd

| Wind farm | State | Installed capacity (MW) |
|---|---|---|
| Gulf Wind Farm | Texas | 283 |
| Hidalgo Wind Farm | Texas | 250 |
| Highland Wind Energy Center | Iowa | 502 |
| Horse Hollow Wind Energy Center | Texas | 736 |
| Ida Grove Wind Farm | Iowa | 301 |
| Javelina Wind Energy Center | Texas | 749 |
| Jumbo Road Wind Farm | Texas | 300 |
| Kay Wind Farm | Oklahoma | 300 |
| Keenan I & II | Oklahoma | 253 |
| King Mountain Wind Farm | Texas | 281 |
| Kingfisher Wind Farm | Oklahoma | 298 |
| Klondike Wind Farm | Oregon | 400 |
| Limon Wind Energy Center | Colorado | 601 |
| Lone Star Wind Farm | Texas | 400 |
| Los Vientos Wind Farm | Texas | 900 |
| Lower Snake River Wind Project | Washington | 343 |
| Lundgren Wind Farm | Iowa | 250 |
| Maple Ridge Wind Farms I & II | New York | 322 |
| Meadow Lake Wind Farm | Indiana | 801 |
| Miami Wind Energy Center | Texas | 289 |
| Milford Wind Corridor Project (I & II) | Utah | 306 |
| Mount Storm Wind Farm | West Virginia | 264 |

TABLE 4.4   List of USA wind farms with a generating capacity of at least 250 MW—cont'd

| Wind farm | State | Installed capacity (MW) |
|---|---|---|
| O'Brien Wind Farm | Iowa | 250 |
| Palo Duro Wind Facility | Texas | 250 |
| Panhandle Wind Farm (I & II) | Texas | 399 |
| Panther Creek Wind Farm | Texas | 458 |
| Papalote Creek Wind Farm | Texas | 380 |
| Peetz Table Wind Energy Center | Colorado | 430 |
| Peñascal Wind Farm | Texas | 605 |
| Pioneer Prairie Wind Farm | Iowa | 293 |
| Pomeroy Wind Farm (1–4) | Iowa | 286 |
| Radford's Run Wind Farm | Illinois | 306 |
| Rattlesnake Creek Wind Farm | Nebraska | 318 |
| Red Dirt Wind Project | Oklahoma | 299 |
| Rock Creek Wind Project | Missouri | 300 |
| Rolling Hills Wind Farm | Iowa | 444 |
| Roosevelt Wind Farm | New Mexico | 250 |
| Roscoe Wind Project | Texas | 781 |
| Rush Creek Wind Project | Colorado | 600 |
| San Gorgonio Pass Wind Farm | California | 619 |
| Santa Rita Wind Energy | Texas | 300 |
| Seiling | Oklahoma | 299 |
| Shepherds Flat Wind Farm | Oregon | 845 |
| Sherbino Wind Farm | Texas | 300 |

*Continued*

**TABLE 4.4** List of USA wind farms with a generating capacity of at least 250 MW—cont'd

| Wind farm | State | Installed capacity (MW) |
|---|---|---|
| Shiloh Wind Farm | California | 300 |
| Smoky Hills Wind Farm | Kansas | 251 |
| South Plains Wind Farm I & II | Texas | 500 |
| Spinning Spur Wind Farm | Texas | 516 |
| Stateline Wind Project | Oregon | 300 |
| Stephens Ranch Wind Farm | Texas | 376 |
| Story County Wind Farm I & II | Iowa | 300 |
| Streator Cayuga Ridge South Wind Farm | Illinois | 300 |
| Sweetwater Wind Farm | Texas | 585 |
| Tahoka Wind Farm | Texas | 300 |
| Tehachapi Pass Wind Farm | California | 690 |
| Thunder Ranch Wind Farm | Oklahoma | 298 |
| Top Crop Wind Farm | Illinois | 300 |
| Tucannon River Wind Farm | Washington | 267 |
| Turtle Creek Wind Farm | Iowa | 200 |
| Twin Groves Wind Farms I & II | Illinois | 396 |
| Wake Wind Farm | Texas | 257 |
| Western Plains Wind Farm | Kansas | 280 |
| Wildcat Wind Farm Phase I | Indiana | 200 |
| Willow Springs Wind Farm | Texas | 250 |
| Windy Point/Windy Flats | Washington | 400 |

*Source: 2020. Wikipedia List of Wind Farms in the United States.*

**TABLE 4.5** List of offshore wind farms in operation in the USA.

| Wind farm | Offshore BOEM wind energy lease area | | | State | Capacity (MW) |
|---|---|---|---|---|---|
| Ocean Wind | Offshore New Jersey OCS-A 0498 (NJWEA South) | 13 nautical miles (24 km) east of Atlantic City (NJ) | 64,940 ha | NJ | 1100 |
| Sunrise Wind | Offshore Massachusetts and Rhode Island OCS-A 0486 (North Lease Area) | 26 nautical miles (48 km) east of Montauk Point, Long Island (NY), and 16.6 nautical miles (31 km) southeast of Block Island (RI) | 39,456 ha | NY | 880 |
| Empire Wind | Offshore New York OCS-A 0512 (Hudson North) | 12 nautical miles (23 km) south of Jones Beach Island, Long Island (NY) at New York Bight | 32,110 ha | NY | 816 |
| Vineyard Wind | Offshore Massachusetts OCS-A 0501 | 13 nautical miles (24 km) west of Marthas Vineyard (MA) | 67,536 ha | MA | 800 |
| Park City Wind | Offshore Connecticut OCS-A 0501 | 23 miles off the coast of Martha's Vineyard (MA) | | CT | 804 |
| Revolution Wind | Offshore Rhode Island OCS-A 0486 (North Lease Area) | Halfway between Montauk Point (NY) and Martha's Vineyard (MA) | 39,456 ha | RI CT | 700 |
| Marwind | Offshore Maryland OCS-A 0490 | 26 nautical miles (48 km) east of Ocean City (MD) | 32,256 ha | MD | 248 |
| South Fork | Massachusetts and Rhode Island OCS-A 0486 (North Lease Area) | 26 nautical miles (48 km) southeast of Montauk Point, Long Island (NY), and 16.6 nautical miles (31 km) southeast of Block Island (RI) | 39,456 ha | RI NY | 130 |
| Skipjack | Offshore Delaware OCS-A 0519 | 16.9 nautical miles (31.4 km) from the coast of Delaware to Maryland state line | 10,656 ha | MD DE | 120 |

*Source: 2020. Wikipedia List of Wind Farms in the United States.*

# References

A Wind Energy Vision for Canada. CanWEA.

Asian, S., Ertek, G., Haksoz, C., Pakter, S., Ulun, S., 2017. Wind turbine accidents: a data mining study. IEEE Syst. J. 11 (3), 1567–1578. https://doi.org/10.1109/JSYST.2016.2565818.

AWEA 3rd Quarter 2019 Public Market Report. American Wind Energy Association (AWEA). January 2020. Retrieved February 20, 2020.

Bates, M., 2020. Offshore Wind Power: Present Challenges and Future Realities. North American Wind Power.

Bidwell, D., 2013. The role of values in public beliefs and attitudes towards commercial wind energy. Energy Policy 58, 189–199.

Bohn, C., Lant, C., 2009. Welcoming the wind? Determinants of wind power development among US States. Prof. Geogr. 61 (1), 87–100. https://doi.org/10.1080/00330120802580271.

Bureau of Ocean Energy Management (BOEM) Draft Report for Vineyard Wind I, n.d. https://www.boem.gov/vineyard-wind.

Canadian Wind Energy Association (CanWEA), 2020. Installed Capacity. CanWEA, Canada.

Corscadden, K., Wile, A., Yiridoe, E., 2012. Social license and consultation criteria for community wind projects. Renew. Energy 44, 392–397. https://doi.org/10.1016/j.renene.2012.02.009.

Danish Wind Industry Association, n.d. Wind Know-how and Wind Energy Reference Manual. Danish Wind Industry Association – windpower.org (xn- -drmstrre-64ad.dk).

Daraniya, N., 2012. Wind Energy Basics. https://www.slideshare.net/NiravDaraniya/wind-energy-basics-11733574.

Davidson, R., 2020. Fishing and offshore industries no closer to finding solutions. Wind Power 2020.

Difference Between Horizontal and Vertical Axis Wind Turbine. Difference Between.net. http://www.differencebetween.net/technology/difference-between-horizontal-and-vertical-axis-wind-turbine/#ixzz6U9ND5ptz;.

Doshi, Y., 2017. Wind Turbine Market by Type of Wind Farm (Onshore and Offshore) and Application (Industrial, Commercial, and Residential); Global Opportunity Analysis and Industry Forecast, 2017-2023. Allied Market Research.

Druzin, R., 2018. Texas Wind Generation Keeps Growing, State Remains at No. 1. Houston Chronicle.

EIA (USA Energy Information Administration), 2016. Department of Energy.

Electrical Power Generation, 2016. Introduction to General Overview. http://machineryequipmentonline.com/electrical-power-generation/introduction-to-general-overview/.

Equinor's Empire Wind—equinor.com, n.d.. www.equinor.com.

Expancel Newsletter, 2018. The Sweeping Winds Across America.

Fast, S., 2015. Social science explanations for host community responses to wind energy. In: Leal (Ed.), Handbook of Renewable Energy. Springer, Berlin, pp. 1–15.

Global Offshore Wind Report 2020. Global Wind Energy Council.

Global Trends in Renewable Energy Investment 2020. Frankfurt School-UNEP Centre/BNEF. http://www.fs-unep-centre.org. (Frankfurt am Main).

Gustavson, M., 1979. Limits to wind power utilization. Science 204 (4388), 13–17.

GWEC Global Wind Report 2019, 2020. Global Wind Energy Council. gwec.net/global-wind-report-2019/.

Holm, L.M., 2017. What Is the Public's Opinion on Renewables? Survey Results on Acceptance Worldwide. German Renewable Energies Agency.

How Do Wind Turbine Works? Office of Energy Efficiency & Renewable Energy, USA Department of Energy (USA DoE); Energy. Gov.

Impact Assessment, 2010. Accompanying document to the Communication from the Commission to the European Parliament, the Council, the European Economic and Social Committee and the Committee of the Regions = Energy infrastructure priorities for 2020 and beyond—A Blueprint for an integrated European energy network;.

International Energy Outlook. EIA.

IRENA, 2018. Renewable Power Generation Costs in 2017. International Renewable Energy Agency, Abu Dhabi.

IRENA, 2019. Future of Wind: Deployment, Investment, Technology, Grid Integration, and Socio-Economic Aspects (A Global Energy Transformation Paper). International Renewable Energy Agency, Abu Dhabi.

IRENA, 2020. Renewable Energy Statistics 2020. The International Renewable Energy Agency, Abu Dhabi.

Kennedy, B., 2017. Two-Thirds of Americans Give Priority to Developing Alternative Energy Over Fossil Fuels. Fact Tanks; Pew Research Center.

Krohn, S., Morthorst, P.-E., Awerbuch, S., 2009. The Economics of Wind Energy. A Report by the European Wind Energy Association.

Lantz, E., Flowers, L., 2011. Social Acceptance of Wind Energy Projects: Country Report of United States. IEA Wind Task 28.

Lee, J., Zhao, F., Dutton, A., Backwell, B., Qiao, L., Lim, S., Lathigaralead, A., Liang, E., 2020. Global Offshore Wind Report 2020. Global Wind Energy Council.

Malnick, E., Mendick, R., 2011. 1,500 Accidents and Incidents on UK Wind Farms. The Telegraph. Dec. 2011. Available: http://www.telegraph.co.uk/news/uknews/8948363/1500-accidents-and-incidents-on-UK-windfarms.html.

McNulty, M., 2018. North America Wind Industry Faces Rising Costs. IHS Markit.

Michigan Wind Farm Accidents (2020) and Windpower Engineering and Development (2020).

Mishra, P., 2017. Types of Wind Turbines—Horizontal Axis and Vertical Axis Wind Turbines. Mechanical Booster.

Morales Pedraza, J., 2016. Power options: energy alternatives for the future. In: Advances in Energy Research. vol. 22. Nova Science Publishers, ISBN: 978-1-63483-230-4 (Chapter 1).

National Offshore Wind Strategy. USA Department of Energy (DoE).

Pasqualetti, M.J., 2001. Wind energy landscapes: society and technology in the California desert. Soc. Nat. Resour. 14 (8), 689–699.

Penn State, 2018. Making Wind Farms More Efficient. Science Daily.

Phadke, R., 2010. Steel forests or smokestacks: the politics of visualization in the cape wind controversy. Environ. Polit. 19 (1), 1–20.

Piramal, A., 2020. 27 Pros and Cons of Wind Energy You Need to Know. Environmental Conscience.

Quick Facts Ren Energy.

Renewables Share of North America Electricity Mix Expected to Rise. Clean Power Professionals Group; US Department of Energy (DoE).

Savin, M., 2018. Advantages and Disadvantages of Wind Energy. Alternative Energies.

Shaw, K., Hill, S.D., Boyd, A.D., Monk, L., Reid, J., Einsiedel, E.F., 2015. Conflicted or constructive? Exploring community responses to new energy developments in Canada. Energy Res. Soc. Sci. 8, 41–51. https://doi.org/10.1016/j.erss.2015.04.003.

Sovacool, B.K., 2009. Rejecting renewables: the socio-technical impediments to renewable electricity in the United States. Energy Policy 37 (11), 4500–4513. https://doi.org/10.1016/j.enpol.2009.05.073.

Special Initiative on Offshore Wind (SIOW), n.d. University of Delaware, USA.

USA Wind Vision Report. US Department of Energy's (DoE's) Wind Energy Technologies Office.

Wikipedia Wind Power in Canada.

Wind Energy. National Center for Appropriate Technology.

Wind Energy Facts. FACTS. NET.

Wind Energy Factsheet. Center for Sustainable Systems. Wind Energy_CSS07-09_e2019.pdf.

Wind Power in Ohio (2020); Ohio Citizen Action; https://www.ohiocitizen.org/wind_power_in_ohio.

Renewable Energy World. Wind Power Technology.

Windpower, 2020. G + Reports Growth in Drop Incidents on Offshore wind Farms. Windpower Engineering and Development.

Wiser, R., Bolinger, M., 2016. 2015 Wind Technologies Market Report. Lawrence Berkeley National Laboratory, Berkeley, CA.

Wiser, R., Jenni, K., Seel, J., Baker, E., Hand, M., Lantz, E., Smith, A., 2016. Expert elicitation survey on future wind energy costs. Nat. Energy 1, 16135.

World Energy Council Issues Monitor. World Energy Council.

Wüstenhagen, R., Wolsink, M., Burer, M.J., 2007. Social acceptance of renewable energy innovation: an introduction to the concept. Energy Policy 35, 2683–2691.

# 5

# The use of geothermal energy for electricity generation

## Introduction

Geothermal energy is a type of renewable energy source independent of the Sun having natural heat located inside the Earth (see Fig. 5.1). In other words, the geothermal energy source is the thermal energy generated and stored inside the Earth from the hot rocks present in it. The geothermal energy of the Earth's crust originates from the original formation of the planet and the radioactive decay of materials. "The adjective *geothermal* originates from the Greek roots γῆ (*gê*), meaning earth, and θερμός (*thermós*), meaning hot" (Wikipedia Geothermal Energy, 2020). Why is that so hot? Because it is full of radioactive elements and metals like uranium and potassium. These materials are naturally decaying, and the process creates a lot of heat inside the Earth (Davis, 2017).

Without a doubt, geothermal power is a cost-effective, reliable, sustainable, and environmentally friendly energy source (Glassley, 2010) but has historically been limited to areas near tectonic plate boundaries. Recent technological advances have dramatically expanded the range and size of viable geothermal resources, especially for home heating applications, opening a potential for widespread exploitation of this type of renewable energy source.[a]

Is it a green energy source? The answer is no. Geothermal wells release greenhouse gases trapped deep within the Earth. Nevertheless, these emissions are much lower per energy unit than those released by fossil fuels used for electricity generation and heating.

It is important to stress that geothermal resources are theoretically more than adequate to satisfy humanity's energy needs, but only a very small fraction could be profitably exploited. Drilling and exploration for deep geothermal resources are very expensive. Forecasts for the future of geothermal power depend on assumptions about technology development, energy

---

[a] Geothermal power plants are most common in the Philippines and Iceland. The USA, Costa Rica, Lithuania, China, Japan, Mexico, France, Italy, Nicaragua, Russia, and Indonesia use geothermal power for heating, for generating electricity, or both. At least 70 countries use geothermal energy for heating.

**FIG. 5.1**   Geothermal power plant in Usulután Department, El Salvador, Central America. *Source: Courtesy of 2020. Mburch Wikipedia Geothermal Power. https://commons.wikimedia.org/wiki/File:Central_Geot%C3%A9rmica_de_Berl%C3%ADn_05.JPG.*

prices, energy subsidies, plate boundary movement, and interest rates. Pilot programs like EWEB's Customer Opt-in Green Power Program show that customers would be willing to pay a little more to use geothermal energy sources to reduce the level of contamination of cities and regions.

## Sources of geothermal energy

There are different sources of geothermal energy, depending on the structure of the Earth. However, these sources can be classified, according to their properties, as follows:

- Mechanical properties. Considering this type of property, the Earth can be divided into the lithosphere, asthenosphere, mesospheric mantle, outer core, and inner core;
- Chemical properties. Based on this type of property, the Earth can be divided into the crust, upper mantle, lower mantle, outer core, and inner core.

The geologic layers of the Earth are at the following depths below the surface (see Table 5.1) (Montagner, 2011).

Geothermal energy can be harvested from any of the layers mentioned above (see Fig. 5.2). The different layers of the Earth are the following:

- **The crust**. It is the Earth's outermost layer, what every living thing walks, slithers, or crawls every day. "The Earth's crust is an extremely thin layer of rock that makes up the outermost solid shell of our planet. In relative terms, it is thickness like that of the skin of an apple. It amounts to less than half of 1% of the planet's total mass but plays a vital role in most of Earth's natural cycles. The crust consists of many different rocks that fall into three main categories: igneous, metamorphic, and sedimentary. However, most of those rocks originated as either granite or basalt. The mantle beneath is made of peridotite. Bridgmanite, the most common mineral on Earth, is found in the deep mantle" (Alden, 2019). The crust forms the

**TABLE 5.1** Depths of Earth's mechanical and chemical properties.

| Depth (km) | Chemical layer | Depth (km) | Mechanical layer | Depth (km) |
|---|---|---|---|---|
| | | 0–80[a] | Lithosphere | 0–80[a] |
| 0–35[b] | Crust | | | |
| | | | | 0–10 |
| | | | | 10–20 |
| | | | | 20–80 |
| 35–670 | Upper mantle | | | |
| | | 80–220 | Asthenosphere | |
| | | 220–2890 | Mesospheric mantle | |
| | | | | 220–410 |
| | | | | 400–600 |
| | | | | 600–670 |
| 670–2890 | Lower mantle | | | |
| | | | | 670–770 |
| | | | | 770–2740 |
| | | | | 2740–2890 |
| 2890–5150 | Outer core | 2890–5150 | Outer core | 2890–5150 |
| 5150–6370 | Inner core | 5150–6370 | Inner core | 5150–6370 |

[a] Depth varies locally between 5 and 200 km.
[b] Depth varies locally between 5 and 70 km.
Source: 2020. Wikipedia Structure of the Earth.

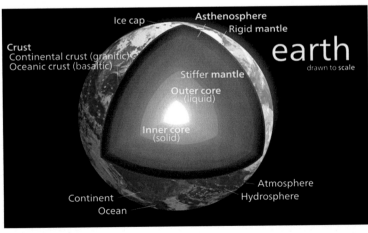

**FIG. 5.2** Layers of the Earth. *Source: Courtesy Kelvinsong https://commons.wikimedia.org/wiki/File:Earth_poster.svg.*

continents and ocean floors.[b] It can be between 4.8 km and 8 km thick under the oceans and between 24.2 km and 56.5 km thick on the continents;

- **The mantle (Upper and Lower mantle).** The mantle is divided into the Upper and Lower mantle (Evers, 2015), separated by the transition zone (Yu et al., 2018). The lowest part of the mantle next to the core-mantle boundary is known as the D″ (pronounced dee-double-prime) layer (Krieger, 2004). As there is intense and increasing pressure as one travels deeper into the mantle, the Lower mantle flows less easily than the Upper mantle. The Earth's mantle is a layer of silicate rock between the crust and the Outer core. Its mass is 67% of the mass of the Earth. It has a thickness of 2900 km, making up about 84% of Earth's volume (Katharina and Fegley, 1998). It is predominantly solid, but in geological time it behaves as a viscous fluid. Partial melting of the mantle at mid-ocean ridges produces oceanic crust, and the partial melting of the mantle at subduction zones produces continental crust, according to the paper entitled What is the Earth's Mantle Made Of? (2016). This layer surrounds the core and is about 2900 miles thick. It is made up of magma and solid rock. Heat is produced in the mantle due to the presence of magma[c];
- **The cores (Outer and Inner cores).** The Earth has two cores: the Outer and Inner cores. The Inner core is solid iron, while the Outer core is made up of insanely hot magma, possibly even hotter than the Sun.

## Types of geothermal power plants

There are three types of geothermal power plants (see Fig. 5.3). These are the following:

1. Liquid Dominated Geothermal Power Plants;
2. Vapor Dominated Geothermal Power Plants;
3. Binary Cycle Geothermal Power Plants.

Liquid Dominated Geothermal Power Plants are built upon liquid reservoirs within the Earth's surface. This liquid is sent through one or more separators or exchangers to lower the water's pressure, creating steam. The steam then propels a turbine generator, triggering it to produce electricity. The steam is then condensed back into a liquid and placed back into the liquid reservoir. This type of geothermal power plant is very common and provides a sustainable, reusable form of energy.

Liquid Dominated Geothermal Power Plant is also known as "Flash Steam Geothermal Power Plant." This type of geothermal power plant conducts flash steam by pressurizing hot water from the Earth's surface. Such geothermal power plants operate using water reservoirs with high temperatures. Liquid-dominated reservoirs are more common than other types of

---

[b] "Oceanic crust covers about 60% of the Earth's surface. Oceanic crust is thin and young – no more than about 20 km thick and no older than about 180 million years. Continental crust is thick and old – on average about 50 km thick and about 2 billion years old – and it covers about 40% of the planet. Whereas almost all of the oceanic crust is underwater, most of the continental crust is exposed to the air" (Alden, 2019).

[c] The amount of heat just within 10,000 m of the Earth's surface contains 50,000 times more energy than all the oil and natural gas resources located an all regions of the world.

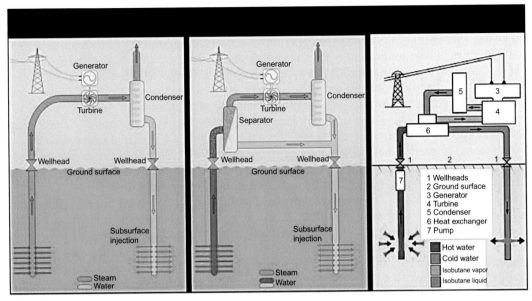

**FIG. 5.3** Conventional geothermal power plant types. *Source: Courtesy of Goran tek-en, Wikipedia, 2019. https:// en.wikipedia.org/wiki/Geothermal_power.*

geothermal power plants, and for this reason, they generate more electricity than others. This reservoir type is found in specific locations, including rift zones, mantle hot spots, and near new volcanoes in the Pacific Ocean.

Another type of geothermal power plant is the Vapor Dominated Geothermal Power Plant. This type of geothermal power plant is very rare but is incredibly efficient in electricity generation. A Vapor Dominated Geothermal Power Plant, also known as "Dry Steam Geothermal Power Plant," works similarly to a Liquid Dominated Geothermal Power Plant. After getting steam from the well, it is filtered and used for the rotation of a turbine. Then the steam reaches the condenser, where it is converted into a liquid. This liquid is further transferred to Earth by another well with the help of a pump. The generator is connected to the turbine, and electricity is produced. It is a vapor-dominated system. This system works well when the Earth's temperature is very high (Ambersariya, 2019).

It is important to mention that the Vapor Dominated Geothermal Power Plants are so rare that they exist in only two US locations: the Geysers in California and the famous dry steam reservoir held at Yellowstone National Park. However, the Yellowstone site is legally protected from geothermal development, so in reality, only one Vapor Dominated Geothermal Power Plant is used in the USA for electricity generation.

A Binary Cycle Geothermal Power Plant is used when the water in a reservoir is not hot enough to be transformed into steam. This lower temperature of the water is instead used to heat a liquid that expands when heated. This fluid increases the pressure around a generator, causing a turbine to turn, producing electricity. The fluid is recycled and used again to form a reusable energy source. This type of geothermal power plant is the most readily available geothermal resource as it does not require specific liquid or steam reservoirs. Due to the abundance of cold water reservoirs on the Earth's surface, Binary Cycle Geothermal Power

Plants have made up most geothermal power plants in generating electricity in the USA since 1960. In the case of Canada. The country does not currently produce electricity from geothermal power plants due to potentially higher upfront costs, competition from existing investments in other power generation types, and a lack of regulatory and policy support.

It is important to note that the Binary Cycle Geothermal Power Plants also create minimal air emissions due to the constant separation between the water from the Earth's surface and the working fluids used during the operation. This type of geothermal power plant requires two liquid water (low heated) and secondary liquid (with a low boiling point). That is why this type of geothermal power plant is called the Binary Cycle Geothermal Power Plant.

The choice of geothermal power plant design depends on the type of geothermal energy source found. For example, if the water comes out of the ground as steam already, it can be applied directly using the Vapor Dominated Geothermal Power Plant design.

Besides the three types of geothermal power plant designs mentioned above, a few other methods are used to capture geothermal heat energy not for electricity generation but for other purposes. One of the most common methods used to capture geothermal heat energy is the direct use of the Earth's heat. The geothermal hot water can heat buildings, raise plants in greenhouses, dry out fish and crops, de-ice roads, improve oil recovery, aid in industrial processes, and heat water for spas or fish farms. "Ancient Roman, Chinese, and Native American cultures used hot mineral springs for bathing, cooking, and heating. Today, many hot springs are still used for bathing, and many people believe the hot, mineral-rich waters have health benefits" (EIA Geothermal Explained, 2020).

Another common way to utilize geothermal energy is by using heat pumps. These pumps are primarily used for the heating or cooling of buildings. These are called "Ground-source heat pumps." They take advantage of the constant year-round temperature, which is just a few feet underground. In this case, air or antifreeze liquid is pumped through the pipes buried beneath the ground and recirculated throughout the building. It moves heat from one building to the ground in the summer, while it does the opposite in the winter.

There are also some industrial applications of geothermal energy sources. These applications include food dehydration (drying), gold mining, and milk pasteurizing (EIA Geothermal Explained, 2020).

## New developments and future achievements in the geothermal energy sector

Geothermal energy for electricity generation and heating is a renewable and diverse energy solution for the USA. It provides clean, reliable, and flexible electricity generation and delivers unique technology solutions for the country's heating and cooling demands. However, despite the benefits of geothermal energy and its ability to meet some of the nation's most pressing energy needs, the USA has tapped only a fraction of its abundant geothermal energy resources.

Without a doubt, harnessing US geothermal energy resources' full potential will strengthen domestic energy security and will allow the country to continue its leadership in energy innovation at the world level. For example, through regulatory reforms alone, geothermal capacity already installed in the country could increase significantly. "Streamlined regulations and permitting requirements can be achieved through a variety of mechanisms to shorten development timelines, which can—in turn—reduce financing costs during construction. For example, the analysis showed that placing geothermal regulatory and permitting requirements on a level similar to oil and gas and other energy industries could allow the geothermal industry to discover and develop additional resources and reduce costs. The GeoVision analysis demonstrated that optimizing permitting alone could increase installed geothermal electricity generation capacity to 13 GWe by 2050. With technology improvements that focus on exploring, discovering, developing, and managing geothermal resources, geothermal electric power generation could increase nearly 26-fold from today, representing 60 GWe of always-on, flexible electricity generation capacity by 2050. This capacity makes up 3.7% of total US installed capacity in 2050, and it generates 8.5% of all electricity generation" (GeoVision: Harnessing the Heat Beneath Our Feet, 2019).

Besides, it is important to know that geothermal energy sources can be integrated into all types of conventional and non-conventional power plants located in small, isolated towns, villages, or autonomous buildings and houses. The expansion in the use of geothermal energy for electricity generation, heating, and cooling in the North American region, particularly in the USA, will create job opportunities in urban and rural communities. The development of a strong residential and commercial geothermal heat pump industry could also expand the US geothermal workforce significantly.

Since geothermal energy sources typically provide baseload electric generation, integrating new geothermal power plants into existing power systems does not present a major challenge. Indeed, in some configurations, geothermal energy can provide valuable flexibility, such as the ability to increase or decrease production or startup/shut down as required. In some cases, however, the location dependence of geothermal resources requires new transmission infrastructure investments in order to deliver geothermal electricity to load centers.

The Technology Roadmap for Geothermal Heat and Power report (2011) offers a strategic plan to maximize this energy resource' deployment by 2050. According to the report mentioned above, 1400 TWh of electricity per year could come from geothermal energy sources by 2050. Separate studies carried out by the National Renewable Energy Laboratory (Geothermal Energy Basics, 2019) and the Massachusetts Institute of Technology (MIT) (Tester, 2006) concluded that over 100,000 MW could feasibly be reached in the next 15–50 years with a reasonable, sustained investment in research and development.

No integration problems have been observed for the direct use of geothermal energy sources for heating and cooling. The use of geothermal energy sources for heating and cooling is already widespread at the domestic, community, and district scales. District heating networks usually offer flexibility with regard to the primary energy source and can, therefore, use low-temperature geothermal resources or cascaded geothermal heat. The existing geothermal power installed capacity is about 16.8 GWth (Lund and Boyd, 2016) and is equivalent to geothermal heat pump installations in about two million households.

The GeoVision analysis determined that the market potential for geothermal heat pump technologies in the residential sector is equivalent to supplying heating and cooling solutions to 28 million households, or 14 times greater than the existing geothermal installed capacity. This potential is foreseen to represents about 23% of the total residential heating and cooling market share by 2050.

It is important to note that the economic potential of district heating systems that use geothermal resources directly combined with advances in EGS technology could be very high (more than 17,500 installations throughout the country). That is a very high figure compared to the total of 21 district heating systems currently installed in the USA in 2017 (Snyder et al., 2017). These district heating installations could satisfy the demand of about 45 million households (US Energy Information Administration, 2015; McCabe et al., 2019; Liu et al., 2019).

In the geothermal power sector, the main new developments that can be expected to occur in the future are the following:

- The improved energy conversion efficiency of geothermal power plants adapted to the reservoir temperatures on-site for conventional turbines. "The conversion efficiency of geothermal power developments is generally lower than that of conventional thermal power plants. The conversion efficiency is of significant importance when calculating the power potential of newly drilled geothermal wells and for resource estimation studies. The conversion efficiency is the ratio of net electric power generated (MWe) to the geothermal heat produced/extracted from the reservoir (MWth)"[d] (Moon and Zarrouk, 2012).
- Successful demonstration of the EGS on key sites and the dissemination of the technology used in these sites to other places and regions;
- Increased overall efficiency in geothermal combined heat and power plants. The electricity generation from a given geothermal fluid source depends on two elements. These elements are a) the fraction of the heat extracted from the geothermal fluid; b) the efficiency of converting the extracted heat to electricity. Different studies indicate that less than 40% of the ideal efficiency is realized in actual geothermal practice, and the ratio is as low as 30% for low reservoir or high ambient temperatures. In contrast, any other modern power technology can enjoy around 70% of its ideal efficiency limit. The geothermal energy practice has room to improve (Gurgency, 2009)[e];
- Improvement of exploration methods, installation technologies, and system components (pumps, pipes, turbines, etc.). In exploration, research and development are required to locate hidden geothermal energy sources and EGS prospects. Geothermal exploration is the search, in the Earth's subsurface, of potential sites to build a geothermal power plant. This power plant will use hot fluids from the Earth's subsurface to move turbines and produce electricity (Manzella, 2009).
  Geothermal exploration includes a broad range of geology, geophysics, geochemistry, and engineering methods (Hulen and Wright, 2001). Four geothermal elements characterize geothermal regions with adequate heat flow to fuel geothermal power plants. These elements are the following:
  - **Heat Source**. Shallow magmatic body, decaying radioactive elements or ambient heat from high pressures;
  - **Reservoir**. Collection of hot rocks from which heat can be drawn;

---

[d] It is commonly assumed that only 10% of the energy from the produced geothermal fluid can be converted to electricity (IEA Electricity Information 2007, 2007). Another study suggests that the power conversion efficiency from geothermal steam ranges from 10% to 17% (Barbier, 2002). However, it is important to stress that, according to different experts opinion, each geothermal power plant has its own conversion efficiency. For this reason, this percentage should be used with this element in mind.

[e] Without any doubt, one of the most important source for efficiency loss in geothermal power plants is the irreversibility in the heat exchangers.

- **Geothermal Fluid**. Gas, vapor, and water found within the reservoir;
- **Recharge Area**. The area surrounding the reservoir that rehydrates the geothermal system. The wider use of rapid reconnaissance tools, such as satellite-based systems, and radar sensors, could make geothermal exploration more effective and less costly.

Special research is needed to improve penetration rate when drilling hard rock and develop advanced slim-hole technologies and large-diameter drilling through ductile, creeping, or swelling formations. Drilling must minimize formation damage resulting from a complex interaction of the drilling fluid (chemical, filtrate, and particulate) with the reservoir fluid and formation. The objectives of new-generation geothermal drilling and well construction technologies are to reduce the cost and increase the useful life of geothermal production facilities through an integrated effort.

- Improvement and innovations in deep drilling are expected due to the deep drilling projects whose aims are to penetrate super-critical geothermal fluids, which can be a potential source of high-grade geothermal energy. The concept behind it is to flow super-critical fluid to the surface to change directly to super-heated (> 450 °C) hot steam at sub-critical pressures.

The future development of the geothermal heating sector includes the following:

- Improved site assessment, exploration, and installation of shallow systems;
- Further increase in the efficiency of ground-source heat pumps, optimized system concepts, application of advanced control systems, and improved components and materials such as compressors, refrigerants, pipes, etc.;
- Construction of new district-heating networks and optimization of the existing networks and plants;
- Increased application of innovative concepts related to the use of geothermal energy in agriculture, aquaculture, and industrial drying processes, etc.;
- The increased use of geothermal energy sources for de-icing and snow melting of roads and airport runways and the seawater desalination process.

## Geothermal energy installed capacity and electricity generation in the North American region

In 2018, the USA was the biggest electricity producer by geothermal energy at the world level[f] with 18,773 GWh, followed by Indonesia with 13,296 GWh and the Philippines with 10,435 GWh to BP. Statistical Review of World Energy 2020 report.

The USA's first geothermal power plant, The Geyser Complex, is located in northern California (San Francisco) and is still the world's largest geothermal power plant. In the beginning, it had 22 geothermal production plants with 350 wells with a capacity installed of 1590 MW. Today, the number of geothermal power plants currently operating within the region is 18, and the present combined generation capacity of these plants is 900 MW, enough to power 900,000 homes.

Most US's potential geothermal areas are located in Western states and Hawaii, where geothermal energy resources are close to the Earth's surface (see Fig. 5.4). The majority of

---

[f] Geothermal power plants generated in 2018 a total of 88,408 GWh.

### Geothermal resources of the United States

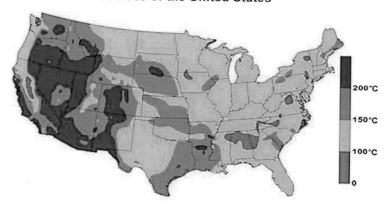

**FIG. 5.4**  Geothermal resources of the United States. *Source: Geothermal Resources in the United States (Map), n.d. US Department of Energy (DoE), Office of Energy Efficiency & Renewable Energy.*

Western states in the USA are rich in geothermal energy resources. There are many 'hot spots' throughout the region where the Earth's crust is thin, allowing the hot water to rise. That re-sults in natural hot springs and geysers, like those in Yellowstone National Park (Davis, 2017). The California state generates most of the electricity produced from the geothermal power plants operating in the country. As has been said before, Canada has no geothermal power plants in operation, but it has some regions where geothermal energy sources could be found and used for electricity generation and heating (see Fig. 5.5).

Fig. 5.4 shows Oregon, California, and Washington state having relatively high sub-surface temperatures, seemingly making the region suitable for geothermal development.

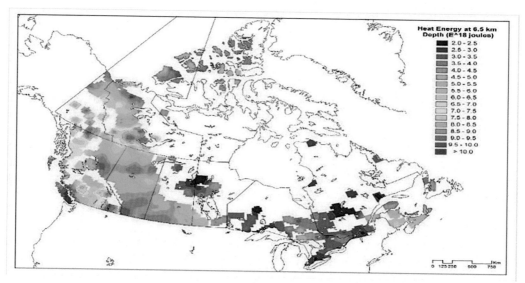

**FIG. 5.5**  Geothermal energy source in Canada. *Source: Geothermal Energy: A No-Brainer for Canada? Pembina Institute. https://www.pembina.org/blog/geothermal-energy-a-no-brainer-for-canada.*

According to the US EIA document entitled "California Profile Analysis (2020)," California is the top producer of electricity from geothermal resources in the USA and accounts for almost three-fourths of the nation's utility-scale electricity generation from geothermal energy sources. About 2730 MW of geothermal power plant capacity is installed at 43 operating power plants located in the state (California Geothermal Energy Statistics and Data, 2019; EIA Electric Power Monthly Table 1.16.B, 2019).

Four areas of California have substantial geothermal energy resources. These areas are the coastal mountain ranges north of San Francisco, volcanic areas of north-central California, near the Salton Sea in southern California, and along the state's eastern border with Nevada (California Known Geothermal Resource Areas, 2015). The Geysers, located in the Mayacamas Mountains north of San Francisco, is the largest complex of geothermal power plants globally (The Geysers, 2019).

Although Oregon has nearly no geothermal development, the US DoE ranked the state as the third-highest potential state for geothermal development in the country. This position is due to its tremendous potential development located east of the Cascade Mountains, according to the US EIA document entitled "Oregon Profile Analysis (2019)."

In 2019, the geothermal power plant capacity installed at the world level reached 13,909 MW, most of them installed in Asia (33%), North America (18.4%), and Eurasia (11.4%)[g] (see Fig. 5.6).

It is important to stress that in 2019 the world geothermal power plant installed capacity (13,909 MW) represents only 0.5% of the world's renewable energy installed capacity (2,532,866 MW), far below the world's wind installed capacity (622,408 MW) and the world's solar installed capacity (584,842 MW). According to industry estimates, geothermal energy sources' electrical potential is between 10 and 100 times what is currently being generated, not to mention the potential heating applications that could displace the use of fossil fuels

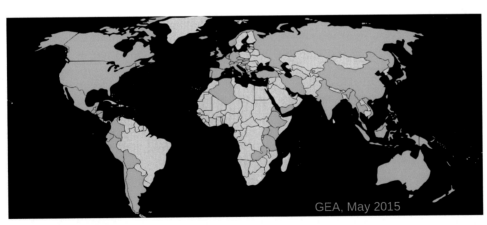

**FIG. 5.6**   Countries with geothermal power plants in operation or development in 2015. *Source: De Rfassbind, own work report May 2015, Public domain, https://commons.wikimedia.org/w/index.php?curid=40277287.*

[g] Twenty-seven countries were, in 2019, producing electricity by geothermal power plants. By regions the number of countries producing electricity using geothermal power plants were the following: Africa (2); Asia (4); Central America and the Caribbean (6); Eurasia (2); Europe (8); North America, including Mexico (2); Oceania (2); and South America (1).

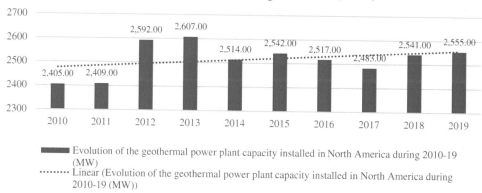

FIG. 5.7  Evolution of the geothermal energy capacity installed in the North America region (USA) during the period 2010–19. *Source: 2020. IRENA Renewable Energy Statistics 2020. IRENA.*

for this purpose. There is loads of room for growth in the geothermal energy sector. For this reason, it is important to single out that "geothermal would not be resource-constrained for at least 50 years, and the International Energy Agency predicts a growth of 28% or 4 GW over the next five years based on planned projects, half of which will happen in the Asia Pacific" (Nanalyze, 2019). The total geothermal power plant capacity installed in the North American region, without Mexico,[h] reached 2555 MW, all of them installed in the USA.

The evolution of the geothermal power plant capacity installed in the North America region (USA) during the period 2010–19 is shown in Fig. 5.7.

Geothermal power plants for electricity generation operating in the North American region are located only in the USA. The biggest geothermal power plant in the USA is located in California.[i] According to the data included in Fig. 5.7, the evolution of the geothermal power plant capacity installed in the region during the period 2010–19 increased by 6.2%, rising from 2405 MW in 2010 to 2555 MW in 2019. It is expected that this trend will continue in the future but at a slow pace. Changes in the annual geothermal installed capacity within the period under consideration are due to the number of wells in operation each year.

The total electricity generated by geothermal power plants at the world level in 2018 reached 88,408 GWh. The electricity generated by the geothermal power plants at the world level represents only 1.3% of the total electricity generated by renewable energy sources (6,586,124 GWh). In the North American region, the electricity generated in the same year reached 18,773 GWh or 21.2% of the total. The regions with the highest electricity generation

---

[h] Mexico has, in 2019, a geothermal energy capacity installed of 936 MW.

[i] In Canada, the first geothermal power plant is under construction in Saskatchewan. The US$50 million facility will power about 5000 homes and offset 27,000 tons of carbon dioxide per year—taking the equivalent of 7400 cars of yearly emissions out of the atmosphere. But in Alberta, despite its strong geothermal potential and energy sector expertise, the renewable power source has yet to take off, possibly due to the province's complex regulatory framework in force.

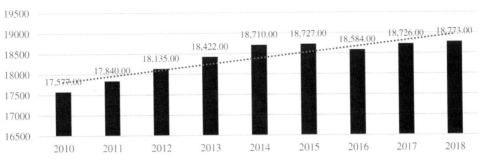

FIG. 5.8   Evolution of the electricity generated by the geothermal power plant in operation in the region during the period 2010–18. *Source: 2020. IRENA Renewable Energy Statistics 2020. IRENA.*

by geothermal power plants at the world level in 2018 were Asia with 29.8% of the total, followed by North America (without Mexico), with 21.2%, and Europe with 14.3%.

The evolution of the electricity generated by the geothermal power plant in operation in the region during the period 2010–18 is shown in Fig. 5.8.

According to Fig. 5.8, the following can be stated: the electricity generated by the only geothermal power plant in operation in the region increased by 6.8% during the period 2010–18, rising from 17,577 GWh in 2010 to 18,773 GWh in 2018. It is expected that a similar trend will be recorded during the coming years.

## Geothermal energy investment costs in the North American region

Without a doubt, geothermal energy is a renewable energy source with very high potential but is currently underrated for its electricity and heating/cooling energy generation. The fuel of geothermal energy, the heat emitted from the Earth's crust, is free. So once the geothermal power plant has been built, geothermal energy costs day-to-day operation is very low. According to Davis (2017), operation and maintenance costs associated with the functioning of a geothermal power plant ranges from US$0.01 to US$0.03 per kWh for an average geothermal power plant in the USA.[j]

---

[j] It is important to single out the following: despite it high upfront capital costs, geothermal power is cost competitive today, compared with other energy sources. Generating electricity from a geothermal heat source costs from US$0.05 to US$0.07 per kilowatt hour, which is comparable to what power from a new fossil fuel plant would cost. However, geothermal electricity generation costs is not competitive with wind and solar energies costs.

However, the upfront capital costs associated with geothermal power plants are high. Besides high upfront capital costs and long payback periods, geothermal power plants have the largest commercially available minimum power units. While solar PV parks and wind farms can be built in smaller modules, geothermal power plants only come online in large capacity, translating to larger investments and more commitment capital upfront. That, combined with difficulties in assessing the resources available in reservoirs, makes geothermal energy sources a less popular investment worldwide, even in places with lots of volcanic activity.

The installed costs of geothermal power plants can be divided into five main components. These components are the following:

- Exploration and assessment of resources of the site;
- Well drilling;
- Field infrastructure;
- Construction of production facilities;
- Grid connection.

Zion Market Research published in 2018 a new report titled "Geothermal Energy Market by Type (Direct Dry Steam Plant, Flash Plant, and Binary Plant) and by Application (Residential, Industrial, and Commercial): Global Industry Perspective, Comprehensive Analysis, and Forecast, 2018–2025". According to the report mentioned above, the world geothermal energy market was valued at approximately US$4 billion in 2018. It is expected to generate around US$9 billion by 2025 (an average increase of around 11% between 2019 and 2025).

Geothermal energy is a renewable energy source, and one of its biggest advantages is its constant availability. The continuous flow of heat from the Earth ensures an unlimited supply of energy for billions and trillions of years to come. The global geothermal energy market is likely to experience significant growth in the future, owing to the rapid depletion of conventional energy resources, such as oil, natural gas, and coal, and the negative impact on the environment and population that has the use of coal and oil for electricity generation. The overutilization of these energy resources has led to the scarcity of fossil fuels and their extraction at the world level (Zion Market Research, 2019).

Additionally, fossil fuels lead to increased carbon emissions, which contribute to air pollution and global warming. To reduce the current level of world pollution, governments and regulatory and civic bodies have formulated stringent rules and regulations regarding sustainability and eco-friendliness. Without a doubt, these norms and regulations will further stimulate the use of renewable energy resources for electricity generation and heating and reduce fossil fuel participation in many countries' energy mix in the future.

Besides, the increased use of geothermal energy for heating and cooling purposes of large infrastructural buildings "is also contributing to the geothermal energy market growth. However, the high installation costs of geothermal energy may hamper the market growth in the future" (Zion Market Research, 2019).

The geothermal flash power plant segment is expected to dominate, in the coming years, the global geothermal energy market due to the high demand for clean and sustainable energy.

North America held an extensive share of the global geothermal energy market in 2018 and is expected to continue occupying this position over the upcoming years, particularly in the USA. This regional geothermal energy role can be attributed to the rising clean energy demand and increasing emission reduction measures adopted by the US and Canadian governments. Moreover, the growing demand for cooling and heating systems will positively impact this regional geothermal market growth (Zion Market Research, 2019).

# Geothermal energy construction and generation costs in the North American region

According to the US DoE, the initial costs for the land and construction of a geothermal power plant in the USA is roughly US$2500 per installed kW. There is only one geothermal power plant in operation in the North American region (in the USA), with a total capacity, in 2019, of 2555 MW. However, according to Nanalyze (2019), an additional geothermal power plant with 1.2 GW capacity is under development in the USA. When this new geothermal power plant enters into operation, the geothermal capacity installed in the USA will increase to around 3.8 GW.

It is important to single out that for a geothermal well to be commercially viable, it needs to produce power at an LCOE lower than today's typical power price level of US$0.06 to US$0.07 per kWh.

Developers experience the effects of geothermal energy's high initial capital investment before drilling even begins. The most immediate costs are identification costs. These costs take the form of exploration and assessment of the site resources. The first step in creating a geothermal well costs between 6% and 8% of the overall capital investment associated with the construction and operation of a geothermal power plant. Still, according to the US DoE report entitled "Energy Efficiency and Renewable Energy Geothermal Program. Geothermal Risk Mitigation Strategies Report (2008)", it has an extremely low sucess rate. Why? In large part, the failure rate is high because there is very little reliable and widely available information to those developers who wish to explore the site.

However, the major area of capital investment risk associated with developing a geothermal energy project is drilling wells. Drilling at any stage in the development process poses a prominent area of risk to developers. Forecasting this risk level for any specific site is generally a matter of speculation or an educated guess based on empirical data from similar projects. Additionally, often drilling a high-grade well field accounts for around 30% of total capital investment. In comparison, low-grade wells can have a drilling cost that accounts for 60% or more of the project's total capital investment (Tester, 2006).

The risk associated with drilling a well comes from various failures in the drilling or exploration process. Circumstances that constitute well failure may consist of little to no self-flow (also known as dry-hole). A well with this symptom cannot be pumped with an external water source because the casing's internal diameter is too narrow to accommodate a pump., and the heated fluid temperature exceeds the pump's limit. Additionally, a well that produces fluid to cool for commercial use would also be considered unsuccessful (Tester, 2006).

It is important to single out that high upfront capital costs, high-grade wells failures, and exploratory failures registered during the exploration and assessment of the geothermal resources phase of development create gaps in available financing for geothermal projects. This gap creates two significant challenges for a developer:

- First, there is no risk-sharing mechanism. Therefore, the developer must bear a disproportionate share of project risk compared to other competing energy investments[k];
- Second could be a money gap created as a result of the risk associated with the well construction. In this case, developers are not getting enough money needed to push the project past the early drilling phases, according to the US DoE document entitled "Energy Efficiency and Renewable Energy Geothermal Program. Geothermal Risk Mitigation Strategies Report (2008)."

These financial issues effectively multiply the impact of failure for those risking development.

What to do to reduce the above risks to the minimum? One remedy is to gain more information about their specific drilling site. As the quality and quantity of site-specific information increases, future wells will become more accurate in their risk assessments and increase successful wells' development. According to the experience gained by geothermal energy developers, the drilling success rate fluctuates heavily, around 33% in the exploration phase. It increases to a more consistent 70% average rate of success within the operations phase. The development stage's drilling success rate (after five wells) is expected to be between 60% and 100%. After the operational stage (40 wells), the rate is expected to be about 90% (Tester, 2006).

Without a doubt, the USA has been slow to include geothermal power into its energy mix in the past.[l] This slow progress is due, in part, to the high initial capital investment and high risk during each stage of the development process associated with the use of geothermal energy for electricity generation and heating. "The result of this high risk and high-cost dilemma has caused only the most secure and high-grade geothermal energy projects to be undertaken, leaving numerous productive geothermal regions untapped along the Pacific Coast of the USA" (Rannestad, 2017).

## The efficiency of the geothermal power plants in the North American region

Geothermal energy is emerging as one of the most reliable renewable energy sources. It is gaining prominence over conventional and non-renewable energy sources because of its eco-friendly nature and constant availability. Geothermal reservoirs are replenished naturally and have massive power generation potential, meeting the baseload demand for energy by generating up to 2TWh of power globally. Without a doubt, the use of geothermal energy resources in electricity generation and heating will continue to increase due to the high-level

---

[k] There are seven risks involved in geothermal steam production business. These risks are the following: "resource risk, facility risk, environmental risk, natural disaster risk, sales risk, location risk, and industrial accident risk." (Adachi, 2011).

[l] Geothermal development within the USA lags significantly behind other types of renewable energy sources used for electricity generation. Geothermal energy's growth rate in the USA stands at a modest 5%. Other forms of renewable energy such as wind and solar PV have growth rates in the range of 20–40%.

efficiency (above 70%)[m] and low maintenance requirements. Thus, the advantages of geothermal energy are expected to drive market growth during the coming years.

When analyzing a power plant's overall efficiency, the capacity factor is often used to understand better how well the power plant operates. The capacity factor of a power plant is defined as the ratio between the electricity produced in a given period and the maximum, ideal amount of electricity that could be produced in the same period. A power plant's efficiency is an easily measured quantity, making it an effective tool to compare different power plants.

It is important to mention that geothermal power developments' conversion efficiency is generally lower than conventional thermal power plants' conversion efficiency. The conversion efficiency is of significant importance when calculating newly drilled geothermal wells' power potential and resource estimation studies (Moon and Zarrouk, 2012). In geothermal power plants, the energy input can be defined as the total mass of fluid (kg/s) multiplied by the average enthalpy (kJ/kg) as shown below:

$$\eta act(\%) = W / \dot{m} \times h \times 100$$

Where W is the running capacity (kWe), $\dot{m}$ is the total mass flow rate (kg/s), and h is the reservoir enthalpy (kJ/kg).

Several factors affect the conversion efficiency of geothermal power plants. These factors are, among others, the following:

- "System design;
- Non-condensable gas content;
- Heat loss from equipment, turbines, and generators efficiencies;
- Parasitic load;
- Weather" (Moon and Zarrouk, 2012).

## Uses of geothermal energy in the North American region

Geothermal energy has two primary applications: heating/cooling and electricity generation, according to the US DoE Geothermal Energy Basics (2019). Ground source heat pumps for heating and cooling use 75% less energy than traditional heating and cooling systems, according to Geothermal Benefits (2019). Unfortunately, geothermal energy dependence on locations with specific conditions, sometimes far from cities and towns, considerably limits its use as a heating and cooling source.

Geothermal heat pumps (GHPs) are the primary method for the direct use of geothermal energy. GHPs use the shallow ground as an energy reservoir because it maintains a nearly constant temperature. GHPs transfer heat from a building to the ground during the cooling season and from the ground into a building during the heating season, according to Geothermal Heat Pump Basics (2019) (see Fig. 5.9).

---

[m] Compared to a coal fired power plant with a capacity factor typically averaging around 60% or a natural gas plant at 45%, it is apparent that geothermal power plants can be more efficient than traditionally fueled power plants. Having a higher capacity factor means that the plant performs closer to its ideal, peak level of performance.

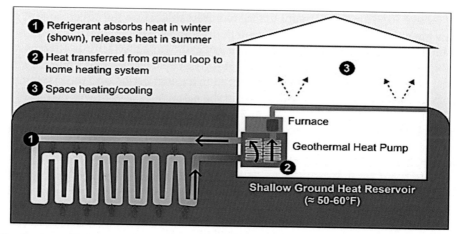

FIG. 5.9   Geothermal heat pump in a residential heating application. *Source: Geothermal Exchange Organization, Inc., 2010. Home Heating With GeoExchange. Geothermal Energy Factsheet. Pub. No. CSS10-10. Center for Sustainable Systems, University of Michigan, USA.*

According to Geothermal Heating and Cooling Technologies (2019) paper, other geothermal energy direct-use applications include space and district heating, greenhouses, aquaculture, and commercial and industrial processes.

## Advantages and disadvantages in the use of geothermal power plants in the North American region

The use of geothermal energy sources for electricity generation has certain advantages and disadvantages like any other energy source.

A-Advantages

The main advantages are, according to Sparrow (2012), TWI (2020), and Miller (2020), the following:

- Geothermal energy can be extracted without burning fossil fuels like coal, gas, or oil;
- Geothermal energy is a reliable energy source. Energy generated from this type of energy source is easy to calculate since it does not fluctuate in the same way as other renewable energy sources, such as solar and wind energies. For this reason, the power output from a geothermal power plant can be predicted with a high degree of accuracy (TWI, 2020);
- While solar and wind energies are most useful when the Sun is shining and the wind is blowing, geothermal energy is always available 100% of the time. In other words, a geothermal power plant can generate electricity 24 h a day, seven days a week. That makes geothermal power a viable and renewable source of baseload power. Geothermal provides a reliable energy source compared to other renewable resources such as wind and solar power. That is because the resource is always available to be tapped into, unlike wind or solar energy (TWI, 2020);

- Geothermal energy for electricity generation is relatively inexpensive during operation. Savings from direct use of the geothermal energy source can be as much as 80% over the cost of the electricity generated by fossil fuel power plants. Also, a home geothermal pump can cut energy bills by 30–40%;
- Geothermal energy is free and abundant. The constant flow of heat from the Earth makes this resource inexhaustible and limitless for the next four billion years;
- Geothermal energy is a non-polluting energy source and environment-friendly as no harmful gases are emitted from geothermal power plant operation. In the case of Binary Cycle Geothermal Power Plants essentially release no emissions. Also, no residue or by-product is generated during the operation of a geothermal power plant. The fields in which geothermal power plants are stationed only produce about 1/6th of the carbon dioxide that a clean natural gas-fueled power plant would produce;
- Geothermal power plants are highly sophisticated and involve large-scale research before installation. This research activity generates employment for skilled and unskilled workers at a very large scale at each production stage and management of the geothermal power plant;
- In cold countries, geothermal energy can be used directly for melting ice on the roads, heating houses in winters and cooling in summers, greenhouses, public baths, etc. Although the initial cost of installation is very high, the cost for maintenance and repair is small;
- Geothermal power offers a significantly longer lifespan. Even if a geothermal power plant's installation cost is at the end of the cost curve, a break-even point at some point can be reached. The indoor components of residential and commercial systems have a rated lifespan of at least 25 years (Miller, 2020);
- Geothermal energy sources can be used to ensure food and water security;
- Geothermal power plants do not produce waste or generated by-products during operation (Sparrow, 2012);
- Geothermal power plants do not occupy too much space and help protect the natural environment (Sparrow, 2012);
- There is a great deal of exploration into geothermal energy at the moment, meaning that new technologies can be developed to improve the geothermal energy process. There is an increasing number of projects to improve and grow the geothermal energy industry sector. With this rapid evolution, many of the current disadvantages of geothermal energy will be mitigated (TWI, 2020);
- There are no cost fluctuations like what happens with gas or oil (Miller, 2020).

B-Disadvantages

The main disadvantages are, according to Sparrow (2012), Max (2020), and Miller (2020), the following:

- Unlike fossil fuels, geothermal energy cannot be transported easily. Once the tapped energy is harnessed, it can only be used efficiently in nearby areas. Also, there are chances of the emission of toxic gases getting released into the atmosphere with the transmission;
- The installation of geothermal power plants to get steam from deep under the Earth requires a huge investment in material and human resources;

- Before setting up a geothermal power plant, extensive research is required. The reason is simple: sites can run out of steam over time due to a drop in the temperature due to excessive or irregular inlet water supply or if the fluid is removed faster than replaced;
- The source of geothermal energy is available in limited regions, some of which are highly inaccessible, such as high-rise mountains and rocky terrains, which renders the process economically unfeasible in some regions;
- Geothermal energy resources are present deep under the Earth, so drilling may release highly toxic gases into the environment near the geothermal sites, which sometimes prove fatal to the workforce involved in the process. The biggest concern is hydrogen sulfide release, a gas that smells like rotten eggs at low concentrations. While not as damaging to the environment as other gases,[n] the smell is pretty bad. Another issue is the disposal of some geothermal fluids, which may contain low levels of toxic materials;
- Most geothermal sites are far from markets or cities, where it needs to be consumed (Sparrow, 2012);
- To install a geothermal power plant, a certain amount of land is needed. For this reason, it is difficult to incorporate geothermal systems in major urban areas for homeowners. In such cases, a heat pump is needed from a vertical ground source (Max, 2020);
- It is impossible to achieve carbon neutrality using geothermal power for electricity generation and heating/cooling. The goal of geothermal power is to save energy compared to what gets spent on a traditional fossil fuel-based power plant. A geothermal power plant provides that outcome exceptionally well, but it would also be incorrect to call this technology a carbon-neutral system. Closed-loop systems still require electricity to provide heating and cooling benefits to a structure, and an external source should provide this electricity. Water must be pumped (and sometimes antifreeze) through the equipment to transfer energy efficiently, and this action also requires the supply of electricity provided by an external source (Miller, 2020);
- There is no assurance that a geothermal power plant's energy amount will justify the initial high capital investment involved (Sparrow, 2012);
- The possibility of earthquakes is one of the disadvantages of using geothermal energy for electricity generation and heating/cooling. This possibility is attributed to changes in the Earth's structure arising from digging. Advanced geothermal power, which pushes water into the Earth's crust to fissure further using the resource, presents a greater challenge. "But because the bulk of geothermal plants are situated away from populated centers, the impact of these earthquakes is limited" (Max, 2020).

## Environmental impact of the use of geothermal energy source for electricity generation in the North American region

The environmental impact of geothermal energy is minimal, especially compared to fossil fuel power plants used for electricity generation. When sited and constructed carefully, geothermal power plants can be reliable renewable and environmentally friendly electricity sources. According to the EIA source, the environmental effects of using geothermal energy

---

[n] These greenhouse gases are carbon dioxide, hydrogen sulfide, methane, and ammonia (Max, 2020).

sources for electricity generation, heating, and cooling depend on how geothermal energy is used or converted to useful energy. Direct use applications and geothermal heat pumps have almost no negative effects on the environment. On the contrary, they can have a positive effect by reducing the use of other energy sources that may have negative effects on the environment.

Why? The answer is very simple. Geothermal power is a clean and renewable energy source because it does not burn fuel to generate electricity, so the levels of air pollutants they emit are low. "Geothermal power plants emit 97% less acid rain-causing sulfur compounds and about 99% less carbon dioxide than fossil fuel power plants of similar size. Geothermal power plants use scrubbers to remove the hydrogen sulfide naturally found in geothermal reservoirs. Most geothermal power plants inject the geothermal steam and water that they use back into the Earth. This recycling helps to renew the geothermal resource" (EIA Geothermal Energy and the Environment, 2019).

One way to compare the environmental impact of different electricity generation technologies is to analyze their life-cycle greenhouse gas emissions. That means considering the system's full life.

The Intergovernmental Panel on Climate Change (IPCC) released a report in 2014 covering climate change mitigation. The IPCC report includes Fig. 5.10 with the life-cycle emissions by electricity-generating source.

For all of the energy technologies included in Fig. 5.10, at least five studies are reviewed. The empirical basis for estimating the emissions associated with geothermal energy is much weaker than others. The Special Report on Renewable Energy Sources and Climate Change Mitigation reported between 6 and 79 $gCO_2eq/kWh$ for geothermal power (Sathaye et al., 2011). According to the IPCC report mentioned above, geothermal power plants have life-cycle emissions of 38 $gCOeq/kWh$, one of the highest among renewable energy sources. But it is 95% less than coal (820 $gCOeq/kWh$) and 92% less than gas (490 $gCOeq/kWh$).

Geothermal power plant emissions compared to other renewable sources are shown in Table 5.2. It is important to single out that, according to Table 5.2, geothermal power plants occupy the fourth place within the five renewable energy sources included in the mentioned table, only below solar PV. The life-cycle emissions for renewable technologies are included in Table 5.2.

One of the main environmental concerns associated with the use of geothermal power plants for electricity generation is surface instability. During the operation of a geothermal power plant, it removes water and steam from reservoirs within the Earth to generate electricity. For this reason, the land above those reservoirs could sink slowly over time if no action is taken to avoid that this happens. One of the measures is to re-inject used water into the Earth via an injection well. This measure significantly reduces the risk of land subsidence.

An additional problem that can arise during the operation of geothermal power plants is the occurrence of earthquakes. Geothermal power plants are usually built near fault zones or geological "hot spots" characterized by instability and different earthquakes. Besides, drilling wells deep into the Earth and removing water and steam to ensure a geothermal power plant's operation could trigger small earthquakes that could affect the power plant's functioning.

However, regardless of what has been said above, there should be no doubt, increased geothermal deployment could improve air quality and reduce $CO_2$ emissions. The GeoVision analysis "indicates opportunities for improved air quality resulting from reductions in sulfur

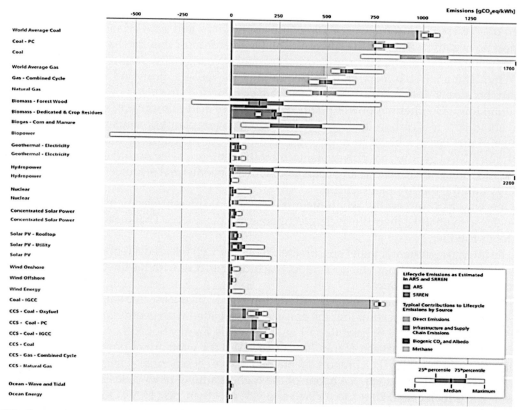

**FIG. 5.10** Comparative life-cycle greenhouse gas emissions from electricity supplied by commercially available technologies (fossil fuels, renewable, and nuclear power) and projected emissions of future commercial plants of currently pre-commercial technologies (advanced fossil systems with CCS and ocean energy). Abbreviations: AR5—IPCC WG III Fifth Assessment Report, CCS—$CO_2$ capture and storage, IGCC—integrated coal gasification combined cycle, PC—pulverized hard coal, PV—photovoltaic, SRREN—IPCC Special Report on Renewable Energy Sources and Climate Change Mitigation. *Sources: Bruckner, T., Bashmakov, I.A., Mulugetta, Y., Chum, H., de la Vega Navarro, A., Edmonds, J., Faaij, A., Fungtammasan, B., Garg, A., Hertwich, E., Honnery, D., Infield, D., Kainuma, M., Khennas, S., Kim, S., Nimir, H.B., Riahi, K., Strachan, N., Wiser, R., Zhang, X., 2014. Energy systems. In: Edenhofer, O., Pichs-Madruga, R., Sokona, Y., Farahani, E., Kadner, S., Seyboth, K., Adler, A., Baum, I., Brunner, S., Eickemeier, P., Kriemann, B., Savolainen, J., Schlömer, S., von Stechow, C., Zwickel, T., Minx, J.C. (Eds.), Climate Change 2014: Mitigation of Climate Change. Contribution of Working Group III to the Fifth Assessment Report of the Intergovernmental Panel on Climate Change. Cambridge University Press, Cambridge, and New York, NY. https://www.ipcc.ch/site/assets/uploads/2018/02/ipcc_wg3_ar5_full.pdf. SRREN (IPCC, 2012. Renewable Energy Sources and Climate Change Mitigation. Special Report of the Intergovernmental Panel on Climate Change (2012). Technical Support Unit Working Group III. Potsdam Institute for Climate Impact Research (PIK). Cambridge University Press, Cambridge.), geothermal power (Sathaye, J., Lucon, O., Rahman, A., Christensen, J., Denton, F., Fujino, F., Heath, G., Kadner, S., Mirza, M., Rudnick, H., Schlaepfer, A., Shmakin, A., 2011. Renewable energy in the context of sustainable development. In: Edenhofer, O., Pichs-Madruga, R., Sokona, Y., Seyboth, K., Matschoss, P., Kadner, S., Zwickel, T., Eickemeier, P., Hansen, G., Schlömer, S., von Stechow, C. (Eds.), IPCC Special Report on Renewable Energy Sources and Climate Change Mitigation, Prepared by Working Group III of the Intergovernmental Panel on Climate Change. Cambridge University Press, Cambridge, and New York, NY).*

TABLE 5.2    Geothermal power plant emissions compared to other renewable sources.

| Technology | Lifecycle emissions (gCO$_2$eq/kWh) |
| --- | --- |
| Wind | 11 |
| Hydropower | 24[a] |
| Concentrated solar | 27 |
| **Geothermal** | **38** |
| Solar PV | 48 |

[a] *Hydropower has several environmental consequences that are still being studied, and the actual life-cycle emissions for a hydropower plant have the potential to be significantly higher than 24 gCO$_2$eq/kWh.*
*Source: Bruckner, T., Bashmakov, I.A., Mulugetta, Y., Chum, H., de la Vega Navarro, A., Edmonds, J., Faaij, A., Fungtammasan, B., Garg, A., Hertwich, E., Honnery, D., Infield, D., Kainuma, M., Khennas, S., Kim, S., Nimir, H.B., Riahi, K., Strachan, N., Wiser, R., Zhang, X., 2014. Energy systems. In: Edenhofer, O., Pichs-Madruga, R., Sokona, Y., Farahani, E., Kadner, S., Seyboth, K., Adler, A., Baum, I., Brunner, S., Eickemeier, P., Kriemann, B., Savolainen, J., Schlömer, S., von Stechow, C., Zwickel, T., Minx, J.C. (Eds.), Climate Change 2014: Mitigation of Climate Change. Contribution of Working Group III to the Fifth Assessment Report of the Intergovernmental Panel on Climate Change. Cambridge University Press, Cambridge, and New York, NY. https://www.ipcc.ch/site/assets/uploads/2018/02/ipcc_wg3_ar5_full.pdf.*

dioxide (SO$_2$), nitrogen oxides (NOx), and fine particulate matter (PM$_{2.5}$) emissions." The report mentioned above "further identifies opportunities for reduced carbon dioxide emissions. For the electric sector, this could cumulatively result in up to 516 million metric tons of avoided carbon dioxide equivalent (CO$_2$) emissions through 2050. The heating and cooling sector's impacts through 2050 could cumulatively include up to 1281 million metric tons of CO$_2$ emissions avoided. By 2050, the combined CO$_2$ reductions for the two sectors is equivalent to removing about 26 million cars from the road annually" (GeoVision: Harnessing the Heat Beneath Our Feet, 2019).

Finally, it is important to stress the following: on average, a coal power plant emits roughly 35 times more CO$_2$ per kWh electricity generated than a geothermal power plant in the USA. Each year, US geothermal electricity offsets the emission of 4.1 million tons of CO$_2$, 80 thousand tons of nitrogen oxides, and 110 thousand tons of particulate matter from coal-powered plants. The US DoE is actively funding research into combining carbon capture and storage with geothermal energy production. However, the risks of long-term and high-volume geologic carbon sequestration are uncertain.

Some geothermal facilities produce solid waste that must be disposed of in approved sites, though some by-products can be recovered and recycled.

## The future of geothermal energy in the North American region

There are many new and exciting developments in the geothermal energy sector in the North American region. The Enhanced Geothermal System (EGS) will greatly expand the use of geothermal energy sources for electricity generation and heating to areas where this type of energy source has not yet been exploited as it should be. Without a doubt, EGS has gained traction in the geothermal spotlight in the past few years. The EGS operates in the following manner: drill two wells into an ultra-hot basement or volcanic rock. One of the wells (injection well) injects cold water at high rates, causing the surrounding rock to fracture. This process

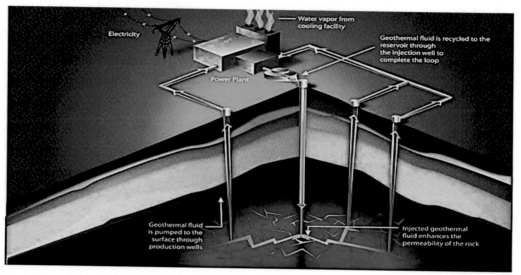

**FIG. 5.11**    Enhanced geothermal system schematic. *Source: Office of Energy Efficiency and Renewables Energy. https://www.energy.gov/eere/articles/geothermal-energy-glance-back-and-leap-forward.*

is known as "hydroshearing." If the hydroshearing fractures develop according to plan, the fracture network created will communicate with the second well (the producer well). The water injected heats up within the induced fractures and is produced from the producer well. By continuing to inject water, the fracture network will remain open (see Fig. 5.11). It is important to stress that the development of the EGS "requires a strong understanding of the geomechanical properties and stresses of the rock to estimate the direction of growth of the fracture network" (Hirschmiller, 2019).

Also, low-temperature and co-production of geothermal electricity can be produced from oil and gas wells. There are also significant amounts of high-temperature water or suitable high-pressure conditions to harvest this energy in areas with large oil or gas reservoirs. Oil and gas reservoirs also contain high amounts of geothermal fluid, which can be utilized in direct-use applications.

A report released in 2019 by the US DoE[o] suggests that the country's geothermal power capacity could increase by more than 26-fold by 2050, reaching a total installed capacity of 60 GW, thanks to accelerated technological development and adoption. The report mentioned above also demonstrates the benefits of geothermal power for residential and industrial heating. Energy Secretary Rick Perry announced his Department had provided funding for a US$140 million research facility at the University of Utah on human-made geothermal energy (McCombs, 2019).

At present, Canada remains the only major country in the Pacific Rim that is not producing electricity from its geothermal resources (IRENA Renewable Energy Statistics 2020, 2020). Although the colder it is outside, the more electricity a geothermal power plant can produce. For this reason, it is important to single out that the larger the temperature differentials

---

[o] See the US DoE report entitled "New Study Highlighting the Untapped Potential of Geothermal Energy in the United States" (2019).

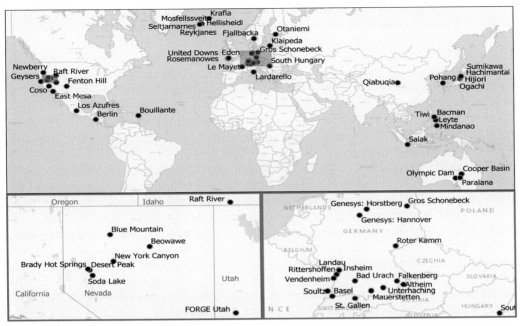

**FIG. 5.12**    Map of 64 EGS projects around the world. *Source: By Ahinoam Pollack – Own work, CC BY-SA 4.0, https://commons.wikimedia.org/w/index.php?curid=92207180.*

between the geothermal resource and the ambient air temperature, the more efficiently geothermal plants operate. That is why geothermal power is an ideal energy source for electricity generation and heating for cold countries.

Beyond question, the future outlook for expanded production from conventional geothermal energy sources and EGS in the North American region is positive as new technologies promise increased growth in locations previously not considered for the construction of a geothermal power plant (see Fig. 5.12).

According to Fig. 5.12, the following can be stated: from the 64 EGS projects worldwide, ten are located in the USA (15.6%).

## Geothermal energy capacity in Canada

There is no geothermal power plant operating in Canada in 2020, but this situation could change in the coming years. According to Canada's First Geothermal Production and Injection Well Test Exceeds Expectations—First 20 MW Facility in Design Phase (2020) paper, prepared by DEEP Earth Energy Production Corp., "the spring/summer flow testing program indicates the temperature and flow rates from the geothermal reservoir in the Deadwood Formation are sufficient to support multiple geothermal power plants." The construction of these facilities commenced with drilling the deepest horizontal well-to-date in Saskatchewan's history, allowing for installing a large diameter submersible pump. "This initial horizontal well will be the first step in constructing the first geothermal power facility in Canada. Early power generation technology is being sourced for immediate self-generation, supplying power for drilling, testing, and construction activities for the Williston Basin's first 20 MW geothermal power plant, which can supply enough power for approximately 20,000 households".

Undoubtedly, Canada is given high priority to implementing its new energy policy to reduce polluting gases. To achieve that goal, it is indispensable that the country increases the use of all available renewable energy sources, particularly geothermal energy sources, for electricity generation and heating/cooling.

It is important to stress that Canada has substantial potential for geothermal energy development (Grasby et al., 2012). To date, the development of geothermal energy in the country has all been for heating applications.

Even though Canada currently has no commercial geothermal electricity production, it has geothermal potential from a reservoir perspective. The Western Canadian Sedimentary Basin and the Williston Basin in southern Saskatchewan are vast areas with numerous porous and permeable rocks at temperatures greater than 100 °C. In other parts of Canada, such as the interior British Columbia and the Yukon area, plutonic rocks have heated groundwater, which has been utilized in commercial hot springs. Areas around these hot springs may be potential targets if sedimentary strata reside nearby.

Summing up, the following can be stated: many Canadian locations are home to natural conditions required to extract geothermal energy. These locations are British Columbia, Alberta, Saskatchewan, Yukon, and the Northwest Territories.

According to Wikipedia Geothermal power in Canada (2020), Rudyk (2020), and Hirschmiller (2019), several projects are underway in varying states of completion. These projects are the following:

- Alberta No. 1 Geothermal Project. In August 2019, Alberta's government announced the first conventional geothermal power plant construction near Grande Prairie. It will have a 5 MW capacity and will provide power and heat to a nearby industrial park. Terrapin Geothermics is leading the project in partnership with the Municipal District of Greenview and PCL Construction. The heat value of the geofluid is equivalent to the energy released from burning 350,000 barrels of oil per year;
- Razor Energy Corp. A second Alberta geothermal power plant with a 21 MW capacity will operate under the concept of co-production (Wikipedia Geothermal Power in Canada, 2020), where hot geofluids are pumped to the surface during oil and gas production. These hot fluids are being tested for power generation;
- Deep Earth Energy Production Corp. (DEEP), Saskatchewan. In January 2019, the Canadian government announced it would provide C$25.6 million in funding to develop a 5 MW geothermal power plant near Estevan, Saskatchewan. DEEP leads the project, and it is expected to be completed in early 2022. The project is set to power 5000 homes a year while offsetting about 40,000 tonnes of carbon emissions, the equivalent of taking about 8000 cars off the road;
- Eavor-Lite Project, Alberta. This is a pilot project in Rocky Mountain House, Alberta, with a capacity of 10 MW. Canadian company Eavor Technologies Ltd. has drilled a pilot well with a new geothermal technology near Rocky Mountain House, Alberta. "They have utilized oil and gas horizontal drilling technology to design a closed-loop system that uses conduction to heat fluid running through the wells rather than producing hot formation water. The system does not produce any brine or rely on hydraulic fracturing.

The fluid in the wells are closed off to the reservoir, and the only interaction between the reservoir is heat through conduction" (Hirschmiller, 2019);

- In January 2020, Eavor Technologies and the Little Salmon Carmacks First Nation entered into a partnership to develop a 3 MW closed-loop geothermal system the company calls an "Eavor-Loop" near Little Salmon Carmacks, Yukon (Rudyk, 2020);

- Clarke Lake, British Columbia. This will be the first project applied for a permit to explore the opportunities to convert an aging gas field into a geothermal resource;

- Lakelse Lake, British Columbia. In January 2020, the British Columbia government awarded a permit to the Fort Nelson First Nation to develop a geothermal project for the Fort Nelson area called the "Clark Lake Geothermal Project." This is the first national project to develop the geothermal potential near Lakelse Lake, south of Terrace, British Columbia. The first phase of the project, the Kitselas geothermal power plant, is estimated to create 87 jobs with 17 permanent positions and save more than 20,000 tonnes of carbon emissions each year. Future phases of the project are projected to create an additional 800 jobs and save a further 300,000 tonnes of carbon emissions yearly;

- Canoe Reach, British Columbia. In June 2018, the British Columbia Oil and Gas Commission issued its first-ever permit authorizing the construction of a geothermal energy project within the province. Borealis GeoPower plans, in collaboration with the Village of Valemount, will drill four geothermal wells near Valemount. It is expected that the geothermal power plant will generate electricity and supply district heating primarily from fractured igneous rocks. The project is estimated to create at least 22 jobs in its first phase with four permanent positions while reducing around 4000 tonnes of carbon emissions each year. Valemount has been reported to have among the worst winter air quality in British Columbia. Developing a geothermal energy district heating project has the potential to increase air and health quality positively.

Regardless of what has been said above, the geothermal energy sources available in the country have not been exploited as they should, but this situation is beginning to change. For this reason, it can be stated that today the geothermal industry is less developed in Canada than in comparable countries around the world. However, with the large availability of sub-surface data in the Western Canadian Sedimentary Basin, from oil and gas exploration and development, there is the ability to map and understand the potential for geothermal energy sources available in the area and reduce exploration risks to the minimum. Drilling costs are significantly less than in many countries worldwide, making Canada attractive for exploiting its geothermal energy resources. "The Intermontane basins in British Columbia and the Yukon Territory hold potential due to the proximity of hot springs. In contrast, the Western Canadian Sedimentary Basin has the potential for conventional geothermal, co-production with oil and gas activities, and the implementation of new technologies to develop this potentially extensive geothermal resource" (Hirschmiller, 2019).

## Geothermal energy electricity generation in Canada

Until September 2020, there was no geothermal power plant generating electricity in Canada. As previously mentioned, this situation will change in 2022, when the first geothermal power plant in the country begins to generate electricity.

## Geothermal energy capacity in the United States

Geothermal energy in the USA was first used for electric power production in 1960. The Geysers in Sonoma and Lake counties, California, was developed into what is now the largest geothermal steam power plant in the world, with a capacity of 1590 MW. Other geothermal steam fields are located in the Western USA and Alaska. The list of 31 operational geothermal power plants in the USA in 2018 is shown in Table 5.3.

**TABLE 5.3**  List of 31 geothermal power plants in the USA in 2018.

| Name | State | Type | Year |
|---|---|---|---|
| Beowawe | Nevada | Flash steam (87%)<br>Binary cycle (13%) | 1985 |
| Blue Mountain | Nevada | Binary cycle | 2009 |
| Blundell | Utah | | 2007 |
| Brady | Nevada | | 1992 |
| Coso | California | Flash steam | 1987 |
| Cove Fort | Utah | | 2014 |
| Desert Peak | Nevada | Flash steam (71.7%)<br>Binary cycle (28.3%) | 1985 |
| Dixie Valley | Nevada | Flash steam (99%)<br>Binary cycle (1%) | 1988 |
| Don A. Campbell | Nevada | | 2013 |
| The Geysers | California | Dry steam | 1960 |
| Heber | California | Binary cycle | 1985 |
| Imperial Valley | California | Dry steam | 1982 |
| Jersey Valley | Nevada | | 2011 |
| Lightning Dock | New Mexico | Binary cycle | 2014 |
| Mammoth | California | Binary cycle | 1984 |
| McGinness Hills | Nevada | Binary cycle | 2012 |
| Neal Hot Springs | Oregon | | 2012 |
| North Brawley | California | Binary cycle | 2010 |
| Ormesa | California | Binary cycle | 1987 |
| Patua | Nevada | Binary cycle (85.7%)<br>Solar PV (14.3%) | 2017 |
| Puna | Hawaii | Binary cycle | 1992 |
| Raft River | Idaho | Binary cycle | 2008 |

**TABLE 5.3**   List of 31 geothermal power plants in the USA in 2018—cont'd

| Name | State | Type | Year |
|---|---|---|---|
| Salt Wells | Nevada | Binary cycle | 2012 |
| San Emidio | Nevada | | 2012 |
| Soda Lake | Nevada | | 1987 |
| Steamboat | Nevada | Binary cycle (80.3%) Flash steam (19.7%) | 1988 |
| Stillwater | Nevada | Binary cycle (65.3%) Solar PV (34.7%) | 2009 |
| Thermo 1 | Utah | | 2013 |
| Tungsten Mountain | Nevada | | 2017 |
| Tuscarora | Nevada | | 2012 |
| Wabuska | Nevada | Binary cycle | 1984 |

In 2018, according to IRENA Renewable Energy Statistics 2020 (2020), the total geothermal power capacity installed in the USA was 2541 MW (2555 MW in 2019).
*Source: 2020. Wikipedia List of Geothermal Power Plants in the United States.*

**TABLE 5.4**   List of geothermal power plants proposed in the USA.

| Name | State | Capacity (MW) | Owner | Type | Year |
|---|---|---|---|---|---|
| Hell's Kitchen | California | 140 | Controlled Thermal Resources | | 2023 |
| Casa Diablo IV | California | 30 | Ormat | Binary cycle | 2021 |

*Source: 2020. Wikipedia List of Geothermal Power Plants in the United States.*

The list of geothermal power plants proposed in the USA is shown in Table 5.4.

According to Table 5.4, two new geothermal power plants are proposed in the USA to enter operation in 2021 and 2023 with a total capacity of 170 MW. Both geothermal power plants are located in California.

The evolution of the US's geothermal power plant capacity during the period 2010–19 is shown in Fig. 5.13.

According to Fig. 5.13, the US's geothermal power plant capacity increased 6.2% during the period 2010–19, rising from 2405 MW in 2010 to 2555 MW in 2019. It is expected that the geothermal power plant capacity in the country will increase further during the coming years with the addition of new capacities in at least three geothermal power plants.

The use of geothermal energy sources for electricity generation and heating/cooling is not free of environmental impact. The environmental impact includes hydrogen sulfide emissions, corrosive or saline chemicals discharged in wastewater, possible seismic effects from water injection into rock formations, waste heat, and noise.

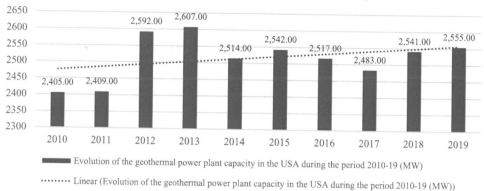

FIG. 5.13    Evolution of the geothermal power plant capacity in the USA during the period 2010–19. *Source: 2020. IRENA Renewable Energy Statistics 2020. IRENA.*

## Geothermal energy electricity generation in the United States

The USA has 8359 geothermal energy zones on public land managed by the Bureau of Land Management. Other data related to geothermal power plants in the USA are the following:

- There are 557 authorized geothermal energy zones;
- There are 227 pending geothermal energy zones for approval;
- There are 16 canceled geothermal energy zones. From this total, 12 were due to the expiration of their licenses, and four were withdrawn;
- There are 22 relinquished geothermal energy zones;
- A total of 18 geothermal energy proposed zones were rejected;
- There are 2,810,915 authorized geothermal energy acres and 916,635 geothermal energy acres waiting for approval;
- There are 7503 closed geothermal energy zones

The USA leads the world in the amount of geothermal electricity generation. In 2019, the total electricity generated by geothermal power plants in the USA reached 24,148 GWh, representing around 0.05% of US utility-scale electricity generation. In 2019, a total of 31 geothermal power plants were operating in seven states in the USA (see Table 5.5). Electricity generated from geothermal plants is projected to increase up to 52.2 billion kWh in 2050, according to the US Energy Information Administration (2020) and Annual Energy Outlook 2020 (2020).

In the USA, exploitation of geothermal energy has hardly begun, despite the numbers reported above. The US Energy Information Administration (EIA) projects that geothermal electricity generation could more than quadruple by 2040, increasing to over 67,000 GWh and helping California and other states satisfy their mandated renewable generation requirements.

**TABLE 5.5**  The US states with geothermal power plants in 2019.

|  | State share of total US geothermal electricity generation | Geothermal share of total state electricity generation |
|---|---|---|
| California | 71.2% | 5.6% |
| Nevada | 23.5% | 9.5% |
| Utah | 2.8% | 1.1% |
| Oregon | 0.9% | 0.2% |
| Hawaii | 0.7% | 1.1% |
| Idaho | 0.5% | 0.4% |
| New Mexico | 0.4% | 0.2% |

*Source: US Energy Information Administration, 2020. Geothermal Explained. Use of Geothermal Energy.*

# References

Adachi, M., 2011. Risk control for disaster at the geothermal field. In: Proceedings of the 9th Asian Geothermal Symposium; November 7-9, 2011.

Alden, A., 2019. Why the Earth's Crust Is so Important. Thought Co.

Ambersariya, D., 2019. What Is Geothermal Energy? Types of Geothermal Power Plant. Invention Sky.

Anon., 2020. Annual Energy Outlook 2020. US EIA.

Anon., 2019. California Geothermal Energy Statistics and Data. California Energy Commission (Accessed 5 December 2019).

Anon., 2015. California Known Geothermal Resource Areas. California Energy Commission. Updated February 25, 2015.

Anon., 2020. California Profile Analysis. US Energy Information Administration.

Anon., 2020. Canada's First Geothermal Production and Injection Well Test Exceeds Expectations—First 20 MW Facility in Design Phase. DEEP Earth Energy Production Corp. September 2020.

Barbier, E., 2002. Geothermal energy technology and current status: an overview. Renew. Sustain. Energy Rev. 6, 3–65.

Davis, M., 2017. Geothermal Energy Costs—Breaking Down Its True Price. Understand Solar.

Anon., February 2019. EIA Electric Power Monthly Table 1.16.B.

Anon., 2019. EIA Geothermal Energy and the Environment. US Energy Information Administration.

Anon., 2020. EIA Geothermal Explained. US Energy Information Administration.

Anon., 2008. Energy Efficiency and Renewable Energy Geothermal Program. Geothermal Risk Mitigation Strategies Report. US DoE.

Evers, J., 2015. Mantle; National Geographic. National Geographic Society. Retrieved June 28, 2019.

Anon., 2019. Geothermal Benefits. Geothermal Exchange Organization.

Anon., 2019. Geothermal Energy Basics. US Department of Energy (DoE), National Renewable Energy Laboratory (NREL).

Anon., 2018. Geothermal Energy Market by Type (Direct Dry Steam Plant, Flash Plant, and Binary Plant) and by Application (Residential, Industrial, and Commercial): Global Industry Perspective, Comprehensive Analysis, and Forecast, 2018–2025. Zion Market Research.

Anon., 2019. Geothermal Heat Pump Basics. US DoE, NREL.

Anon., 2019. Geothermal Heating and Cooling Technologies. US EPA.

Anon., 2019. GeoVision: Harnessing the Heat Beneath Our Feet. US Department of Energy (DoE), Office of Scientific and Technical Information.

Glassley, W.E., 2010. Geothermal Energy: Renewable Energy and the Environment. CRC Press, ISBN: 9781420075700.

Grasby, S.E., et al., 2012. Geothermal Energy Resource Potential of Canada (PDF). Geological Survey of Canada.

Gurgency, H., 2009. How to Increase Geothermal Power Conversion Efficiencies. Queensland Geothermal Energy Centre, School of Mech. and Mining Engineering, The University of Queensland, Canada. 2012.

Hirschmiller, J., 2019. What Is Geothermal Energy? What Is Canada's Potential? GLJ Petroleum Consultants.

Hulen, J.B., Wright, P.M., 2001. Geothermal Energy—Clean Sustainable Energy for the Benefit of Humanity and Environment. US Department of Energy (DoE).

Anon., 2007. IEA Electricity Information 2007. OECD Publishing (International Energy Agency).

Intergovernmental Panel on Climate Change (ICPP), 2014. Climate Change 2014 Mitigation of Climate Change. Working Group III Contribution to the Fifth Assessment Report of the ICPP.

Anon., 2020. IRENA Renewable Energy Statistics 2020. IRENA.

Katharina, L., Fegley, B., 1998. The Planetary Scientist's Companion. Oxford University Press, New York, ISBN: 978-1423759836. OCLC 65171709.

Krieger, K., 2004. D Layer Demystified; Science News. American Association for the Advancement of Science. Retrieved November 5, 2016.

Liu, X., Hughes, P., McCabe, K., Spitler, J., Southard, L., 2019. GeoVision Analysis Supporting Task Force Report: Thermal Applications—Geothermal Heat Pumps. ORNL/TM-2019/502, Oak Ridge National Laboratory, Oak Ridge, TN (Accessed 28 April 2019) https://info.ornl.gov/sites/publications/Files/Pub103860.pdf.

Lund, J.W., Boyd, T.L., 2016. Direct utilization of geothermal energy 2015 worldwide review. Geothermics 60, 66–93. https://doi.org/10.1016/j.geothermics.2015.11.004.

Manzella, A., 2009. Geophysical Methods in Geothermal Exploration. Italian National Research Council, International Institute for Geothermal Research, Pisa. https://hendragrandis.files.wordpress.com/2009/04/a_manzella.pdf.

Max, 2020. Disadvantages of Geothermal Energy: Dominating Threats to Use This Energy. Linquip Tech News. https://www.linquip.com/blog/disadvantages-of-geothermal-energy/.

McCabe, K., Beckers, K., Young, K.R., Blair, N., 2019. GeoVision Analysis Supporting Task Force Report: Thermal Applications—Quantifying Technical, Economic, and Market Potential of Geothermal District Heating Systems in the United States. NREL/TP-6A20-71715, National Renewable Energy Laboratory, Golden, CO. https://www.nrel.gov/docs/fy19osti/71715.pdf.

McCombs, B., 2019. Trump Administration Doubles Down on Fossil Fuels. Associated Press (via LA Times). Retrieved June 3, 2019.

Miller, B., 2020. 23 Big Advantages and Disadvantages of Geothermal Power. Green Garage.

Montagner, J.-P., 2011. Earth's structure, global. In: Gupta, H. (Ed.), Encyclopedia of Solid Earth Geophysics. Springer Science & Business Media, ISBN: 9789048187010, pp. 134–154.

Moon, H., Zarrouk, S.J., 2012. Efficiency of Geothermal Power Plants: A Worldwide Review. Department of Engineering Science, University of Auckland, New Zealand.

Nanalyze, 2019. A Geothermal Energy Stock for Investors.

Anon., 2019. New Study Highlighting the Untapped Potential of Geothermal Energy in the United States. US Department of Energy (DoE). May 30, 2019; Retrieved June 3, 2019.

Anon., 2019. Oregon Profile Analysis. US EIA.

Rannestad, E., 2017. Costs and risks of geothermal energy and the US pacific coast geothermal potential. In: Environmental Economics. Whitman College, p. 461. November 2017.

Rudyk, M., 2020. Yukon First Nation Signs Deal for New Geothermal Project. Retrieved February 4, 2020.

Sathaye, J., Lucon, O., Rahman, A., Christensen, J., Denton, F., Fujino, F., Heath, G., Kadner, S., Mirza, M., Rudnick, H., Schlaepfer, A., Shmakin, A., 2011. Renewable energy in the context of sustainable development. In: Edenhofer, O., Pichs-Madruga, R., Sokona, Y., Seyboth, K., Matschoss, P., Kadner, S., von Stechow, C. (Eds.), IPCC Special Report on Renewable Energy Sources and Climate Change Mitigation; Prepared by Working Group III of the Intergovernmental Panel on Climate Change. Cambridge University Press, Cambridge, and New York, NY.

Snyder, D., Beckers, K., Young, K., 2017. Update on geothermal direct-use installations in the United States. In: 42nd Workshop on Geothermal Reservoir Engineering Proceedings; February 13–15, 2017; Stanford, California. National Renewable Energy Laboratory, Golden, CO, pp. 1–7. SGP-TR-212. (Accessed 28 March 2019) https://pangea.stanford.edu/ERE/db/IGAstandard/record_detail.php?id=27985.

Sparrow, P., 2012. Geothermal Energy: Advantages and Disadvantages. https://www.ianswer4u.com/2012/02/geo-thermal-energy-advantages-and.html#:~:text=%20Geothermal%20Energy%20%3A%20Advantages%20And%20Disadvantages%20,geothermal%20power%20plants%20is%20very%20less.%20More%20.

Anon., 2011. Technology Roadmap-Geothermal Heat and Power. IEA.

Tester, J.W., 2006. The Future of Geothermal Energy. Massachusetts Institute of Technology (MIT), USA.

The Geysers, 2019. About Geothermal Energy, Welcome to the Geysers. Calpine (Accessed 6 December 2019).

TWI, 2020. What Are the Advantages and Disadvantages of Geothermal Energy? https://www.twi-global.com/technical-knowledge/faqs/geothermal-energy/pros-and-cons.

US Energy Information Administration, 2015. Residential Energy Consumption Survey (RECS); 2015 Household Characteristics Technical Documentation Summary. US Energy Information Administration, Washington, DC. https://www.eia.gov/consumption/residential/reports/2015/methodology/pdf/RECSmethodology2015.pdf.

Anon., 2020. Monthly Energy Review. US Energy Information Administration (April 2020).

Anon., March 26, 2016. What Is the Earth's Mantle Made Of? Universe Today. Retrieved November 34, 2018.

Anon., 2020. Wikipedia Geothermal Energy.

Anon., 2020. Wikipedia Geothermal Power in Canada.

Yu, C., Day, E.A., de Hoop, M.V., Campillo, M., Goes, S., Blythe, R.A., van der Hilst, R.D., 2018. Compositional heterogeneity near the base of the mantle transition zone beneath Hawaii. Nat. Commun. 9 (9), 1266. https://doi.org/10.1038/s41467-018-03654-6. Bibcode: 2018NatCo...9.1266Y. 5872023 29593266.

Zion Market Research, 2019. Global Geothermal Energy Market Will Reach US$9 Billion by 2025. Zion Market Research.

# The use of bioenergy for electricity generation

## Introduction

Bioenergy exists in three basic forms. These forms are in solid fuel (biomass[a]), liquids (biofuels), or gases (biogas or biomethane). Bioenergy can be used "to produce heat for cooking and heating residential spaces and water, either in traditional stoves or in modern appliances such as pellet-fed central heating boilers. At a larger scale, bioenergy can provide heat for public and commercial buildings as well as for industry. Bioenergy can also co-generate electricity and heat via combined heat and power (CHP) systems to serve residential, commercial, and industrial buildings, either on-site or distributed from larger production facilities via district heating and cooling systems" (Renewables 2019 Global Status Report, 2019).

In other words, bioenergy involves the use of a wide range of biological materials for energy purposes. These can be converted into thermal energy, electricity, and fuels for transportation (biofuels) through several different processes. Many well-established bioenergy pathways exist that are technically proven and for which systems are available at a commercial level worldwide. Besides, new technologies are at the earlier stages of development, demonstration, and commercialization.[b] Given the potential environmental, social, and economic implications of using biomass materials for energy production, bioenergy's sustainable manufacture and use are key issues to ensure a country's economic development.[c]

---

[a] Biomass is the organic by-product "left over from plants, animals, and waste products from organic sources, such as garbage, wood, crops, landfill gas, and alcohol fuels. When burned, the chemical energy in biomass is released as heat, which then can be converted into biofuels and/or biogas and finally into useable energy such as fuels, electricity, or heat" (McFarland, 2017).

[b] The traditional use of biomass for heating involves the burning of woody biomass or charcoal as well as dung and other agricultural residues in simple and inefficient devices in developing and emerging economies.

[c] Modern bioenergy is any production and use of bioenergy that is not classified as traditional use of biomass.

According to McFarland (2017), agriculture, forestry, municipalities, colleges, universities, scientific and research institutions, food processor plants, hospitals, medical centers, and many more industries, facilities, and institutions create billions of tons of waste. There were very few safe, cost-effective ways to use waste productively or dispose of it in the past. Today, however, the situation is quite different. Renewable energy technology makes waste a potential biomass energy resource that many countries can be used for electricity generation and heating.

The basic process of biomass energy, or bioenergy, has to do with photosynthesis. Plants contain chlorophyll, which absorbs carbon dioxide from the atmosphere and water from the ground. When animals eat the plants, some energy is transferred, producing what is known today as bioenergy. The carbon dioxide and water it contains are then released back into the atmosphere and grow more plants and crops to start the cycle again. Research carried out has proven that using bioenergy as an energy source has several benefits over fossil fuel use as a power source and reduces greenhouse gas emissions. For it to be produced, the organic material must undergo a bioenergy conversion process, of which several technologies can be used. Today's main uses of bioenergy are electricity production through driving turbines and providing biofuel for transportation (biodiesel and ethanol) (see Fig. 6.1) (McFarland, 2017).

Primary energy consumption worldwide rose by 1.3% in 2019. This percentage is "below its 10-year average rate of 1.6% per year and much weaker than the 2.8% growth seen in 2018" (BP Statistical Review of World Energy 2020, 2020). By region, biomass energy consumption fell in North America, Europe, and CIS, and growth was below average in South and Central America. Demand growth in Africa, the Middle East, and Asia was roughly in line with historical averages. In the specific case of the North American region, primary energy consumption fell by 1.1%. The use of renewables for electricity generation in North America in 2019 increased by 5.8% with respect to 2018. Looking at energy by fuel, 2019 growth was driven by

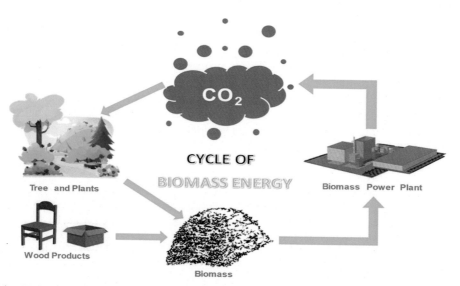

**FIG. 6.1**  Cycle of biomass energy. *Source: Agico Group. http://www.agicosolution.com/biomass-energy-in-china/ and Author own work.*

renewable energy sources for electricity generation, followed by natural gas, which together contributed over 75% of the net increase. "The share of both renewables and natural gas in primary energy increased to record highs. Meanwhile, coal consumption declined, with its share in the energy mix falling to its lowest level since 2003" (BP Statistical Review of World Energy 2020, 2020).

The growth in biomass use for electricity generation in the North American region is driven by higher demand for energy and power among the American and Canadian populations. This growth in the population's energy consumption and changing climatic conditions due to global warming raise the region's power demand. Despite US energy policy supporting fossil fuel use, particularly coal and shale gas, renewable energy sources have become higher in demand in the USA to tackle high energy needs with sustainable energy solutions.

In recent years, renewable energy has grown faster than other energy sources, with solar power generation estimated to become the fastest-growing electricity source by 2050. "Renewables provided the largest increment to power generation (340 TWh), followed by natural gas (220 TWh). These gains came partially at the expense of coal generation, which fell sharply (− 270 TWh), causing the share of coal in power generation to fall by 1.5% points to 36.4% – the lowest since 1985. Despite this reduction, coal remained the single largest source of power generation in 2019. Meanwhile, the share of renewables in electricity generation increased from 9.3% to 10.4%, surpassing nuclear generation for the first time" (BP Statistical Review of World Energy 2020, 2020).

Overall, installed renewable energy capacity was enough to provide an estimated 27.3% of global electricity generation at the end of 2019. Hydropower still made up the majority (58%) of this estimated generation share, followed by wind power (22%), solar PV (10%), and bioenergy (8%) (Renewables 2020 Global Status Report, 2020).

According to the BP Statistical Review of World Energy 2020 report, in 2019, about 15.9% of the electricity generated across sectors in the North American region was from renewable energy sources below natural gas, coal, and nuclear energy. It is foreseen that this percentage could grow up to 25% by 2030 in the USA.[d] Most of the power generation had arisen from hydro, wind, and biomass.

The transport sector saw a considerable rise in the use of biofuels such as ethanol and biodiesel during the past decade. Ethanol transportation fuel (E85) "is a type of renewable energy fuel which is anticipated to grow at the fastest rate at an average annual rate of 9.7% over the next 25 years, even if it starts from a very low base. On the other hand, biomass constitutes about 98% of renewable energy use that comprises nearly 60% of biomass wood, 33% from biofuels, and nearly 8% from biomass waste" (Crifax, 2020).

[d] Considering cumulative renewable energy capacity, China remained the global leader with 789 GW at year's end, followed by the USA with 282 GW, Brazil with 144 GW, India with 137 GW, and Germany with 124 GW. China also led the world in capacity added during the year at 67 GW, followed by the USA with 22 GW, India with 13 GW and Brazil and Japan, each with around 8 GW (Renewables 2020 Global Status Report, 2020).

## Types of bioenergy

Bioenergy can adopt the form of biomass, biofuels, or biogas. Biomass energy is the energy derived from the organic matter of plants and animals. Biomass energy in the form of dead plants, trees, grass, leaves, crops, manure, garbage, and animal waste can be a great source of alternative fuels that can replace fossil fuels in electricity generation (see Fig. 6.2). Plants use a process called "photosynthesis" that converts energy from the Sun into chemical energy.

Bioenergy use falls into two main categories: traditional and modern. According to IRENA Bioenergy (2020), traditional bioenergy use refers to biomass combustion in such forms as wood, animal waste, and traditional charcoal. On the other hand, modern bioenergy technologies include the following:

- "Liquid biofuels produced from bagasse and other plants;
- Bio-refineries;
- Biogas produced through anaerobic digestion of residues;
- Wood pellet heating systems, and other technologies" (IRENA Bioenergy, 2020).

FIG. 6.2   Biomass energy. *Source: Photograph © Andrew Dunn, http://www.andrewdunnphoto.com. Wikimedia Commons: Compost Heap.jpg.*

The traditional use of biomass to supply energy for cooking and heating is simple. Biomass can be used to provide heat in several different markets. "Traditional use of biomass is still the largest sector, an important energy source for industry and buildings, with the heat either provided directly at the site where it is to be used or distributed via district heating systems. The patterns of use have changed relatively slowly in recent years" (IEA Renewables 2018, 2018).

"Given the serious negative health impacts of traditional biomass use, the effects on local air quality, and the unsustainable nature of much of the supply of this biomass, efforts are being made to reduce the use of traditional biomass in the push to improve access to clean fuels" (Renewables 2019 Global Status Report, 2019).

For the reasons mentioned above, the amount of biomass used in traditional applications has decreased slightly in recent years. The decline is due, in addition to what has been said in the previous paragraph, to the negative effects of biomass burning on local air quality, the associated health impacts, and the unsustainable nature of much of the biomass supply for these uses (IEA et al., 2019).

Modern bioenergy, which excludes the traditional use of biomass, provided an estimated 5.1% of total global final energy demand in 2018, accounting for about half of all renewable energy in final energy consumption (Renewables 2020 Global Status Report, 2020). Modern bioenergy also provides for heating around 8.6% of the global energy supply used for this specific purpose, 3.1% of transport energy needs, and 2.1% of the global electricity supply (IEA Renewables 2018, 2018).

It is important to single out that modern bioenergy use has grown most rapidly in the electricity sector – at approximately 6.7% per year over the last five years – compared to around 4.4% in the transport sector and around 1.1% in the energy sector (heating) (Renewables 2020 Global Status Report, 2020).

Without a doubt, modern bioenergy can provide heat more efficiently and cleanly for industry, agriculture, residential, public, and commercial buildings than other types of energy sources. According to the Renewables 2020 Global Status Report, this is a summary of the contribution of modern bioenergy in providing heat for the industry, agriculture, residential, public, and commercial buildings:

- Bioenergy heat can be produced and used directly where it is created, including through the co-generation of electricity and heat using combined CHP systems. Bioenergy heat demand in the industry and agriculture sectors "grew 1.8% annually between 2013 and 2018, and bioenergy heat met 9.3% of the sectors' heat requirements in 2018. Industries that handle biomass – notably paper and board, sugar and other food products, and wood-based industries – often use their residues for energy purposes. For example, in the paper and board sector, 40% of energy use is derived from biomass sources, including the "black liquor" produced in paper manufacture" (Renewables 2020 Global Status Report, 2020). Regrettably, bioenergy is not yet widely used in other industries.
  "Biomass and waste fuels met around 6% of the cement industry's energy needs in 2017, mainly in Europe, where they provided around 25% of the energy used in cement making" (Renewables 2020 Global Status Report, 2020).
  Modern bioenergy provided around 4.6% of total heat demand in the buildings sector but fell by around 1% annually between 2013 and 2018. Biomass's share of heat in buildings also

declined during the same period. "Biomass can produce heat for residential buildings by burning wood logs, chips, or pellets produced from wood or agricultural residues. The informal use of wood and other biomass to heat individual residences is prevalent in developing and emerging economies as well as in more developed economies" (Renewables 2020 Global Status Report, 2020).

It is important to stress that bioenergy heat can be a source of local air pollution if inefficient appliances and/or poor-quality fuels are used. New technologies that allow for high emissions reduction "from biomass combustion are commercially available, triggered by stringent national regulations for small combustion facilities in some countries" (Renewables 2020 Global Status Report, 2020). It is important to stress that it is generally easier to economically meet efficiency and air quality goals at larger operation scales.

According to Vis and van den Berg (2010), the different biomass types can be divided into four categories (see Table 6.1). These categories are the following:

- "Forest biomass and forestry residues;
- Energy crops;
- Agricultural residues;
- Organic waste."

**TABLE 6.1**    Biomass types.

| Main type | Sub-type | Examples |
|---|---|---|
| Forestry | Primary forest products | Stem wood and thinning |
|  | Primary forestry residues | Leftovers from harvesting activities: twigs, branches, stumps, etc. |
|  | Secondary forestry residues | Residues resulting from any processing step: sawdust, bark, black liquor, etc. |
| Energy crops | Oil, sugar, and starch crops | Jatropha, rapeseed, sunflower seed, sugar cane, cereals (wheat, barley, etc.), maize, etc. |
|  | Energy grasses | *Miscanthus*, switchgrass, etc. |
|  | Short rotation coppice | Poplar, eucalyptus, etc. |
| Agricultural residues | Primary or harvesting residues, a by-product of cultivation and harvesting activities | Wheat straw, etc. |
|  | Secondary processing residues of agricultural products, e.g., for food or feed production | Rice husks, peanut shells, oil cakes, etc. |
|  | Manure | Pig manure, chicken manure, cow manure, etc. |
| Organic waste | Tertiary residues, released after the use phase of products | Biodegradable municipal waste, landfill gas, demolition wood, sewage gas, and sewage sludge |

*Source: Vis, M.W., van den Berg, D., 2010. Best Practices and Methods Handbook, vol. I, Harmonization of Biomass Resource Assessments. Biomass Energy Europe; file:///C:/Users/LEX ~ 1/AppData/Local/Temp/BEEBestPracticesandmethodshandbook-1.pdf; https://www.researchgate.net/publication/268388401_Harmonization_of_biomass_resource_assessments_Volume_I_Best_Practices_and_Methods_Handbook.*

About 75% of the world's renewable energy use involves bioenergy, with more than half of traditional biomass use. "Biomass has significant potential to boost energy supplies in populous nations with rising demand, such as Brazil, India, and China. It can be directly burned for heating or power generation or converted into oil or gas substitutes. Liquid biofuels, a convenient renewable substitute for gasoline, are mostly used in the transport sector" (Alternative Energy Today, 2020).

## Forest biomass

In the context of biomass energy, "forest biomass includes several types of raw woody materials derived from forests or from the processing of timber that can be used for energy generation" (see Table 6.1) (Alternative Energy Today, 2020).

According to Morales Pedraza (2015) and Alternative Energy Today (2020), the types of raw woody materials derived from the forests are the following:

- Stem wood (see Fig. 6.3);
- Primary forestry residues;
- Secondary forestry residues;
- Woody biomass;
- Trees outside of forests.

Without a doubt, the most commonly used type of biomass for electricity generation and heating is wood, either round wood or wood waste, from industrial activities. "Wood and wood waste can be combusted to produce heat used for industrial purposes, space and water heating, or to produce steam for electricity generation. Through anaerobic digestion, methane can be produced from solid landfill waste or other biomass materials such as sewage, manure, and agricultural waste. Sugars can be extracted from agricultural crops, and, through distillation, alcohols can be produced for use as transportation fuels. Numerous other technologies exist or are being developed to take advantage of other biomass feedstock" (About Renewable Energy, 2017).

**FIG. 6.3**   Stemp wood. *Source: https://pixabay.com/photos/wood-trunks-cross-section-view-1405527/.*

It is important to single out that "most of the biomass used for heating is wood-based fuel, but liquid and gaseous biofuels also are used – including biomethane, which can be injected into natural gas distribution systems" (Renewables 2020 Global Status Report, 2020).

In 2018, modern bioenergy applications provided an estimated increase of 9.5% compared to the level registered in 2010 (Renewables 2020 Global Status Report, 2020). Biomass contributes the highest share to the global energy supply of all renewable energy resources. It provides energy not only for heating and transport but also for electricity generation, based on IEA PVPS Trends in Photovoltaic Applications 2019 and IEA PVPS Snapshot of Global PV Markets 2020 reports (IEA PVPS, 2019, 2020). According to the same sources mentioned above, biomass and bioenergy's traditional use contributed an estimated 12% to total final energy consumption in 2018.

## Energy crops

According to Morales Pedraza (2015), there are five main types of energy crops (see Fig. 6.4; Ahlberg, 2014):

- Oil containing crops;
- Sugar crops;
- Starch crops;
- Woody crops;
- Grassy crops.

**FIG. 6.4** Bioenergy crops like Miscanthus, pictured here, can store more carbon in their soil than corn or soybean crops. *Source: Ahlberg, L., 2014. Bioenergy Crops Could Store More Carbon in the Soil. University of Illinois at Urbana-Champaign.*

## Agricultural residues

According to the Electrical Power Generation report entitled "Introduction to the Current Situation and Perspectives on the Use of Biomass in the Generation of Electricity" (2016), agricultural residues are the by-products of agricultural practice. A distinction should be made "between primary or harvest residues (like a straw) produced in the fields and secondary residues from the processing of the harvested product (such as bagasse and rice husks) (see Fig. 6.5) produced at a processing facility. Manure is included as a separate category. By-products from further processing of agricultural products, such as molasses and vinasse, are regarded as residues from the food industry and are not included in this group."

In the specific case of bagasse, it is important to stress that there is no capacity installed for using this type of bioenergy for electricity generation in the North American region. There are also no plans for the use of bagasse for this purpose in the region in the future.

**FIG. 6.5** Sugarcane bagasse. *Source: Courtesy of Keechuan;* photo://www.dreamstime.com/stock-photo-sugarcane-bagasse-can-be-recycled-as-paper-fuel-renewable-ener-juice-product-pulp-named-energy-image93735213">93735213</a>©<a href="https://www.dreamstime.com/keechuan_info"itemprop="author">Keechuan</a>- <a href="https://www.dreamstime.com/">Dreamstime.com</a>.

## Organic waste

According to the Electrical Power Generation report entitled "Introduction to the Current Situation and Perspectives on the Use of Biomass in the Generation of Electricity" (2016), "organic waste includes biodegradable waste from households, industry, and trade activities. Organic waste includes biodegradable municipal waste, construction and demolition of wood, and sewage sludge. Biogases from sewage treatment plants as well as landfill gas are also included as energy carriers from organic waste" (see Fig. 6.6).

It is important to single out that cheap, high-quality biomass (e.g., wood waste) for power generation may become scarce in the future because it can also be used for heat production and in the pulp and paper industry.

New biomass feedstock based on energy crops has a larger potential for electricity generation and heating but is more expensive. Technologies and cost of power and heating generation from biomass "depend on feedstock quality, availability, and transportation cost, power plant size, and conversion into biogas, if any. If sufficient biomass is available, biomass power plants are a clean and reliable power source suitable for baseload service" (Electrical Power Generation, 2016).

Bioenergy can be produced from a variety of biomass feedstock. These biomass feedstocks can be directly used to produce electricity or heating or can be used to generate gaseous, liquid, or solid fuels through various processes (Electrical Power Generation, 2016).

On the other hand, it is important to know that "the range of bioenergy technologies is broad, and technical maturity varies substantially. Some commercially available technologies include small- and large-scale boilers, domestic pellet-based heating systems, and ethanol production from sugar and starch. Advanced biomass-integrated gasification combined-cycle power plants and lignocellulose-based transport fuels are examples of technologies that are

FIG. 6.6 Organic waste on the compost. *Source: Animaflora. Dreamstime.com; https://www.dreamstime.com/ organic-waste-compost-organic-waste-compost-image195208609.*

at a pre-commercial stage while liquid biofuel production from algae. Some other biological conversion approaches are in the research and development phase" (Electrical Power
Generation, 2016).

Bioenergy is a special source of renewable energy in several ways. These are the following:

- First, it can directly provide all three types of energy carriers;
- Second, it is easily storable and dispatchable, and when there is not enough Sun or wind,
  biomass-fired generators can be ramped up as needed;
- Third, the major drawback is the following: biomass requires strict management to be
  sustainable.

No matter how many solar panels are installed in a solar park, the Sun cannot be used
faster, nor the wind on Earth can be reduced if more wind turbines are installed. But with
biomass, countries should "avoid resource depletion, prevent monocultures from reducing
biodiversity, and ensure that the energy needs of rich countries are not met at the expense of
food needs in developing countries" (Electrical Power Generation, 2016).

## Bioenergy installed capacity and electricity generation in the North American region

Bioenergy is the second-largest source of renewable energy in the world. The combustion
of organic plant or animal waste has a calorific value that can be harnessed to produce heat
or electricity. The North American region's bioenergy power generation market is witnessing
significant growth due to the ever-increasing demand for energy. The bioenergy generation
market in the North American region is anticipated to rise exponentially in the coming years
because bioenergy only comprises approximately 5% of the USA's total primary energy generation. In 2017, Canada's bioenergy electricity generation share reached 1.4% within all renewable energy sources (About Renewable Energy, 2017).

In 2019, the bioenergy capacity installed at the world level reached 124,026 MW, generating
522,552 GWh in 2018. In the North American region, the bioenergy capacity installed in 2019
reached 15,825 MW, representing 12.8% of the world total and generating 78,546 GWh in 2018.
The bioenergy capacity installed in the USA reached in 2019 a total of 12,450 MW, representing 78.7% of the total at the regional level, and generating 67,885 GWh in 2018. The bioenergy
capacity installed in Canada is much lower (3376 MW in 2019), reaching 21.3% of the total at
the same level and generating 10,661 GWh in 2018.

The evolution of the bioenergy capacity installed in the North American region during the
period 2010–19 is shown in Fig. 6.7.

According to Fig. 6.7, the bioenergy capacity installed in the North American region increased by 31.9% during the period 2010–19, rising from 11,997 MW in 2010 to 15,825 MW in
2019. However, it is important to single out that despite the high increase in the bioenergy capacity installed in the region registered during the period 2010–19, during 2018 and 2019, the
bioenergy capacity installed decreased by 2.2%, falling from 16,172 MW in 2017 to 15,825 MW
in 2019. It is expected that the trend registered in the last two years will change during the
coming years, and more bioenergy capacity will be installed in the region with the aim of
increasing its role in the regional energy mix.

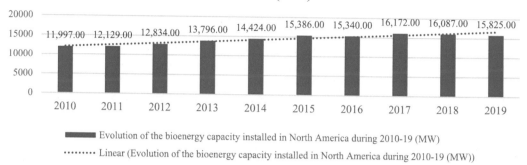

**FIG. 6.7** Evolution of the bioenergy capacity installed in the North American region during the period 2010–19. *Source: 2020. IRENA Renewable Energy Statistics 2020. IRENA and Author own calculations.*

On the other hand, in 2018, the world electricity generation using bioenergy reached 522,552 GWh. In the same year in the North American region, bioenergy electricity generation reached 78,546 GWh, representing 15.4% of the world total. The USA generated 67,885 GWh (86.4%) of the total electricity generated in the North American region in 2018. Canada, in the same year, generated only 10,661 GWh or 13.6% of the regional total.

The evolution of the electricity generation using bioenergy in the North American region during the period 2010–18 is shown in Fig. 6.8.

According to Fig. 6.8, the electricity generation using bioenergy in the North American region increased by 17.1% during the period 2010–18, rising from 67,052 GWh in 2010 to 78,546 GWh in 2018. However, it is important to single out that in the last three years of the period under consideration, the electricity generated by bioenergy power plants in North America decreased by 5.8%, falling from 83,355 GWh in 2015, the highest generation reached within the period, to 78,546 GWh in 2018. It is expected that the trend registered in the last

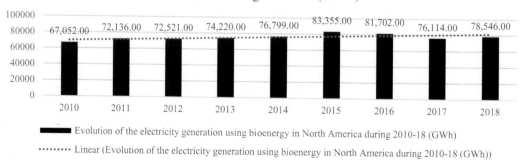

**FIG. 6.8** Evolution of the bioenergy electricity generation in the North American region during the period 2010–18. *Source: 2020. IRENA Renewable Energy Statistics 2020. IRENA and Author own calculations.*

three years will change during the coming years, and the bioenergy role within the regional energy mix will be higher than it is today.

In solid biofuels and renewable waste, the world capacity installed in 2019 reached 101,426 MW. In the North American region, the same year's installed capacity reached 12,277 MW, representing 12.1% of the world's capacity installed, generating 63,964 GWh in 2018. In 2019, the USA's solid biofuels and renewable waste installed capacity reached 9917 MW, representing 77.6% of the total at the regional level, and generating 54,275 GWh in 2018. In the same year, Canada's solid biofuels and renewable waste installed capacity reached 2360 MW, representing 22.4% of the total at the regional level, and generating 9689 GWh in 2018.

The evolution of the solid biofuels and renewable waste capacity installed in the North American region during the period 2010–19 is shown in Fig. 6.9.

According to Fig. 6.9, the following can be stated: the solid biofuels and renewable waste installed capacity in the North American region increased by 22.3% during the period 2010–19, rising from 10,042 MW in 2010 to 12,277 MW in 2019. However, it is important to single out that since 2016, when the solid biofuels and renewable waste installed capacity reached its peak within the period under consideration, it decreased by 4.2%, falling from 12,805 MW in 2016 to 12,277 MW in 2019. One of the primary reasons for the decrease of solid fuels and renewable waste capacity in the USA is their cost. New solid biofuels and renewable waste plant's construction cost often exceeds US$100 million, and larger plants require double or triple that figure to build. In addition to that, the economic benefits of the investment are not immediately noticeable (Folk, 2020).

On the other hand, the use of solid biofuels and renewable waste for electricity generation in the North American region reached, in 2019, a total of 63,964 GWh. In 2019, the USA electricity generated using solid fuels and renewable waste reached 54,275 GWh or 84.9% of the

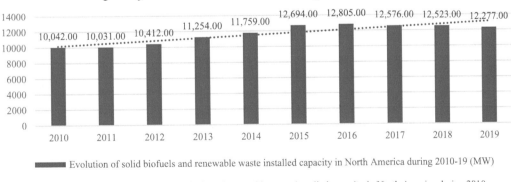

FIG. 6.9    Evolution of the solid biofuels and renewable waste capacity installed in the North American region during the period 2010–19. *Source: 2020. IRENA Renewable Energy Statistics 2020. IRENA and Author own calculations.*

regional total. In the same year, Canada's electricity generated using solid fuels and renewable waste reached 9689 GWh representing 15.1% of the regional total.

The evolution of the electricity generated using solid biofuels and renewable waste in the North American region during the period 2010–18 is shown in Fig. 6.10.

According to Fig. 6.10, the following can be stated: the electricity generated using solid biofuels and renewable waste in North America increased by 4.1% during the period 2010–18, rising from 61,444 GWh in 2010 to 63,964 GWh in 2018. Looking at Fig. 6.10, the period under consideration can be divided into three parts. During the first part covering the period 2011–14, the electricity generated using solid fuels and renewable waste increased by 12.9%, rising from 61,444 GWh in 2011 to 68,589 GWh in 2014, the highest peak electricity generation within the whole period under consideration. During the second part covering the period 2014–17, the electricity generated by this type of bioenergy fell 10.8%, decreasing from 68,589 GWh in 2014 to 61,202 GWh in 2017. During the third part covering the years 2017 and 2018, the electricity generated by solid fuels and renewable waste increased, once again, by 4.5%, rising from 61,202 GWh in 2017 to 63,964 GWh in 2018. It is expected that the use of solid fuels and renewable waste for electricity generation in the North American region will continue to be an important component of the regional energy mix in the future, but it is difficult to confirm what will be the level of this role.

In renewable municipal waste, the capacity installed in North America is relatively small compared to Europe and Asia. In 2019, the region's capacity reached 1133 MW, representing only 7.8% of the world total (14,518 MW), and generating 8554 GWh in 2018. In 2019, the USA installed capacity of renewable municipal waste was 1095 MW, representing 96.6% of the total capacity installed of this type of bioenergy source in the North American region, but only 7.5% of the world level. Using this type of energy source in the USA, the electricity generation reached, in 2018, 8382 GWh. Canada has little renewable municipal waste capacity installed (38 MW or 3.4% of the regional total), generating only 172 GWh in 2018 (2% of the regional total).

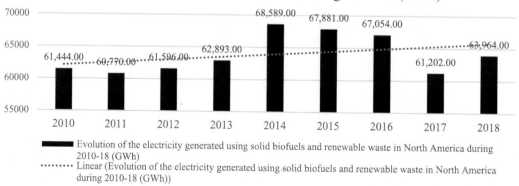

FIG. 6.10  Evolution of the electricity generated using solid biofuels and renewable waste in the North American region during the period 2010–18. *Source: 2020. IRENA Renewable Energy Statistics 2020. IRENA and Author own calculations.*

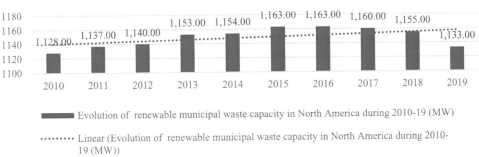

FIG. 6.11    Evolution of the renewable municipal waste in North America during the period 2010–19. *Source: 2020. IRENA Renewable Energy Statistics 2020. IRENA and Author own calculations.*

The evolution of the renewable municipal waste capacity in North America during the period 2010–19 is shown in Fig. 6.11.

According to Fig. 6.11, the following can be stated: the renewable municipal waste capacity in the North American region increased by 0.5% during the period 2010–19, rising from 1128 MW in 2010 to 1133 MW in 2019. The whole period under consideration can be divided into two parts. In the first part, covering the period 2010–2016, the renewable municipal waste capacity increased by 3.1%, reaching the highest peak of 1163 MW in 2015 and 2016. However, in the second part covering the period 2016–2019, the renewable municipal waste capacity decreased by 2.6%, falling from 1163 MW in 2016 to 1133 MW in 2019. This trend reflects the low role that this type of energy source has in the North American region energy mix, particularly in Canada, where this capacity represents only 3.4% of the regional total.

In 2018, the electricity generated by the renewable municipal waste in the North American region reached 8554 GWh, representing 13.8% of the world total (62,148 GWh). In the same year, the USA electricity generated using this type of bioenergy source reached a total of 8382 GWh, representing 98% of the regional total. In 2018, Canada generated a very small amount of electricity using renewable municipal waste (172 GWh), representing 2% of the regional total and 0.3% at the world level.

The evolution of renewable municipal waste electricity generation in North America during the period 2010–18 is shown in Fig. 6.12.

According to Fig. 6.12, the following can be stated: the use of renewable municipal waste for electricity generation in North America decreased by 9.5% during the period 2010–18, falling from 9446 GWh in 2010 to 8554 GWh in 2018. It is expected that this type of bioenergy source will continue to be a component of the regional energy mix in the future, but based on the information available, it is difficult to indicate if its role will be higher or lower than it is today.

In 2019, the liquid fuels installed capacity in the North American region reached 1041 MW, representing 36.6% of the world total (2841 MW). In the USA, the liquid fuels installed capacity in that year reached only 165 MW, representing 15.9% of the regional total and 5.8% at the world level. In 2018, the electricity generated at the regional level using liquid fuels was 351 GWh, representing 100% of the electricity generated using this type of bioenergy at

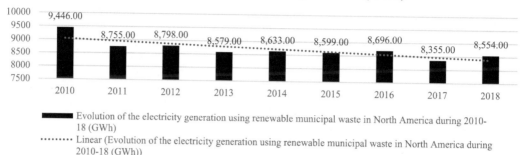

Evolution of the electricity generation using renewable municipal waste in North America during 2010-18 (GWh)

FIG. 6.12  Evolution of the renewable municipal waste electricity generation in North America during the period 2010–18. *Source: 2020. IRENA Renewable Energy Statistics 2020. IRENA and Author own calculations.*

the regional level and 6.1% at the world level (5729 GWh). In 2019, Canada had a liquid fuel installed capacity of 876 MW, representing 84.1% of the regional total and 30.8% worldwide. However, according to IRENA Renewable Energy Statistics 2020, the country is not using this capacity for electricity generation.

North America's evolution in the use of liquid fuels for purposes other than electricity generation is the following: global ethanol production increased by 2% to 114 billion liters in 2019, up from 111 billion liters in 2018. However, large increases in the use of liquid fuels in several countries caused a decreased production of this type of fuel in the USA, the major world ethanol producer. The USA accounted for 50% of world production, followed by Brazil, China, India, Canada, and Thailand (Renewables 2020 Global Status Report, 2020).

In 2019, the USA ethanol production fell 2% and reached 59.7 billion liters (US EIA Monthly Energy Review, 2020). "Key factors behind the decline were reduced domestic demand for ethanol as blending limits were approached and the US Environmental Protection Agency's continued support for small refinery exemptions, both of which reduced domestic demand, lowered prices and led to a scale-back in production," according to the Renewable Fuels Association (2020) report entitled "Global Ethanol Production (2019)".

Besides, ongoing US-China trade negotiations (among other factors) restricted ethanol export opportunities, leading US fuel exports to decline 14% in 2019, to 5.6 billion liters (IEA Renewables 2019, 2019). "In response to the reduction in overall demand, several ethanol production plants cut back production" (Thompson, 2019).

Another important bioenergy type of source in the North American region is biogas.[e] Biogas is the gaseous product of anaerobic digestion, a biological process in which microorganisms break down biodegradable material in the absence of oxygen, comprised primarily of methane (between 50% and 70%) and carbon dioxide (between 30% and 50%), with trace amounts of other particulates and contaminants (Anaerobic Digestion in US—A Large Biogas Opportunity, 2017).

---

[e] The average calorific value of biogas is about 21–23.5 MJ/m$^3$, so that 1 m$^3$ of biogas corresponds to 0.5–0.6 L diesel fuel or about 6 kWh (Fachagentur Nachwachsende Rohstoffe e.V. (FNR), 2009).

Biogas can be produced from various waste sources, including landfill material, animal manure, wastewater, industrial, institutional, and commercial organic waste.[f] Biogas can also be produced from other lignocellulosic biomass, such as crop and forest residues, dedicated energy crops, or through dry fermentation, co-digestion, or thermochemical conversions (e.g., gasification). Biogas can be combusted to provide heat, electricity, or both. It can also be "upgraded to pure methane—also called biomethane or renewable natural gas—by removing water, carbon dioxide, hydrogen sulfide, and other trace elements (see Fig. 6.13). This upgraded biogas is comparable to conventional natural gas and thus can be injected into the pipeline grid or used as a transportation fuel in a compressed or liquefied form" (Anaerobic Digestion in US—A Large Biogas Opportunity, 2017).

Renewable natural gas is considered a "drop-in" fuel for the natural gas vehicles currently on the road. "It can also be a source for renewable hydrogen, which can be used in stationary fuel cells and fuel cell electric vehicles" (NREL Biogas Potential in the United States, 2013).

The process shown in Fig. 6.13 is the following: Organic input substances such as food scraps, fats, or sewage sludge are fed into the biogas plant. Renewable resources such as corn, carrots, or grass serve both as food for animals such as cows and pigs, as well as for microorganisms in the biogas plant. Semi-liquid fertilizers and manure are also fed into the biogas plant. In the hot fermenter for approximately 38–40 °C, microorganisms degrade substrates

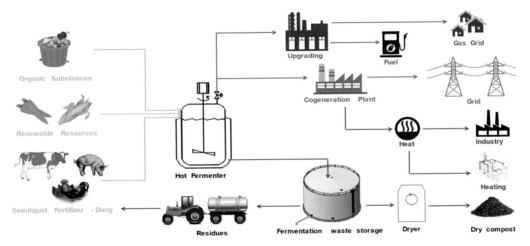

**FIG. 6.13** Schematic of a biogas plant used for power generation. *Source: Weltec Biopower. https://www.weltec-biopower.es/centro-de-informaciones/biogas/como-funciona-una-planta-de-biogas.html and Author own work.*

[f] According to Home Biogas Industrial Biogas in the United States of America (n.d.), "there are three main sources of industrial biogas production currently pursued in the United States: wastewater, landfills, and agricultural organic waste. In addition to replacing reliance on conventional fuels sources, using biogas as an alternative can have auxiliary benefits depending on the source. It is important to single out that the production of biogas from landfills reduces the methane emission of solid waste. Biogas production at wastewater treatment facilities can reduce their net energy consumption. The conversion of agricultural organic waste into biogas can reduce methane emissions from animal waste, assist in odor management, prevent ground water contamination, and produce natural fertilizer."

with light and oxygen exclusion. The end product of this fermentation process is biogas, with methane as the main component. But biogas also contains aggressive hydrogen sulfide. A stainless steel fermenter has the distinct advantage of resisting hydrogen sulfide attack and being usable for decades. Furthermore, with a stainless steel fermenter, the biogas plant also can operate in the thermophilic temperature range (up to 56 °C).

After the substrate fermentation, it is transported to the final fermentation waste store to be removed for use. The residues can be used as high-value fertilizers. The advantage: biogas manure has a lower viscosity and consequently enters the soil more quickly. Furthermore, in many cases, the fermentation residues have a higher fertilizer value and less intense smell. But drying and subsequent use as dry compost are also possible. The biogas generated is stored on the container's roof, and from there, it is transported and burned in the co-generation plant (BHKW) to generate electricity and heat. The electrical current is fed directly to the electrical network. The heat generated can be used to heat buildings or to dry wood or crop products.

The North American region's biogas situation is the following. The total biogas installed capacity in 2019 reached 2507 MW, representing 12.9% of the world total (19,381 MW) and generating 14,231 GWh in 2018. In the USA, the installed capacity in that year reached 2368 MW representing 94.5% of the regional total and 12.2% at the world level and generating 13,259 GWh in 2018. Canada had a biogas installed capacity of 140 MW, representing 5.5% of the regional total and only 0.7% at the world level, generating 972 GWh in 2018.

The evolution of the biogas installed capacity in the North American region during the period 2010–19 is shown in Fig. 6.14.

According to Fig. 6.14, the following can be stated: the biogas installed capacity in North America increased by 41.5% during the period 2010–19, rising from 1772 MW in 2010 to 2507 MW in 2019. Looking at Fig. 6.14, the period under consideration can be divided into two parts. During the first part covering the period 2010–14, the biogas installed capacity increased by 41.6%, rising from 1772 MW in 2010 to 2510 MW in 2014. During the second part covering the period 2014–19, the biogas installed capacity decreased by 0.2%, falling from

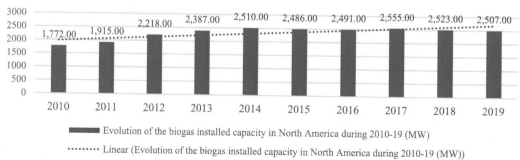

FIG. 6.14   Evolution of the biogas installed capacity in the North American region during the period 2010–19. *Source: 2020. IRENA Renewable Energy Statistics 2020. IRENA and Author own calculations.*

2510 MW in 2014 to 2507 MW in 2019. However, the North American biogas industry is projected to grow exponentially over the period 2020–26 (Bora, 2020).

On the other hand, the electricity generation using biogas in the North American region during the period 2010–18 is shown in Fig. 6.15.

According to Fig. 6.15, the following can be stated: the biogas electricity generation in North America increased by 34.3% during the period 2010–18, rising from 10,598 GWh in 2010 to 14,231 GWh in 2018. Looking at Fig. 6.15, the period under consideration can be divided into two parts. During the first part covering the period 2010–14, the biogas electricity generation increased by 37.4%, rising from 110,598 GWh in 2010 to 14,558 GWh in 2014. During the second part covering the period 2014–18, the electricity generation using biogas decreased by 2.3%, falling from 14,558 GWh in 2014 to 14,231 GWh in 2018. It is expected that the role of this type of energy source for electricity generation in the North American region will continue to be an important component of the regional energy mix, but its level will be lower than it is today.

Besides, it is important to know the following: during 2018, nearly 290 TWh per year of energy was generated from energy crops. It is estimated that nearly 4157 TWh per year of energy could have been produced if 7% of the total agricultural land was utilized.[8] The potential to generate energy from the major feedstock is estimated to be from 10,000 TWh to 14,000 TWh by 2030. The energy produced is expected to satisfy up to 16% of the electricity demand, 32% of the natural gas demand, and displace 28% of the coal usage across the world. Thus, energy crops are expected to be a prominent energy source, which will help the biogas market grow during the forecast period at the world level and in the North American region.

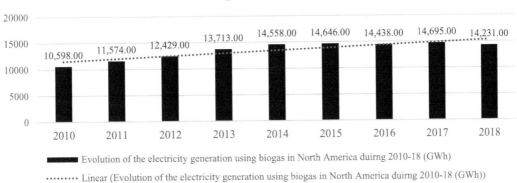

FIG. 6.15   Evolution of the biogas electricity generation in the North American region during the period 2010–18. *Source: 2020. IRENA Renewable Energy Statistics 2020. IRENA and Author own calculations.*

[8] During 2018, it was estimated that around 0.5% of the total agricultural land produced energy crops.

# Bioenergy investment costs in the North American region

According to the IRENA Global Landscape of Renewable Energy Finance 2020 report,[h] world renewable energy investment reached US$322 billion in 2018, with modest growth in 2019. Without a doubt, global investment in renewable energy made significant progress during the period 2013–18, with a cumulative US$1.8 trillion invested. "However, the pace must accelerate considerably for the world to meet internationally agreed climate goals adopted in the Paris agreement on climate change. To ensure a climate-safe future, annual investment in renewables – including various types of power generation, solar heat, and biofuels – would have to almost triple to US$800 billion by 2050. With the onset of the COVID-19 pandemic, renewable energy investments dropped by 34% in the first half of 2020, compared to the same period of 2019" (IRENA and CPI, 2020).

The private sector remains the main capital source for renewables, accounting for 86% of the sector's investments during the period 2013–18. "Project developers provided 46% of private finance, followed by commercial, financial institutions at 22%. Project-level equity was initially the most widely used financial instrument, linked to 35% of the renewables' investments in 2013-16. Since 2017, it has been overtaken by project-level conventional debt, which reached 32% in 2017-18. Public finance, representing 14% of total investments in renewables in 2013-18, came mainly via development finance institutions. Public financing resources, although limited, can be crucial to reduce risks, overcome initial barriers, attract private investors, and bring new markets to maturity" (IRENA and CPI, 2020).

It is important to single out that "East Asia and the Pacific attracted the largest share of renewable energy investments, with 32% of global financial commitments over 2013-18, mainly driven by China. The OECD countries of the Americas – including Chile, Canada, Mexico, and the USA – followed, with 18% of the world's investments in renewables, respectively" (IRENA and CPI, 2020).

Annual investment commitments in renewable energy during the period 2013–18 are shown in Fig. 6.16. According to the data included in that figure, the annual investment commitments in renewables increased by 34.7% during the period 2013–18, rising from US$239 billion in 2013 to US$322 billion in 2018. The peak in annual investment in renewables within the period under consideration was reached in 2017 (US$351 billion).

According to IRENA (2020b) report, the required cumulative investment in the energy sector needs to reach US$60 trillion by 2030 and US$110 trillion by 2050 (see Fig. 6.17) to accomplish the Paris agreement on climate change goals. Particularly, to set the world on a more climate-friendly trajectory by 2050, over US$37 trillion will be required for energy efficiency solutions, US$27 trillion for renewables, US$13 trillion for electrification of end-use sectors (e.g., for electric vehicles and railways), and US$13 trillion for power grids and energy flexibility measures, such as smart meters and energy storage (IRENA, 2020b).

# Bioenergy construction and generation costs in the North American region

According to IRENA Renewable Power Generation Costs in 2019 (IRENA, 2020a), the wide range of bioenergy-fired power generation technologies installed costs, capacity fac-

---

[h] IRENA and CPI (2020), Global Landscape of Renewable Energy Finance, 2020, International Renewable Energy Agency, Abu Dhabi. ISBN 978-92-9260-237-6, 2020.

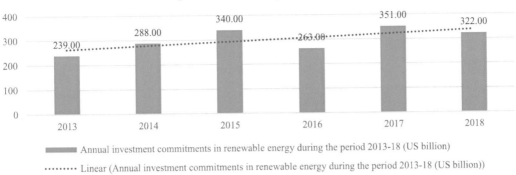

FIG. 6.16    Annual investment commitments in renewable energy during the period 2013–18 (billion US). *Source: IRENA and CPI, 2020. Global Landscape of Renewable Energy Finance 2020. International Renewable Energy Agency, Abu Dhabi. ISBN 978-92-9260-237-6.*

tors, and feedstock costs result in a wide range of observed LCOEs for bioenergy-fired electricity. Bioenergy can provide very competitive electricity where capital costs are relatively low, and low-cost feedstocks are available. Indeed, this technology can provide dispatchable electricity generation with an LCOE as low as around US$0.04/kWh. The most competitive projects use agricultural or forestry residues already available at industrial processing sites, where marginal feedstock costs are minimal or even zero. Where on-site industrial process steam or heat loads are required, bioenergy CHP systems can reduce the LCOE for electricity to as little as US$0.03/kWh, depending on the alternative costs for heat or steam available to the site.

Projects using low-cost feedstocks such as agricultural or forestry residues or the residues from processing agricultural or forestry products tend to have the lowest LCOE. For projects in the IRENA Renewable Cost Database (n.d.), the weighted average project LCOE by feedstock is US$0.06 per kWh or less for projects using black liquor, primary solid bioenergy (typically wood or wood chips), renewable municipal solid waste, and other vegetal and agricultural waste. Projects relying on municipal waste come with high capacity factors and are generally an economical source of electricity.

However, it is important to stress that "the LCOE for projects in the North American region is significantly higher than the average. Given that these projects have been developed mostly to solve waste management issues, though, and not primarily for the competitiveness of their electricity production, this is not necessarily an impediment to their viability" (IRENA, 2020a).

Many of North America's higher-cost projects using municipal solid waste as a feedstock rely on technologies with higher capital costs. More expensive technologies are used to ensure local pollutant emissions are reduced to a minimum level. Excluding these types of projects, which are typically not the largest, reduces the weighted average LCOE in North America by around US$0.01/kWh and narrows the gap with the LCOE of non-OECD regions.

On the other hand, "a range of bioenergy power generation technologies are now mature and represent competitive power generation options wherever low-cost agricultural or forestry waste is available" (IRENA Bioenergy for Power, 2020). Besides, new bioenergy "technologies

## Needed by 2030 (US$ 60 trillion)

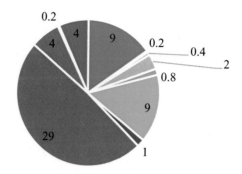

- Fossil fuels-supply
- Fossil fuels power
- Renewables-end-uses
- CCS and others
- Hydrogen
- Biofuels
- Energy efficiency
- Power grid and energy flexibility
- Nuclear
- Renewable power generation
- Electrification

## Needed by 2050 (US$ 110 trillion)

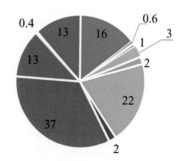

- Fossil fuels-supply
- Fossil fuels power
- Renewables-end-uses
- CCS and others
- Hydrogen
- Biofuels
- Energy efficiency
- Power grid and energy flexibility
- Nuclear
- Renewable power generation
- Electrification

**FIG. 6.17** Cumulative energy-sector investment needed through 2030 and 2050 under IRENA's Transforming Energy Scenario. Note: hydrogen=electrolyzers for hydrogen production; CCS and others=carbon capture and storage for use in industry and material improvements. *Source: IRENA, 2020. Global Renewables Outlook: Energy Transformation 2050 (2020 edition). International Renewable Energy Agency, Abu Dhabi. www.irena.org/-/media/Files/ IRENA/Agency/Publication/2020/Apr/IRENA_Global_Renewables_Outlook_2020.pdf.*

are emerging that show significant potential for further cost reductions" (IRENA Bioenergy for Power, 2020).

Secure, long-term supplies of low-cost, sustainably sourced feedstocks are critical to bioenergy power plants' economics. Feedstock costs can be zero for some types of wastes, including those produced on-site "at industrial installations such as black liquor at pulp and paper mills or bagasse at sugar mills. Their use can sometimes save on disposal costs. Bioenergy for power using sustainably sourced feedstocks can provide dispatchable low-cost electricity and – in many circumstances – heat for industrial processes or heat networks" (IRENA Bioenergy for Power, 2020).

In order to analyze the use of bioenergy power generation, it is important to consider three main factors. These factors are the following:

- Feedstock type and supply;
- The conversion process;
- The power generation technology.

According to IRENA Renewable Power Generation Costs in 2019 (IRENA, 2020a), the economics of bioenergy power generation depends upon the availability of a feedstock supply. This supply should be predictable, sustainably sourced, low-cost, and adequate over the long term. An added complication is that there is a range of cases where electricity generation is not the site's primary activity but is tied to forestry or agricultural processing activities that may impact when and why electricity generation happens. For instance, in electricity generation at pulp and paper plants, a significant proportion of the generated electricity will be used to run their own activities.

Bioenergy feedstocks are very heterogeneous, with the chemical composition highly dependent on the plant species. The cost of feedstock per unit of energy is highly variable too. This cost could range from on-site processing residues that would otherwise cost money to dispose of them to dedicated energy crops that must pay for the land used, harvesting, and logistics of delivery, as well as storage on-site at a dedicated bioenergy power plant. "Examples of low-cost residues combusted for electricity and heat generation are sugarcane bagasse, rice husks, black liquor, and other pulp and paper processing residues, sawmill offcuts and sawdust, and renewable municipal waste streams" (IRENA, 2020a).

In addition to cost, the feedstocks' physical properties matter, as they will differ in ash content, density, particle size, and moisture, with heterogeneity in quality. These factors also impact transportation, pre-treatment, storage costs, and different conversion technologies' appropriateness. Some technologies are relatively robust and can cope with heterogeneous feedstocks, while others require more uniformity (e.g., some gasification processes) (IRENA, 2020a).

A key cost consideration for bioenergy is that most forms have a relatively low energy density. Therefore, collection and transport costs dominate feedstocks' costs derived from forest residues and dedicated energy crops. Consequently, logistical costs increase significantly the further from the power plant the feedstocks need to be sourced. In practical terms, this tends to limit bioenergy power plants' economic size. The lowest electricity cost is achieved once feedstock delivery reaches a certain radius around the bioenergy power plant.

The ordinary share of the feedstock cost in the total LCOE ranges from between 20% and 50% for biomass technologies. However, prices for biomass sourced and consumed locally are difficult to obtain, meaning whatever market indicators are available for feedstock costs must be used as proxies. Alternatively, estimates of feedstock costs from techno-economic analyses that may not necessarily be representative or up-to-date can be used (IRENA, 2015).

According to IRENA Renewable Power Generation Costs in 2019 (IRENA, 2020a), between 2010 and 2019, the global weighted-average LCOE of bioenergy for power projects fell 13.2%, dropping from US$0.076 per kWh to US$0.066 per kWh – a figure at the lower end of the cost of electricity from new fossil fuel-fired projects. Bioenergy for electricity generation offers a suite of options, spanning a wide range of feedstocks and technologies. "Where low-cost feedstocks are available – such as by-products from agricultural or forestry processes on-site – they can provide highly competitive, dispatchable electricity. For bioenergy projects newly commissioned in 2019, the global weighted-average total installed cost was US$2141 per kW, representing an increase of 79.1% in the 2018 weighted-average of US$1693 per kW" (IRENA, 2020a).

Different regions have varying bioenergy power generation costs, with both a technology component and a local cost component to the total. Projects in emerging economies tend to have lower investment costs when compared to projects in the OECD countries, as emerging economies often benefit from lower labor and commodity costs but often also benefit from less stringent environmental regulations, thus allowing lower-cost technologies with reduced emission control investments, albeit in some cases with higher local pollutant emissions (IRENA, 2020a).

Planning, engineering and construction costs, fuel handling and preparation machinery, and other equipment (e.g., the prime mover and fuel conversion system) represent the major categories of total investment costs of a bioenergy power plant. Additional costs are associated with grid connection and infrastructure (e.g., roads). Equipment costs tend to dominate, but specific projects can have high infrastructure and logistics or grid connection costs when located in remote areas. CHP bioenergy installations have higher capital costs than other bioenergy power plants. But due to their higher overall efficiency (between 80% and 85%) and the ability to produce heat and/or steam for industrial processes or space and water heating through district heating networks, the economics associated with this type of bioenergy power plants can improve significantly. Although the pattern of deployment by feedstock varies by country and region, it is clear that total installed costs across feedstocks tend to be higher in Europe and North America and lower in Asia and South America.

For bioenergy projects in North America, costs range from US$591 per kW for landfill gas projects since the technological options used to develop projects are more heterogeneous. The relatively small bioenergy power plants for electricity generation result in low-energy-density bioenergy feedstocks and increasing logistical costs in enlarging the collection area to provide a greater feedstock volume. These are valid reasons to select large-scale bioenergy power plants. The optimal size of a bioenergy power plant to minimize the LCOE of the project, in this context, is a trade-off between the cost benefits of economies of scale and the higher feedstock costs as the average distance to the power plant of the feedstocks sourced grows.

It is important to single out that co-firing bioenergy power plants with coal are "much cheaper since the power plant is already built, and costs are limited to the biomass fuel and its preparation at the plant site. Costs can hover from almost nothing to US$0.03 per kWh for a project where biomass is 10% to 25% of the total fuel input of the power plant. The cost of power from landfill gas can range from US$0.035 to US$0.079 per kWh, depending on the size of the landfill, financing available, distance from the grid or local application, and other factors" (Selasa, 2009).

Concerning biogas power plants, what is the expected cost of constructing this type of power plant? The following factors determine the construction costs of biogas power plants:

- The purchasing costs or opportunity costs for the land for the biogas power plant and slurry storage;
- The model of the biogas power plant;
- The size and dimensions of the biogas power plant;
- The prices of material, labor input, and wages (Anaerobic Digestion Community, 2019).

To gain a rough idea of the typical costs of a simple, unheated biogas power plant, the following figures can be used as a reference, according to Anaerobic Digestion Community (2019) can be used:

- A biogas power plant's total cost, including all vital installations but not including land, is between US$50 and US$75 per $m^3$ capacity.
- Between 35% and 40% of the total costs are for the digester.

Using the figures mentioned above as a reference, the following can be stated: the specific cost of gas production in community biogas power plants or large biogas power plants is generally lower than in small family biogas power plants. The gas distribution costs (mainly piping) usually increase with the size of the biogas power plant. For community bioenergy power plants with several end-users of biogas, the piping costs are high and compensate the degression by "economics of size" partly or wholly. In regions where biogas power plant heating is necessary, "large-scale biogas power plants would be more economical. To keep the construction costs low, labor provided by future biogas users is desirable. Often, the whole excavation work is done without hired labor. On the whole, a reduction of up to 15% of the wages can be affected by user labor. If periods of low farm activities are chosen to construct the biogas power plant, opportunity costs for labor can be kept low" (Anaerobic Digestion Community, 2019).

The main two components of operation and maintenance costs associated with a biogas power plant are wage and material cost for:

- The purchase, collection, and transportation of the substrate;
- Water supply for cleaning the stable and mixing the substrate;
- Feeding and operating of the power plant;
- Supervision, maintenance, and repair of the power plant;
- Storage and disposal of the slurry;
- Gas distribution and utilization;
- Administration (Anaerobic Digestion Community, 2019).

In general, for bioenergy power plants, fixed operations and maintenance costs include labor, insurance, scheduled maintenance, and routine replacement of plant components, such as boilers, gasifiers, feedstock handling equipment, and other items. In total, these operations and maintenance costs account for between 2% and 6% of the total installed costs per year. Large bioenergy power plants tend to have lower per kW fixed operations and maintenance costs due to economies of scale. Variable operations and maintenance costs, at an average of

US$0.005 per kWh, are typically low for bioenergy power plants compared to fixed operations and maintenance costs of other types of power plants. Replacement parts and incremental servicing costs are the main components of variable operations and maintenance costs, including non-biomass fuel costs, such as ash disposal (IRENA, 2020a).

According to Energypedia Electricity Generation from Biogas (2016), economically, biogas' electricity can compete with electricity generation from fossil fuels and other renewable energy sources. "Supporting factors are:

- Rising prices of fossil fuels;
- Low reliability of electricity provision from national grids with the persistent risk of power cuts;
- The vulnerability of hydropower to drought."

Inhibiting factors are, according to Energypedia Electricity Generation from Biogas (2016), the following:

- "Relatively low prices of fossil fuels;
- Need to buy high-quality components from industrialized countries;
- Unfavorable conditions for selling electricity;
- Lack of awareness, capacity, and experience preventing the economic operation of infrastructure components."

A biogas power plant's economic feasibility depends on the entire range of power plant outputs' economic value. These are, according to Energypedia Electricity Generation from Biogas (2016), the following:

- "Electricity or mechanical power;
- Biogas;
- Heat, co-generated by the combustion engine;
- The sanitation effect with chemical and biological oxygen demand reduction in the runoff of agro-industrial settings;
- The slurry is used as fertilizer."

## The efficiency of the bioenergy plants in the North American region

One important factor in increasing the use of bioenergy-fired electricity plants for electricity generation is its extremely high capacity factor – ranging between 85% and 95% – where feedstock availability is uniform over the entire year. That capacity factor is even higher than the capacity factor of nuclear power plants.

According to the report on conversion efficiency of biomass entitled "Biomass Availability and Sustainability Information System" (2015), energy efficiency has become a generic term that "hides various realities and calculation methodologies that can slightly differ from one study to another. In general, efficiency commonly refers to a percentage corresponding to the ratio between the input of energy content and heat and electricity energy output as follow":

*Efficiency = Final energy in the form of heat or electricity / Fuel Input*

The formula mentioned above "is used mostly to obtain the potential efficiency of boilers units. However, in practice, boiler efficiency is directly affected by a wide range of internal and external factors (e.g., climate, scale, the moisture content in biomass, etc.)" (Biomass Availability and Sustainability Information System, 2015).

It can also be "optimized over time (using the produced heat, applying add-ons to increase electricity or heat production, technological improvements of the combustion technology, increasing plant capacity)" (Biomass Availability and Sustainability Information System, 2015). Moreover, if boilers are central elements within bioenergy power plants, "the global efficiency of these plants is also affected by the efficiency of side components and the network in which they are integrated" (Biomass Availability and Sustainability Information System, 2015).

It is well known "that photosynthesis is the process used by plants, algae, and certain bacteria to harness energy from sunlight and turn it into chemical energy. It is essentially the source of the world's food, animal feed, fiber, and timber. But aside from that, photosynthesis is also the source of biomass-based biofuels that are renewable energy sources. To perform photosynthesis, plants, algae, and certain bacteria need three things — carbon dioxide, water, and sunlight — and after this process, starch and sucrose are the main products" (Alessandra, 2019).

The following equation can summarize photosynthesis:

$$CO_2 + H_2O + light\ energy = CH_2O + O_2.$$

It is important to single out that "the maximum efficiency of converting solar energy to biomass energy is estimated to be approximately 4.5% for algae, 4.3% for land plants such as woody, round-leafed plants, and 6% for land plants such as sugarcane, switchgrass, Miscanthus, and sweet sorghum. Additionally, their respiration consumes 30% to 60% of the energy they make from photosynthesis. They also spend half of each day in the dark, thus needing to use previous carbohydrate stores to keep them growing. But even with this low number already, the actual conversion efficiency is even lower than the calculated potential efficiency" (Alessandra, 2019).

On the other hand, capacity factors for bioenergy power plants are very heterogeneous, depending on technology and feedstock availability. "Between 2010 and 2019, the global weighted-average capacity factor for bioenergy projects varied between a low of 65% in 2012 to a high of 86% in 2017. In 2019, the weighted-average LCOE ranged from a low of US$0.057 per kWh in India and US$0.059 per kWh in China to highs of US$0.08 per kWh in Europe and US$0.099 per kWh in North America" (IRENA Bioenergy for Power, 2020), over the last ten years.

However, if the heat is sold, these values would be lower. Individual projects typically generate electricity that costs between US$0.030 and US$0.140 per kWh. Still, higher values exist – up to US$0.250 per kWh – "particularly for waste incineration projects in the OECD where the primary purpose of the process is not electricity generation but waste disposal" (IRENA Bioenergy for Power, 2020).

The availability of feedstock is based on seasonal agricultural harvests. However, capacity factors are typically lower. An emerging issue for bioenergy power plants is the impact climate change may have on feedstock availability and how this might impact its total annual volume and distribution throughout the year. That is an area where the need for research will be ongoing as the climate changes. Biomass power plants that rely on bagasse and landfill gas and other biogases tend to have lower capacity factors, between 50% and 60%. Power plants relying on wood, fuelwood, rice husks, and industrial and renewable municipal waste tend to have weighted-average capacity factors between 60% and 85%. After accounting for feedstock handling, the assumed net electrical efficiency of the prime mover (generator) averages around 30% but varies from a low of 25% to a high of around 36%. CHP plants that produce heat and electricity achieve higher efficiencies, with an overall efficiency between 80% and 85%.

## Uses of bioenergy for electricity generation in the North American region

Bioenergy makes the largest renewable contribution to the global energy supply. Bioenergy is a sustainable and renewable option for reducing fossil fuel demand in the electricity sector. In 2019, a total of 522.6 TWh was produced globally using bioenergy, representing an increase of 27.2 TWh (4%) over the previous year (IRENA Renewable Energy Statistics 2020, 2020).

Bioenergy was, in 2019, the third-largest renewable electricity source after hydropower and wind energy and accounted for 70% of renewable energy consumption. However, the contribution of bioenergy share has decreased by a few percentage points (approximately between 0.5% and 1%) annually, partly due to the decreasing use of traditional biomass sources for electricity generation in several countries.

Without a doubt, in the transport sector, biofuels (bioethanol, biodiesel, etc.) are among the best options for replacing the use of fossil oil. The share of biofuels in the transport sector in 2017 was about 3%.

Considering the traditional use of biomass, in 2017 in the North American region, this type of energy source contributed an estimated 12.4% to final energy consumption. Modern sustainable biomass energy (excluding the traditional use of biomass) provides around half of all renewable energy in final energy consumption (Renewables 2018, 2018).

In 2017, modern biomass energy contributed an estimated 5% to total final energy consumption. It contributed an estimated 5% of the heat total, 3% of the transport total, and 2.1% of the electricity total. "Modern biomass energy use is growing most quickly in the electricity sector at around 9% per year, compared to around 7% in the transport sector; its use for heating is growing more slowly, at around 1.8%" (Renewables 2019 Global Status Report, 2019).

Finally, it is important to single out that "the global bioenergy market size stood at US$ 344.90 billion in 2019 and is projected to reach US$642.1 billion by 2027" (Fortune Business Insights—Power and Renewables/Bioenergy Market, 2021). That represents an increase of 86.2%.

## Advantages and disadvantages in the use of bioenergy plants for electricity generation in the North American region

Bioenergy is, perhaps, one of the most controversial types of renewable energy sources used today for electricity generation and heating in many countries. According to Rinkesh (2020), Ecavo.com (2016), Ames (2018), and McFarland (2017), the main advantages and disadvantages of using bioenergy for electricity generation in the North American region are the following:

A-**Advantages**

- **Bioenergy is a renewable energy source**: Bioenergy is considered a renewable energy source because the organic materials used to produce it are never-ending. The key difference between this and other renewable energy sources, such as solar, wind, and hydro, is the need for replenishment. While plant life is abundant, using it without making efforts to replenish stocks can lead to large amounts of it still being wasted, as can be seen with deforestation (Ecavo.com, 2016);
- **The dependency of bioenergy on fossil fuels is minimal:** By utilizing natural materials for electricity generation and heating, the country's dependency on fossil fuels is minimal. The abundance of bioenergy materials available far outweighs fossil fuels, making it a more readily available fuel source (Ecavo.com, 2016);
- **The use of bioenergy for electricity generation does not generate carbon dioxide**: One of the major advantages of using bioenergy for electricity generation and heating is that it produces fewer harmful greenhouse gases than any other fossil fuel alternatives.
  Bioenergy produces less carbon dioxide than any fossil fuel energy used for electricity generation and heating. The levels of sulfur dioxide, a major component of acid rain, produced by using bioenergy for electricity generation and heating are also extremely low (Ames, 2018).
  It is important to know that bioenergy is a natural part of the carbon cycle. For this reason, the only carbon dioxide emitted to the environment from the use of bioenergy for electricity generation and heating is the amount absorbed by plants in the course of their life cycle. In replenishing the used plant materials, the new ones that spring up absorb an equal quantity of carbon dioxide, hence, developing neutrality that witnesses no new carbon dioxide is generated. This aspect renders bioenergy uniquely clean;
  - **Bioenergy is widely available and easy to store:** Like solar and wind energy, bioenergy sources are abundant in supply. Bioenergy can be found in virtually all countries and regions of the world. In the form of dead leaves, grass, trees, and animal carcasses, organic waste is available in abundance in all countries and can be used to produce bioenergy. However, this abundance must be maintained. While bioenergy will always be available because they are part of the planet's natural life cycle, this does not mean acting in an irresponsible manner concerning their use (Ecavo.com, 2016).
    It is important to single out that bioenergy is not only easy to source but also to stored;
  - **Bioenergy can be used in many forms**: Bioenergy is among the most versatile renewable energy sources available for electricity generation and heating. It can be converted into many different fuel sources, each of which has varied applications, creating different products from different organic matter forms (Ecavo.com, 2016).
    It can be used to produce methane gas, biodiesel, and other biofuels. It can also be used directly as heat or to generate electricity using a steam turbine (Rinkesh, 2020);
  - **Bioenergy helps reduce waste:** "Most waste produced in homes is either plant matter or biodegradable. This kind of waste can be channeled to more profitable use. Bioenergy generation utilizes any waste that would have otherwise found a way into landfills. Bioenergy uses this waste so that it's no longer sitting in landfills. Minimized waste means a reduction of land intended for landfills; hence, more space for human habitats" (Rinkesh, 2020).

By burning solid waste, the amount of garbage dumped in landfills is reduced between 60% and 90%, decreases the cost of landfill disposal and the amount of land required for landfill (McFarland, 2017);

- **Bioenergy is cheaper than fossil fuels**: Producing bioenergy does not involve high upfront capital investment. Compared to drilling for oil or creating gas pipelines, the costs involved in collecting biomass fuels are extremely low. This low cost can be passed onto consumers, "whose energy bills can then no longer be dependent on issues like availability and the decisions of the companies who supply energy. These low costs also make bioenergy more attractive to producers, as they can enjoy higher profits for less output" (Ecavo.com, 2016). Besides, producers of waste can add value by channeling their garbage to create a more profitable use in the form of bioenergy (McFarland, 2017);
- **Bioenergy is a domestic production**: Bioenergy can take energy production out of larger energy companies' hands. That means that people no longer need to be beholden to power companies and their charges. Furthermore, bioenergy means "that practically anybody could produce and use it on a domestic level. While this does take some work, even something as simple as burning wood instead of using a central heating system can save money and have a more beneficial effect on the environment" (Ecavo.com, 2016).

## B-Disadvantages

While the advantages of using bioenergy for electricity generation and heating are many, it is not a perfect energy source. As with all energy sources, the disadvantages of using this type of energy source for electricity generation and heating must also be considered. That is crucial if they are to be confronted properly to allow bioenergy to work for those who need it. Some of these disadvantages are related directly to the use of bioenergy for electricity generation and heating, whereas others are indirect consequences of its production or application.

- **Bioenergy is not totally clean when burned:** The biggest contention against the use of bioenergy as a clean energy source for electricity generation and heating is the pollution created from burning wood and other natural materials. Burning wood and other plant life does create other emissions in addition to carbon. These emissions can pollute the local environment, even if the effects are not as drastic as they may be from the use of fossil fuels for electricity generation and heating (Ecavo.com, 2016).
Some biomass resources used to produce electricity are crops, forest products, agricultural waste, and urban waste. However, the biomass feedstock and the way it is harvested can negatively impact land use and global warming emissions. Furthermore, using tree or tree products to create bioenergy comes with its own set of problems. For example, substantial forest land needs to be cleared to collect enough lumber, which causes topical changes and destroys animal habitats.
For this reason, the use of bioenergy for electricity generation and heating is no complete better off from the environmental point of view compared to other energy sources, particularly renewable energy sources.
- **Bioenergy can lead to deforestation:** Uncontrolled bioenergy production can result in deforestation because wood is a major bioenergy source in many countries. A large amount of wood and other waste products have to be burned to produce a considerable

quantity of bioenergy. The desire to produce bioenergy using wood as an energy source on a large scale can lead to deforestation destroying many plants and animals' habitats.

"In fact, this is the main reason for slowing down the large scale use of biomass fuel in many countries. Governments feel replanting efforts may not match the rate of cutting down of trees" (Rinkesh, 2020);

- **Bioenergy is not as efficient as fossil fuels for electricity generation and heating:** Even though bioenergy is natural in many ways, it does not get close to fossil fuel efficiency. In fact, some renewable energy sources like biofuels are fortified with fossil fuels to increase their efficiency.

    As biodiesel, ethanol is inefficient compared to gasoline. For this reason, it often has to be mixed with some gasoline to make it work properly anyway. On top of that, it is important to single out that ethanol is harmful to combustion engines over long term use;

- **Bioenergy requires a lot of space:** A great amount of land is needed for some biomass crops to be produced and, when they have grown, the product requires a large amount of storage room before being converted into bioenergy (Ames, 2018). Often, bioenergy power plants are located in urban areas, which means that they are causing more traffic and pollution in the area;

- **Bioenergy is expensive:** While the cost of extracting bioenergy is lower than most types of fossil fuels, it still generally exceed those of many other forms of renewable energy. In some cases, bioenergy projects are considered not worth the price of implementing them, especially when solar, wind, and hydro alternatives are available. This cost comes from the need for biomass resources to be maintained and for extracted biomass to be replanted. Furthermore, the cost of the machinery used in biomass extraction is also a factor to be considered as well as its transportation to the bioenergy power plant (Ecavo. com, 2016).

    Transport and resource gathering expenses are high and will be continually needed every day. Transport the fuel to the bioenergy power plant and the carbon emissions and pollution made in doing so could be very high compared with other energy sources, particularly renewable energy sources;

- **Bioenergy requires water:** "An often unseen disadvantage of bioenergy is the amount of water needed in production. All plants need water to live, which means sources must be available at all times. Not only does this lead to increased costs in terms of irrigation, but it may result in water sources becoming less available to people and wildlife. Furthermore, with water itself being an alternative form of energy, which is also far cleaner than bioenergy, it raises the question of why the water is not used for that purpose instead" (Ecavo.com, 2016);

- **Bioenergy can be harmful to human health:** The use of bioenergy for electricity generation and heating can be harmful to human health if burned in an enclosed space without proper ventilation, where airborne toxins from the burning particulate matter produce health risks.

    "Biomass from livestock excrement, which often contains harmful bacteria, can be detrimental to the ecosystem and human health. For example, during a heavy rainfall event, the waste material will run off into nearby waterways that many people use to drink and bath. This exposure can lead to an assortment of illnesses like gastrointestinal diseases and developmental problems such as stunted growth in children" (Greentumble, 2018).

Using animal and human waste to power engines may save on carbon dioxide emissions, but it increases methane gases, which are also harmful to the Earth's ozone layer (Rinkesh, 2020). One important element must be considered when using waste products for electricity generation and heating: smell;

- **Bioenergy is still under development:** More needs to be done to harness bioenergy potential. However, it is held back as a renewable energy source by many of the disadvantages mentioned here. When compared to solar and hydro sources, bioenergy is inefficient and under-researched. Scientists are still working on ways to make bioenergy more efficient. Bioenergy is unlikely to be adopted as a viable alternative energy source on a wide scale until that happens.

- **Bioenergy consumes more fuel than other energy sources**: Using trees and tree products to power machines is inefficient. Not only does it take a lot more fuel to do the same job as using conventional fuels, but it also creates environmental problems of its own. For example, to collect enough lumber to power a country full of vehicles or even a power plant, companies would have to cut considerable forest area. That results in major topological changes and destroys the homes of countless animals and plants.

## The future of bioenergy in the North American region

Bioenergy plays a significant role in North America for achieving the Paris goals of limiting climate change to well under 2 °C, despite the decision to reject the Paris agreement on climate change adopted by the former Trump administration.[i]

However, "several studies have challenged the greenhouse gas accounting, raising the concern that lifecycle emissions have been underestimated and that the 'carbon debt' associated with bioenergy often results in greater near-term emissions than the fossil fuels being replaced" (Reid et al., 2019). In addition, "other work emphasizes the prospect that growth in bioenergy could reduce food production and accelerate biodiversity loss" (DeCicco and Schlesinger, 2018; European Academies Science Advisory Council, 2019; Searchinger et al., 2018).

Without a doubt, bioenergy is a significant part of the energy economy, accounting for 9.5% of the total primary energy supply[j] and some 70% of renewable energy in use today (IEA, 2017b; IEA et al., 2019). More than half of this bioenergy involves the traditional use of biomass, mostly in households for cooking and heating and within small industries such as charcoal and brick kilns. While there is considerable scope for improving the sustainability, efficiency, and health safety associated with the use of traditional biomass for electricity generation and heating in the North American region (Creutzig et al., 2015), the use of modern bioenergy for the same purpose will grow more in the coming decades. Modern bioenergy

---

[i] The new Biden administration has declared that the country will sign the Paris agreement after January 21, 2021, and it is expected that the US Senate ratify it later.
[j] The recent Intergovernmental Panel on Climate Change (IPCC) Special Report on Global Warming found that biomass made up a median of 26% of primary energy in 2050 (range from 10% to 54%), up from 10% in 2020 (Rogelj et al., 2018).

was responsible for half of all renewable energy consumed in 2017, providing four times the contribution of solar PV and wind power combined. That role is expected to be higher in the future.

Most bioenergy delivers heat in buildings and industry, but bioenergy is also expected to account for 3% of electricity production and around 4% of transport energy demand in 2023.

"Production of liquid biofuels for transportation grew at annual rates greater than 10% prior to 2010 but then slowed to 4% annual growth during the period 2010-16. The annual average growth rate of bioenergy electricity capacity was 6.5%" during the same period (IEA, 2017b). Throughout the period 2018–23, "bioenergy (including liquid biofuels) is projected to account for 30% of the growth in renewable energy production". Liquid biofuels are considered one of the most cost-competitive sources of high energy density liquid fuels within the North American region. They have been seen as a good replacement for maritime and aviation fuels and fuel that can help meet demands for any road transportation difficult to electrify. "The IEA has projected that advanced biofuels will comprise 50% of the fuel mix for shipping and 70% of aviation fuel demand in 2060" (IEA, 2017a).

However, the coronavirus pandemic has changed the scenario described above. "Lower transport fuel demand resulting from the Covid-19 crisis reduces biofuel consumption in countries where mandate policies require a set percentage of biofuels to be blended with fossil transport fuels. Overall, global gasoline demand is anticipated to contract by 10%, and diesel by 6% in 2020. Diesel is less affected, as a substantial share of its consumption is for the transport of goods, which the crisis has impacted less than personal mobility" (IEA Renewables 2020, 2020).

Without a doubt, "a lowering of crude oil prices since the start of the coronavirus pandemic has made biofuels less competitive with fossil transport fuels. The average crude oil price for 2020 is currently estimated at around US$40 per barrel, down from US$64 per barrel in 2019. The drop in oil price has reduced unblended ethanol purchases in Brazil (-17% in the first half of 2020) and delayed increases to biofuel blending rates in the Association of Southeast Asian Nations (ASEAN) region" (IEA Renewables 2020, 2020).

This delay is because using biofuels would imply higher costs for governments that subsidize this energy source to ensure their competitiveness with gasoline and fossil diesel. "Lower crude oil prices have also caused biofuel prices to fall, albeit generally to a lesser extent, which challenges production economics for some plants" (IEA Renewables 2020, 2020).

"If fossil transport fuel demand rebounds to close to pre-pandemic levels and policy support in key markets continues to expand, transport biofuel production could reach 162 billion liters in 2021, a return to the 2019 level" (Tavares Kennedy, 2020) if the current situation change. Output in 2022 is anticipated to increase 4% or an increase to 169 billion liters with respect to 2021. During the period 2023–25, the average global output of 182 billion liters is anticipated. The greatest production increases for ethanol are expected in China and Brazil, and biodiesel and Hydrotreated Vegetable Oil in the USA and the ASEAN region. "Biofuels are expected to meet around 5.4% of road transport energy demand in 2025, rising from just under 4.8% in 2019" (IEA Renewables 2020, 2020).

Another important element that should be considered to ensure the correct use of bioenergy in the North American region is land availability. Without a doubt, land availability could be a limiting factor in the increased use of bioenergy in the region. Bioenergy is not expected to be a competitive energy source over the long-term. That is because it requires so much land to generate the raw materials needed for electricity generation and heating, which is a scarce resource at the world level.

The conversion of natural habitats and eco-systems to managed landscapes, agriculture, and urban areas has already undermined eco-system services that humanity depends upon and has placed one million species at the risk of extinction (Díaz et al., 2019; Reid et al., 2005). Thus, there is a need to protect remaining natural ecosystems and, wherever possible, restore lost or degraded ecosystem services from degraded lands or lands retired from food or fiber production. "The need for conservation and restoration is largely incompatible with the large-scale expansion of land-intensive bioenergy" (Reid et al., 2019). The use of land to produce energy by bioenergy sources should be used most efficiently in the future if bioenergy wishes to continue to play an important role within the North American energy matrix.

Even though a relatively inefficient land use and the potential competition with other land uses within the North American region, bioenergy plays a significant role in most energy scenarios past mid-century for three reasons. These reasons are the following:

- First, unlike intermittent energy sources, bioenergy can meet the needs for baseload electrical power, a characteristic thought to be increasingly important as the existing fossil fuel-based thermal capacity is retired[k];
- Second, shipping and aviation applications require fuels with a high energy density, and biofuels can meet this criterion at a relatively low cost[l];
- Third, it can provide a carbon-negative energy source.

In the North American region, particularly in the USA, rather than considering bioenergy as a substitute for baseload power production, it will compete with other energy sources to supply this firm power for inter-day and seasonal loads mid-century balancing. There are several reasons for this:

---

[k] "Global bioenergy electricity capacity expanded 8.5 GW in 2019, the second-highest level of annual additions on record. However, the forecast anticipates a 16% decline in bioenergy capacity additions in 2020. Major deployment of bioenergy power projects is concentrated in relatively few countries, with just ten nations accounting for 90% of new capacity in 2019. Of these, China, Brazil, Japan and the UK have been the most affected by the pandemic, so potential exists for some project delivery delays. "Nevertheless, with forestry activity ongoing and ports operational, widespread supply disruptions of biomass fuels (e.g. wood chips and pellets) for existing projects have not been observed" (IEA Renewables 2020, 2020). According to IRENA Renewable Energy Statistics, the global bioenergy electricity capacity expanded 6.3 GW in 2019 with respect to 2018.

[l] The biofuels industry has been strongly impacted by the Covid-19 pandemic. Global transport biofuel production in 2020 is anticipated to be 144 billion liters, equivalent to 2480 thousand barrels per day. This biofuel production level is 11.6% lower than 2019's record output. It is also the first reduction in annual biofuel production in two decades. While this projection is a slight upward revision from the IEA forecast update in May 2020, it is far below the 3% growth anticipated for 2020 in the pre-pandemic forecast. The greatest year-on-year drops in output are for USA and Brazilian ethanol, and European biodiesel (IEA Renewables 2020, 2020).

- First, because of the low cost of natural gas in the North American region, particularly in the USA, gas infrastructure is more pervasive and growing much faster than bioenergy infrastructure;
- Second, where that infrastructure exists, the least cost option for baseload power will involve using existing natural gas power plants at a reduced capacity factor to avoid stranding those assets;
- Third, because they will be online for limited periods, their emissions will be relatively low;
- Fourth, by mid-century, it is likely that this type of power plant will use CCS or hydrogen. Already the USA provides a tax credit for power plants that capture and store $CO_2$, and investments in new carbon capture technologies are growing.

Undoubtedly, the most attractive use of bioenergy as baseload power may not be land-intensive biofuel but biogas. Thus, it is likely that the USA expand the use of biogas for baseload power in the future. For this reason, the US budget for energy research and development increased by more than 12% in 2018, with notable rises for solar energy, hydrogen, and alternative vehicle technologies. The budget approved for 2019 continues the upward trend but with lower growth rates in most areas. The five leading countries for public spending on energy research and development were the USA, China, Japan, France, and Germany. These five countries accounted for around 70% of all such spending worldwide. Research and development budgets in Denmark and Italy also increased in 2018, while in Germany declined by 2% (IEA World Energy Investment 2019, 2019). This trend reflects recent pledges and commitments adopted by several governments.

"On a larger scale, North America ranks third in the world in the waste-to-energy movement, behind the European countries and the Asia Pacific region" (Folk, 2020).

## Bioenergy capacity in Canada

Bioenergy comprises about 6% of Canada's total energy supply and is the country's second-largest renewable energy source after hydro. According to the Canada Energy Regulator (2020) and IRENA Renewable Energy Statistics 2020, in 2015, there were approximately 70 biomass power plants with a total installed capacity of 2417 MW (see Fig. 6.18). Most of these power plants rely on wood, wood by-products, and landfill gas for electricity generation. In 2015, these power plants' capacity represented 1.7% of the total energy capacity installed in the country and 1.9% of the total electricity generated in that year (12,683 GWh). In 2019, the bioenergy capacity installed in Canada reached 3376 MW, an increase of 39.7% with respect to 2015, generating 10,661 GWh in 2018.

British Columbia, Ontario, Alberta, Quebec, and New Brunswick are the Canadian provinces with the largest bioenergy capacity installed and generation. In addition, some Canadian municipalities, such as Edmonton, Nanaimo, and Vancouver/Burnaby, actively produce energy from waste, either from landfills or large-scale anaerobic digesters. Increasingly, Canadian landfills and waste-to-energy facilities generate electricity for nearby utilities and industries or convert landfill gas to natural gas to be moved in natural gas pipelines.

**FIG. 6.18**  Map of bioenergy power plants in Canada. *Source: NEB publication titled "Canada's Adoption of Renewable Power Sources – Energy Market Analysis," published in May 2017. This figure was adapted from the NRCan map titled "Renewable Electric Power Plants, North America" (June 2016).*

Canada has been privileged in using bioenergy sources due to its abundant biomass products, mainly from the forestry industry. This type of energy source "has been growing within the Canadian industry, providing a variety of new jobs to replace the lost jobs that were formerly reliant on traditional forest-related jobs," according to the Government of Canada Statistics (2016). Furthermore, after the sharp decline in the paper and pulp industry over the past 20 years, bio-energy has become an integral part of Canada's renewable energy sector (Bradburn, 2014).

In 2014, Canada generated 8.7 GWh of electricity using wood, organic municipal solid wastes, and landfill gas as fuel. That was most prominently seen where forestry industries are still prevalent: British Columbia, Ontario, Quebec, Alberta, and New Brunswick, according to the Government of Canada Statistics (2016).

The evolution of the bioenergy installed capacity in Canada during the period 2010–19 is shown in Fig. 6.19.

According to Fig. 6.19, the following can be stated: bioenergy installed capacity in Canada increased by 97.8% during the period 2010–19, rising from 1707 MW in 2010 to 3376 MW in 2019. However, the period under consideration can be divided into two parts: The first part, covering the period 2010–13, the bioenergy installed capacity in Canada decreased 11.5%, falling from 1701 MW in 2010 to 1512 MW in 2013. In the second part, covering the period 2013–19, Canada's bioenergy installed capacity increased by 123%, rising from 1512 MW in 2013 to 3376 MW in 2019. It is expected that this last trend will continue in the future but with a slower face, considering the Canadian government's decision to reduce the use of fossil fuels for electricity generation, especially coal and oil.

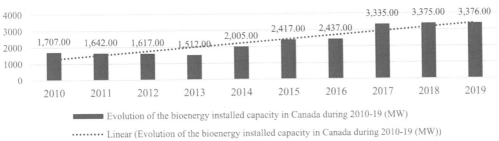

FIG. 6.19  Evolution of the bioenergy installed capacity in Canada during the period 2010–19. *Source: 2020. IRENA Renewable Energy Statistics 2020. IRENA.*

The peak in bioenergy installed capacity in Canada during the period 2009–18 was reached in 2016 with 12,685 MW. The list of the main bioenergy power plants operating in Canada in 2020 is shown in Table 6.2.

In 2019, solid biofuels and renewable waste installed capacity in Canada reached 2360 MW, the same capacity reported in 2018. The evolution of the solid biofuels and renewable waste installed capacity in the country during the period 2010–19 is shown in Fig. 6.20.

According to Fig. 6.20, the following can be stated: solid fuels and renewable waste installed capacity in Canada increased by 50.2% during the period 2010–19, rising from 1571 MW in 2010 to 2360 MW in 2019. However, the period under consideration can be divided into two parts: The first part, covering the period 2010–13, the solid fuels and renewable waste installed capacity in Canada decreased 11.2%, falling from 1571 MW in 2010 to 1396 MW in 2013. In the second part, covering the period 2013–19, Canada's solid fuels and renewable waste installed capacity increased by 69.1%, rising from 1396 MW in 2013 to 2360 MW in 2019. Thus, the evolution of solid fuels and renewable waste installed capacity in Canada is similar to the evolution of the country's bioenergy installed capacity during the same period. The reason is the following: the main bioenergy sources in Canada used for electricity generation are solid fuels and renewable waste representing, in 2019, 90.1% of the different bioenergy sources used in Canada for this specific purpose.

Another type of bioenergy source used in Canada for electricity generation is renewable municipal waste but on a small scale. In 2019, the country's renewable municipal waste installed capacity was only 38 MW, representing 3.4% of the regional total and 0.3% at the world level. In that year, the electricity generated by renewable municipal waste was only 172 GWh, representing 2% of the regional total and 0.3% at the world level.

In 2016, the renewable solid municipal waste generated in Canada was 34.2 million tonnes, 9.2 million tonnes diverted (27%), and 24.9 million tonnes disposed of (73%). From the total renewable municipal waste, 41% is residential, and 59% is non-residential. More than 95% is landfilled, and less than 5% is thermally treated (mostly with energy recovery, although this also includes open burning) (Statistics Canada, 2016).

The evolution of Canada's renewable municipal waste installed capacity during the period 2010–19 is shown in Fig. 6.21.

**TABLE 6.2** Main bioenergy power plants.

| Bioenergy power plants | Capacity (metric tons per year) |
|---|---|
| BioPower | 110,000 |
| Canfor Energy North Ltd. | 78,800 |
| Chetwynd Sawmill | 100,000 |
| Eagle Valley Fuel Pellets | 100,000 |
| Energex Pellet Fuel Inc. | 114,000 |
| Grand River Pellets | 100,000 |
| Granule 777 | 210,000 |
| Granules de la Mauricie | 55,000 |
| Granules LG Inc. | 120,000 |
| Great Northern Pellets | 100,000 |
| Groupe Savoie Inc. | 90,000 |
| Houston Pellet Limited Partnership | 230,000 |
| KD Quality Pellets | 88,000 |
| La Crete Sawmills Ltd. | 140,000 |
| Lauzon Recycled Wood Energy-Saint-Paulin | 60,000 |
| Pacific Bioenergy Pellet Mill | 350,000 |
| Pinnacle Renewable Energy Inc.-Armstrong | 75,000 |
| Pinnacle Renewable Energy Inc.-Burns Lake | 440,000 |
| Pinnacle Renewable Energy Inc.-Meadowbank | 270,000 |
| Pinnacle Renewable Energy Inc.-Williams Lake | 210,000 |
| Pinnacle Renewable Energy-Entwistle | 400,000 |
| Pinnacle Renewable Energy-Lavington | 300,000 |
| Pinnacle Renewable Energy-Smithers | 125,000 |
| Premium Pellet Ltd. | 185,000 |
| Shaw Resources-Belledune | 100,000 |
| Shaw Resources-Shubenacadie | 50,000 |
| Skeena Bioenergy Ltd. | 75,000 |
| Vanderwell Contractors Ltd. | 60,000 |

Total power plants: 28.
*Source: Biomass Magazine.*

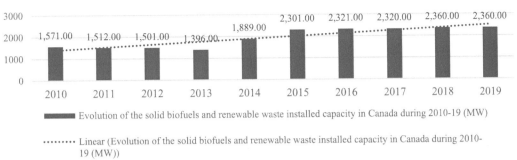

FIG. 6.20   Evolution of the solid biofuels and renewable waste installed capacity in Canada during the period 2010–19. *Source: 2020. IRENA Renewable Energy Statistics 2020. IRENA.*

Based on the data included in Fig. 6.21, the following can be stated: the renewable munici-pal waste in Canada increased by 111% during the period 2010–19, rising from 18 MW in 2010 to 38 MW in 2019. However, the new capacity installed remained almost unchanged after the increase in the renewable municipal waste installed capacity in Canada from 18 MW in 2011 to 39 MW in 2012. Therefore, it is expected that Canada's renewable municipal waste installed capacity will not change significantly, at least during the coming years.

Another type of bioenergy source used by Canada for electricity generation can be classi-fied as "other solid biofuels." In 2019, other solid biofuels installed capacity in Canada reached 2322 MW, representing 20.8% of the regional total and 3.4% at the world level. The evolution of other solid biofuels installed capacity in Canada during the period 2010–19 is shown in Fig. 6.22.

Based on the data included in Fig. 6.22, the following can be stated: other solid biofu-els installed capacity in Canada increased by 49.5% during the period 2010–19, rising from 1553 MW in 2010 to 2322 MW in 2019. The period under consideration can be divided into

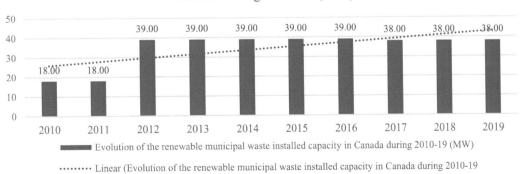

FIG. 6.21   Evolution of the renewable municipal waste installed capacity in Canada during the period 2010–19. *Source: 2020. IRENA Renewable Energy Statistics 2020. IRENA.*

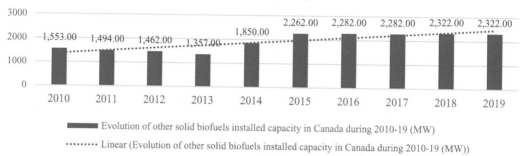

FIG. 6.22   Evolution of other solid biofuels installed capacity in Canada during the period 2010–19. *Source: 2020. IRENA Renewable Energy Statistics 2020. IRENA.*

two parts. In the first part, covering the period 2010–13, other solid biofuels installed capacity decreased by 12.7%, falling from 1553 MW in 2010 to 1357 MW in 2013. In the second part, covering the period 2019–19, other solid fuels installed capacity increased by 71.1%, rising from 1357 MW in 2013 to 2322 MW in 2019. Therefore, it is expected that this type of bioenergy source's installed capacity will not increase significantly, at least during the coming years.

Biofuel is made from biological material. Biomass from agricultural waste and crops is predominantly used to make biofuel. Biofuel may be classified into two categories, dependent on its source. Primary biofuel is made from unprocessed, natural materials such as wood chips. Secondary biofuel is made from processed primary energy sources better to adapt them to a broader range of ethanol applications, according to Renewable Energy in Canada (2019) report.

It is important to single out that Canada is considered a major world biofuel producer. "Canada produces over 250 million liters per year. In recent years, the Canadian government has begun funding the research and development of biofuel production. A 5% biofuel mandate was implemented with coordination between federal and provincial governments back in 2010" (Mabee, 2007).

It is considered that biofuels have many benefits when compared to other types of fuels. One of them is that biofuel reduces greenhouse gas emissions and decreases fossil fuel dependency. Canada's potential growth as a world leader in biofuel production exists thanks to a robust agricultural sector, producing significant biofuel feedstocks. In addition, implementing the Canadian government's Environmental Protection Act of 2008 requires that gasoline in the country consist of at least 5% biofuel (Mukhopadhyay and Thomassin, 2011).

There are additional mandates in place that require at least 2% ethanol in biodiesel and heating oil. That would require that Canada produce at least two hectometers of ethanol per year, opening the door to a significant increase in ethanol production during the coming years. That would equate to an additional 1.9 billion liters of ethanol needed to be produced to meet demand. Ethanol in Canada is produced from cereal grains. Corn and wheat account for virtually all ethanol output (Laan et al., 2009).

The Canadian federal government started the Ethanol Expansion Program back in 2008, intending to kick-start its production across the country. Subsidies were given to producers to persuade them to continue ethanol production while reducing greenhouse gas emissions.

Beyond question, the biofuel sector developments will have major impacts on the Canadian economy, especially the agricultural sector, during the coming years (Laan et al., 2009).

However, it is important to know that there are "environmental concerns about biofuel production in Canada despite its benefits, including water usage, land usage, habitat loss, and nitrogen runoff" (Laan et al., 2009). The biofuel mandate that the government put in place requires that it reviews biofuel production's environmental impacts every two years. There are also concerns that the lack of thorough assessment wastes the multi-billion dollar funding for these environmental reviews. Ethanol and biodiesel are often cited as ways to reduce greenhouse gas emissions. However, several other alternatives can work, as well. "These include the use of electric cars and the development of more efficient engines. Furthermore, R&D tax incentives and carbon taxes can ignite energy innovation in Canada and might demonstrate that biofuel is not the only answer to our sustainable energy needs" (Laan et al., 2009).

In the specific case of liquid biofuels, since 2017, Canada's installed capacity of this type of bioenergy source was stable at 876 MW, representing 84.1% of the regional total and 30.8% worldwide. However, there are no records in the country on the electricity generated by this type of bioenergy source.

In 2019, the biogas installed capacity in Canada reached 140 MW representing 5.2% of the regional total and 0.7% at the world level. In 2018, the electricity generated by the biogas power plants in Canada reached 972 GWh, representing approximately 1.3% of Canada's electricity demand.

The evolution of the biogas installed capacity in Canada during the period 2010–19 is shown in Fig. 6.23.

According to Fig. 6.23, the following can be stated: the biogas installed capacity in Canada increased by 2.9% during the period 2010–19, rising from 136 MW in 2010 to 140 MW in 2019. Analyzing Fig. 6.23, four clear phases can be identified. In the first phase, covering the period 2010–12, the country's biogas installed capacity decreased by 14.8%, falling from 136 MW in 2010 to 116 MW in 2012. In the second phase, covering the period 2012–16, the biogas installed capacity remained stable at 116 MW. In the third phase, covering the years 2016 and 2017, the biogas installed capacity increased by 19.8%, rising from 116 MW in 2016 to 139 MW in 2017. Finally, in the fourth phase, the biogas installed capacity remained stable at 139–140 MW. It is expected that this trend will continue without change during the coming years.

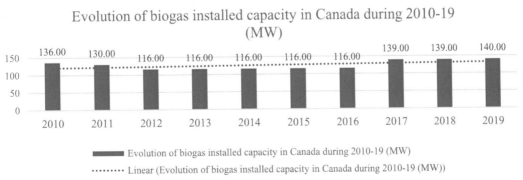

FIG. 6.23 Evolution of the biogas installed capacity in Canada during the period 2010–19. *Source: 2020. IRENA Renewable Energy Statistics 2020. IRENA.*

## Bioenergy electricity generation in Canada

Bioenergy comprises about 6% of Canada's total energy supply and is the country's second-largest renewable energy source after hydro. British Columbia, Ontario, Quebec, New Brunswick, Nova Scotia, and Alberta are the Canadian provinces with the largest bioenergy capacity and generation (see Table 6.3).

Some Canadian municipalities, such as Edmonton, Nanaimo, and Vancouver/Burnaby, actively produce energy from waste, either from landfills or large-scale anaerobic digesters. Increasingly, Canadian landfills and waste-to-energy facilities generate electricity for nearby utilities and industries or convert landfill gas to natural gas to be moved in natural gas pipelines. In Canada, "the most commonly employed biomass is wood; wood waste is used to produce heat for industrial facilities, create steam for electricity production, and water and space heating" (Bradburn, 2014).

In 2018, the electricity generated by the Canadian bioenergy power plants reached 10,661 GWh. The evolution of the electricity generated by the Canadian bioenergy power plants during the period 2009–18 is shown in Fig. 6.24.

According to Fig. 6.24, the following can be stated: bioenergy electricity generation in Canada increased by 37.9% during the period 2009–18, rising from 7733 GWh in 2009 to 10,661 GWh in 2018. However, the period under consideration can be divided into the following manner:

- During the period 2010–13, the bioenergy electricity generation decreased by 7.5%, falling from 10,392 GWh in 2010 to 9860 GWh in 2013;
- After an increase of 27.2% in 2014 concerning 2013, the bioenergy electricity generation in the country remains relatively stable at 12,683 GWh until 2016;

**TABLE 6.3**   Bioenergy generating capacity in Canadian provinces (MW).

| Provinces | Total (MW) |
|---|---|
| Newfoundland and Labrador | 27 |
| Prince Edward Island | 2 |
| Nova Scotia | 66 |
| New Brunswick | 113 |
| Quebec | 205 |
| Ontario | 681 |
| Manitoba | 52 |
| Saskatchewan | 16 |
| Alberta | 55 |
| British Columbia | 827 |
| Canada | 2043 |

According to IRENA Renewable Energy Statistics 2020, in 2016, the bioenergy capacity installed in Canada was 2437 MW.
*Source: Government of Canada Statistics, 2016. (Chapter 1). www.statcan.gc.ca. Retrieved 29,3,2017.*

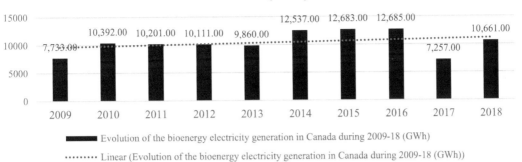

FIG. 6.24 Evolution of the electricity generated by the Canadian bioenergy power plants during the period 2009–18. *Source: 2020. IRENA Renewable Energy Statistics 2020. IRENA.*

- During the years 2016 and 2017, the bioenergy electricity generation in Canada decreased significantly (74.8%), falling from 12,685 GWh in 2016 to 7257 GWh in 2017;
- During the years 2017 and 2018, the country's bioenergy electricity generation increased once again by 46.9%, rising from 7257 GWh in 2017 to 10,661 GWh in 2018.

It is expected that bioenergy electricity generation will increase during the coming years due to Canada's decision to reduce coal and oil's role in the country's energy mix during the coming decades and the closure of some nuclear power plants.

In 2018, Canada's electricity using solid fuels and renewable waste as fuel reached 9689 GWh, a 54.2% increased with respect 2017. The evolution of the electricity generated by solid fuels and renewable waste in Canada during the period 2009–18 is shown in Fig. 6.25.

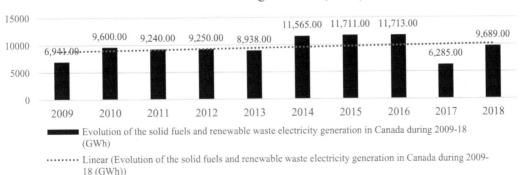

FIG. 6.25 Evolution of the electricity generated by solid fuels and renewable waste in Canada during the period 2009–18. *Source: 2020. IRENA Renewable Energy Statistics 2020. IRENA.*

According to Fig. 6.25, the following can be stated: solid fuels and renewable waste electricity generation in Canada increased by 37.9% during the period 2009–18, rising from 7733 GWh in 2009 to 10,661 GWh in 2018. However, the period under consideration can be divided into the following manner:

- During the period 2010–13, the solid fuels and renewable waste electricity generation decreased by 5.1%, falling from 10,392 GWh in 2010 to 9860 GWh in 2013;
- After an increase of 29.4% in 2014 concerning 2013, the bioenergy electricity generation in the country remain relatively stable at around 11,711 GWh;
- During the years 2016 and 2017, the solid fuels and renewable waste electricity generation in Canada decreased significantly (46.4%), falling from 11,713 GWh in 2016 to 6285 GWh in 2017;
- During the years 2017 and 2018, the country's solid fuels and renewable waste electricity generation increased once again by 54.2%, rising from 6285 GWh in 2017 to 9689 GWh in 2018.

The evolution of solid fuels and renewable waste electricity generation in Canada is similar to the evolution of bioenergy electricity generation in the country in the same period. The reason is the following: the main bioenergy sources in Canada used for electricity generation are solid fuels and renewable waste representing, in 2019, 90.1% of the different bioenergy sources used in Canada for electricity generation.

The peak in solid fuels and renewable waste electricity generation in Canada during the period 2009–18 was reached in 2016 with 11,713 GWh. It is expected that the role of solid fuels and renewable waste in the Canadian energy mix during the coming years will be higher than today due to the Canadian government's decision to reduce the use of coal and oil for electricity generation and the closure of some nuclear power plants.

Another type of bioenergy source used in Canada for electricity generation is renewable municipal waste but on a small scale. In 2019, the country's renewable municipal waste installed capacity was only 38 MW, representing 3.4% of the regional total and 0.3% at the world level. In that year, the electricity generated by renewable municipal waste was only 172 GWh representing 2% of the regional total and 0.3% at the world level.

In 2018, the electricity generated by renewable municipal waste power plants in Canada reached 172 GWh, representing only 2% of the regional total and 0.3% at the world level. The evolution of the electricity generation by renewable municipal waste power plants in Canada during the period 2010–18 is shown in Fig. 6.26.

According to Fig. 6.26, the following can be stated: electricity generated by the renewable municipal waste in Canada increased by 24.6% during the period 2010–18, rising from 138 GWh in 2010 to 172 GWh in 2018. It is important to single out that after an increase in the electricity generated by the renewable municipal waste registered in 2014 of 8.2% with respect to 2013, no further increase was reported during the period 2014–18. It is expected that this trend will not change at least during the coming years.

Another type of bioenergy source used by Canada for electricity generation can be classified as "other solid biofuels." In 2018, the electricity generated by other solid biofuels power plants in Canada reached 9517 GWh, representing 17% of the regional total and 3.1% at the world level. The evolution of the electricity generated by this type of bioenergy power plant in Canada during the period 2010–18 is shown in Fig. 6.27.

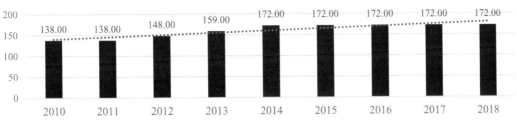

**FIG. 6.26** Evolution of the electricity generation by renewable municipal waste power plants in Canada during the period 2010–18. *Source: 2020. IRENA Renewable Energy Statistics 2020. IRENA.*

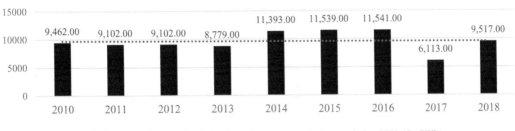

**FIG. 6.27** Evolution of the electricity generated by this type of bioenergy power plant in Canada during the period 2010–18. *Source: 2020. IRENA Renewable Energy Statistics 2020. IRENA.*

Based on the data included in Fig. 6.27, the following can be stated: the use of other solid biofuels for electricity generation in Canada during the period 2010–18 increased very little (0.6%), rising from 9462 GWh in 2010 to 9517 GWh in 2018. However, the period under consideration can be divided into four parts. In the first part, covering the period 2010–13, the electricity generation by other solid biofuels power plants decreased by 7.3%, falling from 9462 GWh in 2010 to 8779 GWh in 2013. In the second part, covering the period 2013–16, the electricity generated by other solid biofuels power plants increased by 17.6%, rising from 8779 GWh in 2013 to 11,541 GWh in 2016. In the third part, covering the years 2016 and 2017, the electricity generated by other solid biofuels decreased by 47.1%, falling from 11,541 GWh in 2016 to 6113 GWh in 2017. Finally, in the last part, covering the years 2017 and 2018, the electricity generated by other solid biofuels power plants increased by 55.7%, rising from 6113 GWh in 2017 to 9517 GWh in 2018.

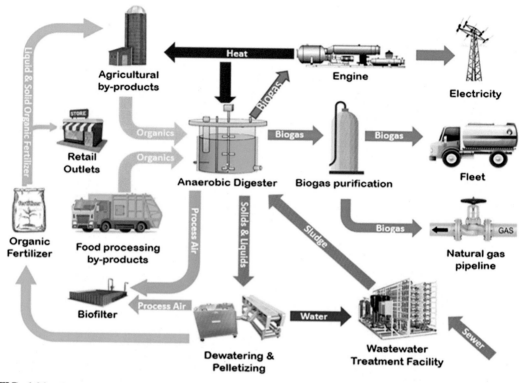

FIG. 6.28   Biogas scheme. *Source: Canadian Biogas Association, 2020. About Biogas. https://biogasassociation.ca/about_ biogas and Author own creation.*

Concerning the use of biogas for electricity generation[m] in Canada (see Fig. 6.28), the situation is as follows. Biogas is getting the government and the energy industry's attention because it is an energy source that could help meet the ever-increasing energy demand in Canada, an environment-friendly energy source, and a smart waste management option. While biogas can be generated from any organic waste source, the most common sources remain organic waste from landfill or municipal waste, including organic food waste from:

- Houses;
- Animal manure from livestock activities;
- Wastewater treatment plants;
- Local farms;
- Industries;
- Public and private institutions;
- Commercial facilities.

[m] Biogas produces renewable heat, electricity, and pipeline quality gas that can be stored in the pipeline and used for transportation, household heating or industrial, commercial and institutional processes. Biogas provides a closed loop opportunity for multiple businesses, extracting energy while recycling valuable nutrients (Canadian Biogas Association, 2020).

According to the Canadian Biogas Association research, there are 61 operational biogas facilities in Canada's agriculture and agri-food sector, and at least five facilities either planned, under construction, or commissioning. In terms of actual utilization, more than 60% of the biogas energy produced is used for electricity generation for sale, followed by a mixed-use of electricity and heating onsite. The Canadian provinces of Ontario, Quebec, and Alberta have the highest number of biogas power plants, with Ontario accounting for 64%. Still, there is huge potential for using biogas for electricity generation and heating in the country not yet exploited.

In 2018, the electricity generated by the biogas power plants in Canada reached 972 GWh, representing around 1.3% of Canada's electricity demand.

The evolution of the electricity generation by biogas power plants in Canada during the period 2010–18 is shown in Fig. 6.29.

Based on the data included in Fig. 6.29, the following can be stated: the electricity generation using biogas as fuel in Canada increased by 22.7% during the period 2010–18, rising from 792 GWh in 2010 to 972 GWh in 2018. However, it is important to single out since 2014, electricity generation using biogas as fuel remained stable at 972 GWh.

Without a doubt, biogas has many environmental benefits. In addition, it improves local economies and creates job opportunities. However, biogas remains less popular than other energy sources, and several factors can explain this situation:

- First, there is still a lack of awareness among the conventional energy generators and the public about its potential benefits, so there is less demand for constructing biogas power plants in the country;
- Second, the high upfront capital cost includes the plant itself's capital cost, feedstock procurement, and the pipeline connection;
- Third pipeline access and distribution;
- Fourth, getting a continuous and large amount of feedstock;
- Fifth, the type of technology used by the biogas industry.

Finally, forest biomass is the dominant energy source used to generate heat and electricity by industries and utilities among different bioenergy sources. Historically, bioenergy

Evolution of biogas electricity generation in Canada during 2010-18 (GWh)

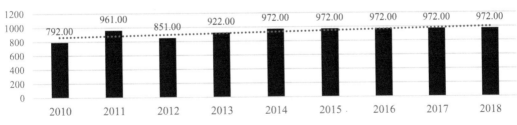

**FIG. 6.29** Evolution of the electricity generation by biogas power plants in Canada during the period 2010–18. *Source: 2020. IRENA Renewable Energy Statistics 2020. IRENA.*

consumption was especially important for home energy use, as Canadians burned wood for heating and cooking in Canada.

A former coal-burning thermal plant in north-western Ontario is now operating on biomass, making it the largest power plant in North America fueled completely by biological material. The conversion of the coal power plant to a biomass power plant cost US$170 million. According to the provincially-own Ontario Power Generation Inc. (OPG), the Atikokan biomass power plant is now burning 100% biomass after being converted from a coal-fired power plant as part of Ontario's plan to eliminate coal power generation. OPG's generating power plant located nearby Thunder Bay burned its last coal in April 2014, making Ontario the first jurisdiction in the North American region to scrap coal for electricity generation (Canadian Manufacturing, 2014).

The plant modifications included constructing a fuel storage and handling system to repower the plant from coal to biomass fuel. Two silos, each of which can store up to 5000 metric tons of wood pellets, and a new truck receiving and transfer infrastructure was built. The boiler of the coal power plant was also modified. OPG chose wood-pellet biomass as fuel because the energy content is similar to the lignite coal that the Atikokan biomass power plant was designed to burn. Much of the existing equipment could be adapted for the use of biomass as fuel. Electricity produced at Atikokan power plant using biomass will provide back-up for OPG's hydroelectric generation and for intermittent renewable power such as solar and wind as needed. Annual fuel consumption and electricity production will depend on the Ontario Power Authority's power purchase agreement and electricity demand. The Atikokan biomass power plant can produce approximately 200 MWh at full capacity.

## Bioenergy capacity in the United States

In the USA, there were, in 2019, a total of 178 bioenergy power plants, according to a Biomass Magazine paper entitled "US Biomass Power Plants (2019)". In 2019, the total bioenergy installed capacity in the USA reached 12,450 MW, according to IRENA Renewable Energy Statistics 2020 report. The biggest US biomass power plant is the Okeelanta Biomass Cogeneration power plant in Florida, with a capacity of 128.90 MW. In addition, there are 86 municipal waste-to-energy facilities in the USA across 25 states for energy recovery (see Fig. 6.30). While several have expanded to manage additional waste, the last new facility opened in 1995. The high investment costs represent a serious barrier to expanding the use of this type of energy source for electricity generation in the USA.

The construction cost of a new municipal waste-to-energy facility often exceeds US$100 million, and larger plants require double or triple that figure to build. In addition to that, the economic benefits of the investment are not immediately noticeable. Presently, the USA processes 14% of its trash in municipal waste-to-energy plants, which is still a substantial amount of refuse given today's consumption rate.

North America followed the EU for bioenergy consumption in buildings. In 2017, more than 2 million US households (2% of the total) used wood or wood pellets as their primary heating fuel, and a further 8% of households used wood as a secondary heat source, according to Winter Fuels Outlook (2018).

**FIG. 6.30** The Palm Beach County Renewable Energy Facility is an RDF-based waste-to-energy (WTE) facility. *Source: Folk, E., 2020. Progress of Waste-to-Energy in the USA. BioEnergy Consult. https://www.bioenergyconsult.com/ waste-to-energy-in-usa/.*

In the industry sector, heat supplied from bioenergy accounted for some 6.1% of all heat consumption. Generally, the use of bioenergy has been concentrated in industries where biomass residues are created as part of the production process – such as pulp and paper (where bioenergy provides 30% of energy needs), food, tobacco, and wood and wood products. Bioenergy can deliver low-temperature heat for heating and drying applications and high-temperature process heat through direct use of the fuel or gasifying the biomass and using the resulting fuel gas. However, little bioenergy is used in the more energy-intensive US industrial sectors where very high-temperature heat is required, such as iron, steel, and chemicals. In these sectors, lower-cost, higher energy density fossil fuels usually are preferred (Renewables 2018, 2018).

"The United States produces mainly biodiesel and ethanol fuel, which uses corn as the main feedstock. The US is the world's largest ethanol producer, producing nearly 16 billion gallons in 2017 alone," according to the Renewable Fuels Association (2020). Together with Brazil, the USA accounted for 85% of world ethanol production (27.05 billion gallons). Biodiesel is commercially available in most oilseed-producing states.

Biofuels are mainly used in combination with fossil fuels and also as additives. It is important to single out that "most light vehicles on the road today in the USA can run on blends of up to 10% ethanol, and motor vehicle manufacturers already produce vehicles designed to run on much higher ethanol blends. The demand for bioethanol fuel in the USA was stimulated by the discovery in the late 90s that methyl tertiary butyl ether (MTBE), an oxygenate additive in gasoline, was contaminating groundwater. Cellulosic biofuels are under development to avoid upward pressure on food prices and land-use changes that would be expected to result from a major increase in the use of food biofuels" (Goettemoeller and Goettemoeller, 2007).

On the other hand, it is important to stress that biofuels are not limited to liquid fuels. "One of the often-overlooked uses of biomass in the USA is in the gasification of biomass. There is a small but growing number of people using wood gas to fuel cars and trucks all across America", according to Woodgas.net (2010). The challenge is to expand the biofuels market beyond the US farm states, where they have been most popular to date. Flex-fuel vehicles assist in this transition because they allow drivers to choose different fuels based on price and availability.

In 2019, the USA's bioenergy installed capacity reached 12,450 MW, representing 78.7% of the regional total and 10% at the world level. The evolution of the bioenergy installed capacity in the USA during the period 2010–19 is shown in Fig. 6.31.

According to Fig. 6.31, the following can be stated: the USA's bioenergy installed capacity increased by 21% during the period 2010–19, rising from 10,290 MW in 2010 to 12,450 MW in 2019. However, the period under consideration can be divided into two parts. In the first part, covering the period 2010–15, the bioenergy installed capacity increased by 26%, rising from 10,290 MW in 2010 to 12,969 MW in 2015, the highest figure within the whole period under consideration. On the other hand, in the second part, covering the period 2015–19, the bioenergy installed capacity decreased by 4.1%, falling from 12,969 MW in 2015 to 12,450 MW in 2019. It is expected that, despite the decrease in bioenergy installed capacity in the USA during 2015–19, the country will continue using bioenergy sources for electricity generation, heating, and other purposes in the coming years. For this reason, it is foreseen an increase in the current bioenergy installed capacity in the country but at a low rate.

The list of the main bioenergy power plants in the USA in 2020 is included in Table 6.4.

In 2019, about 2.3% of total US annual energy consumption was from wood and wood waste—bark, sawdust, wood chips, wood scrap, and paper mill residues. In the same year, 64% of the total British thermal units (TBtu) consumed in the USA (2297 TBtu) was in the industrial sector, 23% (529 TBtu) was in the residential sector, 9% (211 TBtu) in the electrical sector, and 4% (84 TBtu) in the commercial sector.

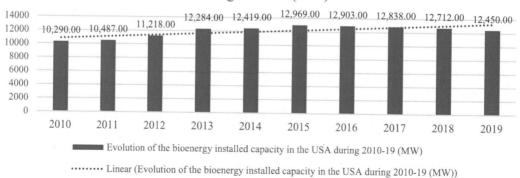

**FIG. 6.31** Evolution of the bioenergy installed capacity in the USA during the period 2010–19. *Source: 2020. IRENA Renewable Energy Statistics 2020. IRENA.*

TABLE 6.4 Bioenergy power plant.

| Bioenergy power plants | Capacity (metric tons per year) |
| --- | --- |
| Allendale White Pellet Plant | 60,000 |
| Amite BioEnergy | 512,500 |
| Appalachian Wood Pellets | 50,000 |
| Appling County Pellets LLC | 200,000 |
| Archer Forest Products LLC | 150,000 |
| Barefoot Pellet Co. | 63,500 |
| Convergent Energy | 108,900 |
| Curran Renewable Energy LLC | 108,900 |
| Dejno's Inc. | 54,400 |
| Dover Resources Inc. | 150,000 |
| Dry Creek Wood Pellets | 77,100 |
| Energex America Inc. | 114,000 |
| Enviva Pellets Ahoskie | 370,000 |
| Enviva Pellets Amory | 120,000 |
| Enviva Pellets Cottondale LLC | 720,000 |
| Enviva Pellets Greenwood LLC | 600,000 |
| Enviva Pellets Hamlet | 600,000 |
| Enviva Pellets Northampton LLC | 750,000 |
| Enviva Pellets Sampson | 750,000 |
| Enviva Pellets Southampton LLC | 750,000 |
| Equustock-Chester | 72,600 |
| Fiber By-Products Corp. | 90,000 |
| Fiber Energy Products AR | 80,000 |
| Fiber Resources Inc. | 108,900 |
| Forest Energy Show Low | 51,000 |
| Georgia Biomass | 750,000 |
| Hamer Pellet Fuel | 70,000 |
| Hazlehurst Wood Pellets LLC | 500,000 |
| Highland Pellets LLC-Pine Bluff | 600,000 |
| Indeck Energy Ladysmith LLC | 81,600 |
| LaSalle BioEnergy | 450,000 |

*Continued*

**TABLE 6.4**   Bioenergy power plant—cont'd

| Bioenergy power plants | Capacity (metric tons per year) |
| --- | --- |
| Lignetics of Idaho Inc. | 72,600 |
| Lignetics of Maine Inc. | 82,000 |
| Lignetics of New England-Allegheny | 55,000 |
| Lignetics of New England-Deposit | 80,000 |
| Lignetics of New England-Jaffrey | 76,200 |
| Lignetics of New England-Schuyler | 70,000 |
| Lignetics of Oregon-Brownsville | 113,400 |
| Lignetics of Virginia Inc. | 82,000 |
| Lignetics of West Virginia Inc. | 113,400 |
| LJR Forest Products | 158,800 |
| Maine Woods Pellet Co. | 100,000 |
| Mallard Creek Inc. | 108,900 |
| Michigan Wood Fuels | 54,400 |
| Morehouse BioEnergy | 512,500 |
| MRE-Crossville | 99,800 |
| MRE-Jasper | 113,400 |
| MRE-Ligonier | 59,000 |
| MRE-Peebles | 59,000 |
| MRE-Quitman | 120,000 |
| Nature's Earth Pellets NC LLC-Laurinburg | 90,700 |
| North Idaho Energy Logs-Hauser | 54,400 |
| Ozark Hardwood Products | 175,000 |
| Pinnacle Renewable Energy- Aliceville | 270,000 |
| Spearfish Pellet Co. | 52,600 |
| Telfair Forest Products LLC | 120,000 |
| Trae Fuels Ltd.-Pellet Plant | 108,900 |
| Varn Wood Pellets | 80,000 |
| Wood Fuel Developers LLC | 100,000 |
| Woodville Pellets LLC | 500,000 |
| Zilkha Biomass-Selma | 275,000 |

Total Plants: 61.
*Source: Biomass Magazine.*

In 2019, solid biofuels and renewable waste installed capacity in the USA reached 9917 MW, representing 80.8% of the regional total and 9.8% at the world level. The evolution of solid fuels and renewable waste installed capacity in the USA during the period 2010–19 is shown in Fig. 6.32.

According to Fig. 6.32, the following can be stated: the USA's solid fuels and renewable waste installed capacity increased by 17.1% during the period 2010–19, rising from 8471 MW in 2010 to 9917 MW in 2019. However, the period under consideration can be divided into two parts. In the first part, covering the period 2010–15, the solid fuels and renewable waste installed capacity increased by 23.6%, rising from 8471 MW in 2010 to 10,444 MW in 2015, the highest figure within the whole period under consideration. On the other hand, in the second part, covering the period 2015–19, the solid fuels and renewable waste installed capacity decreased by 5.1%, falling from 10,444 MW in 2015 to 9917 MW in 2019. It is expected that, despite the decrease in solid fuels and renewable waste installed capacity in the USA during the period 2015–19, the country will continue using this type of bioenergy source for electricity generation, heating, and other purposes during the coming years. For this reason, it is foreseen an increase in the current solid fuels and renewable waste installed capacity in the country but at a lower rate.

Another type of bioenergy source used in the USA for electricity generation and heating is renewable municipal waste. In 2019, the USA's renewable municipal waste installed capacity reached 1095 MW, representing 96.6% of the regional total and 7.5% at the world level. The evolution of renewable municipal waste installed capacity in the USA during the period 2010–19 is shown in Fig. 6.33.

Based on the data included in Fig. 6.33, the following can be stated: the US renewable municipal waste installed capacity decreased by 1.4% during the period 2010–19, falling from 1110 MW in 2010 to 1095 MW in 2019. The period under consideration can be divided into five parts. In the first part, covering the years 2010 and 2011, the country's renewable municipal waste installed capacity increased by 0.8%, rising from 1110 MW in 2010 to 1119 MW in 2011. In the second part, covering the years 2011 and 2012, renewable municipal waste installed

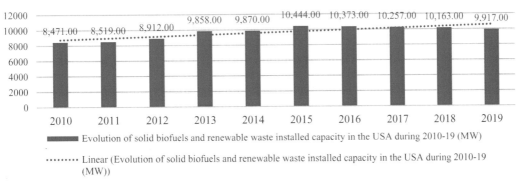

**FIG. 6.32** Evolution of solid fuels and renewable waste installed capacity in the USA during the period 2010–19.
*Source: 2020. IRENA Renewable Energy Statistics 2020. IRENA.*

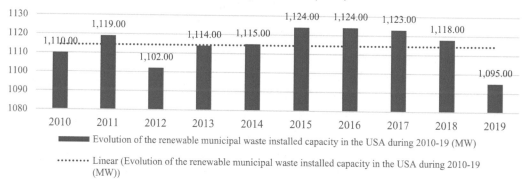

Evolution of the renewable municipal waste installed capacity in the USA during 2010-19 (MW)

Evolution of the renewable municipal waste installed capacity in the USA during 2010-19 (MW)

Linear (Evolution of the renewable municipal waste installed capacity in the USA during 2010-19 (MW))

FIG. 6.33   Evolution of the renewable municipal waste installed capacity in the USA during the period 2010–19. *Source: 2020. IRENA Renewable Energy Statistics 2020. IRENA.*

capacity decreased by 1.6%, falling from 1119 MW in 2011 to 1102 MW in 2012. In the third part, covering the period 2012–15, renewable waste installed capacity increased once again by 2%, rising from 1102 MW in 2012 to 1124 MW in 2015. In the fourth part, covering the period 2015–17, renewable municipal waste installed capacity remained stable at 1124 MW. However, in the fifth part, covering the period 2017–19, renewable municipal waste installed capacity decreased once again by 2.5%, falling from 1123 MW in 2017 to 1095 MW in 2019. Due to ups and downs in renewable municipal waste installed capacity registered in the USA during the period under consideration, it is difficult to foresee whether it would increase or decrease during the coming years.

Another type of bioenergy source used in the USA for electricity generation is the so-called "other solid biofuels." In 2019, the other solid biofuels installed capacity in the USA reached 8822 MW, representing 79.2% of the regional total and 13% at the world level. In the USA, the evolution in using other solid biofuels for electricity generation during the period 2010–19 is shown in Fig. 6.34.

Based on the data included in Fig. 6.34, the following can be stated: The other solid biofuels installed capacity in the USA increased by 19.8% during the period 2010–19, rising from 7361 MW in 2010 to 8822 MW in 2019. Therefore, the whole period under consideration can be divided into three parts. In the first part, covering the period 2010–15, the other solid biofuels installed capacity in the USA increased by 26.6%, rising from 7361 MW in 2010 to 9320 MW in 2015. In the second part, covering the period 2015–18, other solid biofuels' installed capacity remained relatively stable between 9320 MW and 9406 MW. Finally, in the third part, covering the years 2018 and 2019, other solid biofuels installed capacity in the USA decreased by 6.3%, falling from 9406 MW, the highest capacity reached within the whole period under consideration to 8822 MW in 2019. Therefore, it is expected that the other solid biofuels installed capacity in the USA will continue to increase during the coming years but at a low rate (between 1% and 2% as average).

Another type of bioenergy used in the USA for electricity generation and other purposes is liquid biofuels. The two most common types of liquid biofuels are bioethanol

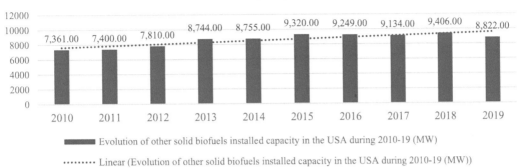

**FIG. 6.34** Evolution in using other solid biofuels for electricity generation in the USA during the period 2010–19. *Source: 2020. IRENA Renewable Energy Statistics 2020. IRENA.*

and biodiesel. "Bioethanol is an alcohol made by fermentation, mostly from carbohydrates produced in sugar or starch crops such as corn, sugarcane, or sweet sorghum. In addition, cellulosic biomass, derived from non-food sources, such as trees and grasses, is also developed as a feedstock for ethanol production. Ethanol can be used as a fuel for vehicles in its pure form, but it is usually used as a gasoline additive to increase octane and improve vehicle emissions. Bioethanol is widely used in the USA and Brazil" (Wikipedia—Biofuel, 2020).

Biodiesel is produced from oils or fats using transesterification, and it is the most common biofuel in Europe. It can be used as a fuel for vehicles in its pure form, or it can be used as a diesel additive to reduce particulates, carbon monoxide, and hydrocarbons from diesel-powered vehicles (Wikipedia—Biofuel, 2020).

In 2019, liquid biofuels installed capacity in the USA reached 165 MW representing 15.9% of the regional total and 5.1% at the world level, and generated 351 GWh in 2018. In the same year, worldwide biofuel production reached 161 billion liters (43 billion gallons), up 6% from 2018, and provided 3% of the world's fuels for road transportation. In the USA, the liquid biofuel production was 35 Mtoe in 2019, and it is expected that the consumption will reach 95 Mtoe by 2030. The IEA wants liquids biofuels to meet more than a quarter of the world's demand for transportation fuels by 2050 to reduce petroleum dependency. However, liquid biofuels' production and consumption are not on track to meet the IEA's sustainable development scenario. From 2020 to 2030, global liquid biofuel output has to increase by 10% each year to reach IEA's goal. However, only 3% of growth annually is expected for the next five years (IEA Transport Biofuels, 2020).

The evolution of the liquid biofuels installed capacity in the USA during the period 2010–19 is shown in Fig. 6.35.

According to Fig. 6.35, the following can be stated: the liquid biofuels installed capacity in the USA decreased by 10% during the period 2010–19, falling from 183 MW in 2010 to 165 MW in 2019. However, it is important to single out that since 2013 the liquid biofuels installed capacity in the country remaining stable between 155 MW and 165 MW. It is expected that this trend will remain without significant change during the coming years unless a radical

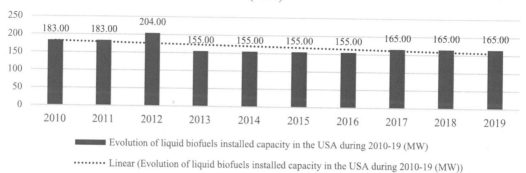

FIG. 6.35   Evolution in liquid biofuels installed capacity in the USA during the period 2010–19. *Source: 2020. IRENA Renewable Energy Statistics 2020. IRENA.*

adjustment in the energy policy is adopted by the new US administration in 2021, prioritizing the use of bioenergy sources for electricity generation, particularly liquids biofuels over coal and oil.

Biogas is the last type of bioenergy source used in the USA for electricity generation to be considered. "The U.S. has over 2,200 sites producing biogas in all 50 states: 250 anaerobic digesters on farms, 1,269 water resource recovery facilities using an anaerobic digester (~ 860 currently use the biogas they produce), 66 stand-alone systems that digest food waste, and 652 landfill gas projects. For comparison, Europe has over 10,000 operating digesters, and some communities are essentially fossil fuel-free because of them." (The American Biogas Council, 2018).

According to NREL Biogas Potential in the United States (2013), "biogas is the gaseous product of anaerobic digestion, a biological process in which microorganisms break down biodegradable material in the absence of oxygen." Biogas comprises methane (between 50% and 70%) and carbon dioxide (between 30% and 50%), with trace amounts of other particulates and contaminants.

Biogas—methane ($CH_4$) derived from biological feedstocks such as wastewater treatment plants or landfills, animal waste, wood chips, and agricultural residues—is one potential renewable fuel with multiple potential uses in the USA. For example, biogas could be captured and used where it is produced to generate electricity, or it could be refined and transported through pipelines to centralized electricity generation facilities, centralized chemical refineries, such as gas-to-liquids power plants, or elsewhere for other energy uses such as the production of heat (Murray et al., 2014).

Biogas can also be upgraded to pure methane, called "biomethane or renewable natural gas," by removing water, carbon dioxide, hydrogen sulfide, and other trace elements. This upgraded biogas is comparable to conventional natural gas. It thus can be injected into the pipeline grid or used as a transportation fuel in a compressed or liquefied form (NREL Biogas Potential in the United States, 2013).

In 2019, the USA's biogas installed capacity reached 2368 MW representing 94.4% of the regional total and 12.2% at the world level. The evolution of the biogas installed capacity in the USA during the period 2010–19 is shown in Fig. 6.36.

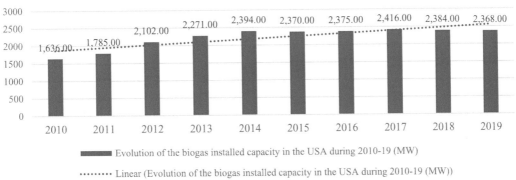

FIG. 6.36   Evolution of the biogas installed capacity in the USA during the period 2010–19. *Source: 2020. IRENA Renewable Energy Statistics 2020. IRENA.*

Based on the data included in Fig. 6.36, the following can be stated: the US's biogas installed capacity increased by 44.7% during the period 2010–19, rising from 1636 MW in 2010 to 2368 MW in 2019. The period under consideration can be divided into two parts. In the first part, covering the period 2010–14, the country's biogas installed capacity increased by 46.3%, rising from 1636 MW in 2010 to 2394 MW in 2014. In the second part, covering the period 2014–19, the USA's biogas installed capacity decreased by 1.1%, falling from 2394 MW in 2014 to 2368 MW in 2019. It is important to single out that the US's biogas installed capacity during the period 2014–19 remained stable between 2368 MW and 2416 MW, the highest biogas installed capacity registered in the country within this period.

Beyond question, the bioenergy sector provides many benefits, allowing communities a method of repurposing their waste. However, it also has negative aspects that are also important to note, as the potential for pollution. Thus, while the bioenergy sector offers solutions, some of them come at a cost that the people and the industry should be ready to assume for environmental and health reasons.

## Bioenergy electricity generation in the United States

According to a US EIA source,[n] "as of December 31, 2019, there were 22,731 electric generators at about 10,346 utility-scale electric power plants in the USA. Utility-scale power plants have a total nameplate electricity generation capacity of at least 1 MW. A power plant may have one or more generators, and some generators may use more than one type of fuel."

The US's share of renewable electricity generation increased from 10.2% to 17.4% between 2009 and 2019 (Global CCS Institute, 2019). In 2018, the electricity generation using renewable energy sources in the USA reached 743,177 GWh. That figure represents an increase of 3.5% with respect to 2018. The electricity generation in the USA in 2018 represented 63.5%

[n] See https://www.eia.gov/toqs/faq.php?id=65=2.

of the regional total and 11.3% at the world level. At least three large US cities, Chicago, Los Angeles, and Philadelphia, committed to 100% renewable electricity by 2045 (IEA Renewables 2019, 2019).

The evolution in the electricity generation using renewables as fuel in the USA during the period 2010–18 is shown in Fig. 6.37.

According to Fig. 6.37, the following can be stated: the use of renewables in electricity generation in the USA increased by 68.6% during the period 2010–18, rising from 440,677 GWh in 2010 to 743,177 GWh in 2018. It is expected that the use of renewables for electricity generation in the USA will continue to increase during the coming years, particularly if the new US administration supports an increased role of renewables energy sources within the US energy mix.

Bioenergy is an important component in the US energy matrix. In 2018, the USA's electricity generation using bioenergy as fuel reached 67,885 GWh, representing 86.4% of the regional total and 13% at the world level. The evolution in the electricity generation using bioenergy as fuel in the USA during the period 2010–18 is shown in Fig. 6.38.

Based on the data included in Fig. 6.38, the following can be stated: the electricity generation using bioenergy in the USA increased by 14.4% during the period 2010–18, rising from 59,319 GWh in 2010 to 67,885 GWh in 2018. However, the period under consideration can be divided into two parts. In the first part, covering the period 2010–15, the use of bioenergy for electricity generation increased by 19.4%, rising from 59,319 GWh in 2010 to 70,818 GWh in 2015, the peak in electricity generation using this type of energy source registered in one single year within the period under consideration. In the second part, covering the period 2015–18, the use of bioenergy for electricity generation decreased by 4.2%, falling from 70,818 GWh in 2015 to 67,885 GWh in 2018. This decrease results from the implementation of President Trump's energy policy that supports the use of conventional energy sources for electricity generation and heating, particularly coal, instead of renewables.

One component of bioenergy in electricity generation in the USA is solid biofuels and renewable waste. According to the US Energy Information Administration data, wood and waste energy accounted for nearly 25% of US renewable energy use in 2019, while biofuels accounted

## Evolution of the electricity generation using renewables as fuel in the USA during 2010-18 (GWh)

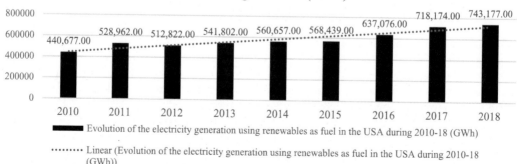

FIG. 6.37 Evolution in the electricity generation using renewables as fuel in the USA during the period 2010–18. *Source: 2020. IRENA Renewable Energy Statistics 2020. IRENA.*

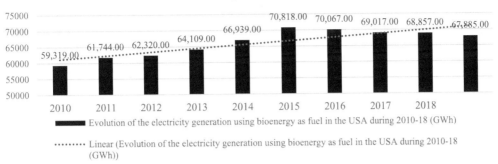

**FIG. 6.38** Evolution in the electricity generation using bioenergy as fuel in the USA during the period 2010–18. *Source: 2020. IRENA Renewable Energy Statistics 2020. IRENA.*

for 20% of renewable energy consumption. According to the same source, the USA consumed a record amount of renewable energy in 2019, at 11.5 quadrillion British thermal units (Btu), or 11% of total US energy consumption. Industrial, commercial, and electric power facilities use wood and waste to generate electricity, produce heat, and manufacture goods. Also, about 2% of US households used wood as their primary source of heat in 2019 (Voegele, 2020).

In 2018, the electricity generation using solid biofuels and renewable waste reached 54,275 GWh, representing 84.9% of the regional total and 12.7% at the world level. The evolution in the use of solid biofuels and renewable waste for electricity generation in the USA during the period 2010–18 is shown in Fig. 6.39.

Based on the data included in Fig. 6.39, the following can be stated: the use of solid biofuels and renewable waste for electricity generation in the USA increased by 8.7% during the period 2010–18, rising from 49,947 GWh in 2010 to 54,275 GWh in 2018. Therefore, the period under consideration can be divided into two parts. In the first part, covering the period 2010–15, the use of solid biofuels and renewable waste for electricity generation increase by 14.2%, rising

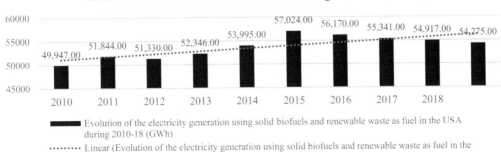

**FIG. 6.39** Evolution in the electricity generation using solid fuels and renewable waste as fuel in the USA during the period 2010–18. *Source: 2020. IRENA Renewable Energy Statistics 2020. IRENA.*

from 49,947 GWh in 2010 to 57,024 GWh in 2015, when the peak in the electricity generation using this type of bioenergy source within the period under consideration was reached. In the second part, covering the period 2015–18, the use of solid biofuels and renewable waste for electricity generation decreased by 4.9%, falling from 57,024 GWh in 2015 to 54,275 GWh in 2018. As indicated before, this decrease results from the implementation of President Trump's energy policy that supports conventional energy sources for electricity generation above renewables, in general, and bioenergy in particular.

Another type of bioenergy used in the USA for electricity generation but on a low scale is renewable municipal waste. In 2018, the USA's electricity generation using renewable municipal waste reached 8382 GWh, representing 98% of the regional total and 13.5% at the world level. The evolution in the use of renewable municipal waste for electricity generation in the USA during the period 2010–18 is shown in Fig. 6.40.

According to Fig. 6.40, the following can be stated: the use of renewable municipal waste for electricity generation in the USA decreased by 10% during the period 2010–18, falling from 9308 GWh in 2010 to 8382 GWh in 2018. However, it is expected that with the new US administration, this trend could change if it decides to support the use of renewable energy sources for electricity generation in the country in the future, particularly bioenergy.

Other solid biofuels are also used in the USA for electricity generation. In 2018, the electricity generated by other solid biofuels reached 45,893 GWh, representing 82.8% of the regional total and 14.8% at the world level. The evolution in the electricity generation using other solid biofuels as fuel in the USA during the period 2010–18 is shown in Fig. 6.41.

Based on the data included in Fig. 6.41, the following can be stated: the use of other solid biofuels for electricity generation in the USA increased by 7.9% during the period 2010–18, rising from 42,536 GWh in 2010 to 45,893 GWh in 2018. The period under consideration can be divided into two parts. In the first part, covering the period 2010–14, the use of solid biofuels for electricity generation increase by 14.2%, rising from 42,536 GWh in 2010 to 48,563 GWh in 2014, when the peak in the electricity generation using this type of bioenergy source was reached within the period under consideration. In the second part, covering the period 2014–18, the use of solid biofuels for electricity generation decreased by 5.5%, falling from 48,563 GWh in 2014 to 45,893 GWh in 2018. As indicated before, this decrease results from the

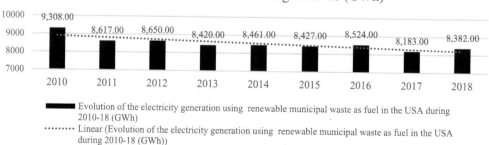

FIG. 6.40 Evolution in the electricity generation using renewable municipal waste as fuel in the USA during the period 2010–18. *Source: 2020. IRENA Renewable Energy Statistics 2020. IRENA.*

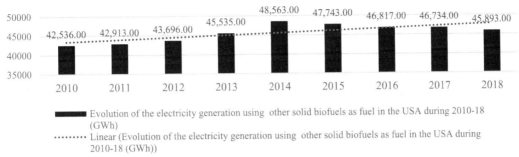

Evolution of the electricity generation using other solid biofuels as fuel in the USA during 2010-18 (GWh)

FIG. 6.41 Evolution in the electricity generation using other solid biofuels as fuel in the USA during the period 2010–18. *Source: 2020. IRENA Renewable Energy Statistics 2020. IRENA.*

implementation of President Trump's energy policy that supports the use of conventional energy sources for electricity generation above the use of renewables for this specific purpose, in general, and bioenergy in particular.

Summing up, the following can be stated regarding biomass electricity generation in the USA in 2019. The role of biomass for electricity generation in the USA by type of energy source is shown in Table 6.5.

According to Table 6.5, the following can be stated: within biomass, wood is the main type of biomass used for electricity generation in the USA in 2019, with 39 billion kWh or 0.9% of the total, followed by landfill gas with 10 billion kWh or 0.3% of the total. It is expected that this trend will not change significantly during the coming years.

Another type of bioenergy source used in the USA for electricity generation, but a low percentage, and in the transportation sector, is liquid biofuels, including fuel ethanol, biodiesel, and other renewable liquid fuels. Liquid fuels accounted for approximately 20% of US renewable energy consumption in 2019, mainly in the transportation sector.

According to EIA sources, biofuels usually are blended with petroleum-based motor gasoline and diesel and are consumed as liquid fuels in automobiles. The US agency also reported that biofuels' industrial consumption accounts for about 3% of US biofuel energy consumption.

TABLE 6.5   Biomass role in the USA for electricity generation in 2019.

| Energy source | Billion kWh | Share of the total (%) |
| --- | --- | --- |
| Biomass (total) | 58 | 1.4 |
| Wood | 39 | 0.9 |
| Landfill gas | 10 | 0.3 |
| Municipal solid waste (biogenic) | 6 | 0.1 |
| Other biomass waste | 3 | 0.1 |

*Source: US EIA, 2020. What Is US Electricity Generation by Energy Source? https://www.eia.gov/tools/faqs/faq.php?id=427&t=2.*

The use of liquid biofuel for electricity generation in the USA is very small. For example, in 2018, the electricity generated by liquid biofuels only reached 351 GWh, representing 100% at the regional total and 6.1% at the world level. The evolution in the use of liquid biofuels for electricity generation in the USA during the period 2010–18 is shown in Fig. 6.42.

According to Fig. 6.42, the following can be stated: the use of liquid fuels for electricity generation in the USA increased 3.9-fold during the period 2010–18, rising from 94 GWh in 2010 to 351 GWh in 2018. Furthermore, it is expected that this trend could continue at a similar rate during the coming years.

The USA is using biogas for electricity generation. Biogas' electricity generation reached in 2018 a total of 13,259 GWh, representing 93.2% of the regional total and 15% at the world level. "Biomass is burned directly in steam-electric power plants, or it can be converted to a gas that can be burned in steam generators, gas turbines, or internal combustion engine generators" (Electricity in the United States, 2020).

The evolution in the use of biogas for electricity generation in the USA during the period 2010–18 is shown in Fig. 6.43.

According to Fig. 6.43, the following can be stated: the use of biogas for electricity generation in the USA increased 35.2% during the period 2010–18, rising from 9806 GWh in 2010 to 13,259 GWh in 2018. It is expected that this trend could continue at the same rate during the coming years, depending on the importance that the new US administration gives to the role of renewables within the country's energy mix, particularly the use of biogas for electricity generation.

Summing up, the following can be stated: aside from a few hydropower plants such as the Grand Coulee Dam in Washington State, which generate 19 million MWh per year – most renewable power plants operating in the USA are on a smaller scale. Generally speaking, renewable energy sources are also more dependent on geography. For example, a geothermal power plant cannot be built in an area with no thermal energy in the ground, or a wind farm cannot be constructed where there is mostly calm weather or where the wind flows slow. For

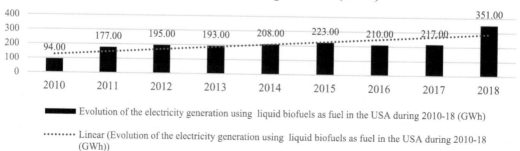

**Evolution of the electricity generation using other solid biofuels as fuel in the USA during 2010-18 (GWh)**

| | 2010 | 2011 | 2012 | 2013 | 2014 | 2015 | 2016 | 2017 | 2018 |
|---|---|---|---|---|---|---|---|---|---|
| | 94.00 | 177.00 | 195.00 | 193.00 | 208.00 | 223.00 | 210.00 | 217.00 | 351.00 |

■ Evolution of the electricity generation using liquid biofuels as fuel in the USA during 2010-18 (GWh)

········ Linear (Evolution of the electricity generation using liquid biofuels as fuel in the USA during 2010-18 (GWh))

**FIG. 6.42** Evolution in the electricity generation using liquid biofuels as fuel in the USA during the period 2010–18. *Source: 2020. IRENA Renewable Energy Statistics 2020. IRENA.*

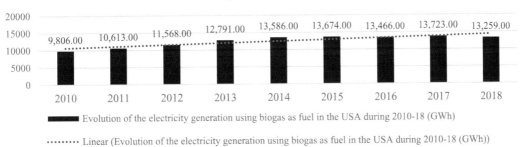

## Evolution of the electricity generation using biogas as fuel in the USA during 2010-18 (GWh)

**FIG. 6.43** Evolution in the electricity generation using biogas as fuel in the USA during the period 2010–18. *Source: 2020. IRENA Renewable Energy Statistics 2020. IRENA.*

this reason, the dispersion of green sources around the country is also quite interesting to look at (see Fig. 6.44) (Desjardins, 2019).

The best location for wind farms is in the country's central states, while the best location for geothermal power plants is in California, Nevada, and Utah. In the case of solar parks, the best location for this type of power plant is in California and several other states around the Atlantic coast, from Massachusetts to Florida. The best location for hydropower plants is in some states in the western part of the country, such as Washington State, Oregon, California, and several other states in the country's eastern part.

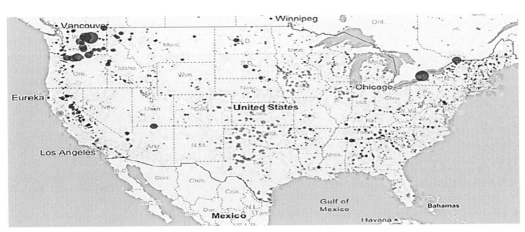

**FIG. 6.44** Mapped: Every power plant in the USA. Note: Hydro (dark blue), wind (light blue), solar (yellow), biomass (brown), and geothermal (green) (2019). *Source: Department of Physics. Weber State University. Figura US Power Plants, 2019. Created by: Daniel V. Schroeder. Email: dschroeder@weber.edu; https://physics.weber.edu/schroeder/energy/ PowerPlantsMap.html.*

# References

About Renewable Energy. Government of Canada.

Ahlberg, L., 2014. Bioenergy Crops could Store More Carbon in Soil. The University of Illinois at Urbana-Champaign.

Alessandra, R., 2019. An In-Depth Comparison: Solar Power Vs. Biomass Energy. Solar Feeds.

Alternative Energy Today, 2020. Bioenergy has Significant Potential.

Ames, H., 2018. The Advantages & Disadvantages of Biomass Energy. Sciencing.

Anaerobic Digestion Community, 2019. What Is a Biogas Power Plant?.

Anaerobic Digestion in US—A Large Biogas Opportunity. Anaerobic-Digestion.com Blog.

Bora, R., 2020. North America Biogas Market Size is Expected to Grow Significantly during 2019-25. Fractovia Market Trending News.

BP Statistical Review of World Energy 2020, sixty-ninth ed. British Petroleum (BP).

Bradburn, K., 2014. 2014 CanBio Report on the Status of Bioenergy in Canada (PDF). Natural Resources Canada.

Canada Energy Regulator, 2020. Canada's Adoption of Renewable Power Sources—Energy Market Analysis. Biomass.

Canadian Biogas Association, 2020. About Biogas. https://biogasassociation.ca/about_biogas.

Canadian Manufacturing, 2014. North America's Largest Biomass Power Plant Operating in Ontario. Cleantech Canada Staff.

Creutzig, F., Ravindranath, N.H., Berndes, G., Bolwig, S., Bright, R., Cherubini, F., Masera, O., 2015. Bioenergy and climate change mitigation: an assessment. GCB Bioenergy 7 (5), 916–944. https://doi.org/10.1111/gcbb.12205.

Crifax, 2020. North America Waste-to-Energy Plants Market Research Report; North America Outlook (US and Canada), Growth Potential, Competitive Industry Size, Share, Growth, Trends, and Forecasts 2019–2027.

DeCicco, J.M., Schlesinger, W.H., 2018. Opinion: reconsidering bioenergy given the urgency of climate protection. Proc. Natl. Acad. Sci. U. S. A. 115 (39), 9642–9645. https://doi.org/10.1073/pnas.1814120115. 2019.

Desjardins, J., 2019. Mapped: Every Power Plant in the United States. https://www.visualcapitalist.com/mapped-every-power-plant-in-the-united-states/.

Díaz, S., Settele, J., Brondízio, E., Ngo, H.T., Guèze, M., Agard, J., Vilá, B., 2019. Summary for Policymakers of the Global Assessment Report on Biodiversity and Ecosystem Services of the Intergovernmental Science-Policy Platform on Biodiversity and Ecosystem Services. Intergovernmental Panel on Biodiversity and Ecosystem Services, Bonn.

Electrical Power Generation, 2016. Introduction to the Current Situation and Perspectives on the Use of Biomass in the Generation of Electricity.

Electricity in the United States, 2020. Electricity Explained. US Energy Information Administration (EIA).

Energypedia Electricity Generation From Biogas.

European Academies Science Advisory Council, 2019. Forest Bioenergy, Carbon Capture, and Storage, and Carbon Dioxide Removal: An Update. Retrieved from https://easac,eu/fileadmin/PDF_s/reports_statements/Negative:Carbon/EASAC:Commentary:Forest_Bioenergy_Feb_2019_FINAL.pdf.

Fachagentur Nachwachsende Rohstoffe e.V. (FNR), 2009. Biogas Basisdaten Deutschland. Stand; Oktober 2008. 7 pp. Very short but comprehensive overview of the biogas situation in Germany.

Folk, E., 2020. Progress of Waste-to-Energy in the USA. BioEnergy Consult.

Fortune Business Insights—Power & Renewables/Bioenergy Market, 2021. Bioenergy Market Size, Share & COVID-19 Impact Analysis, by Product Type (Solid Biomass, Liquid Biofuel, Biogas, and Others), By Feedstock (Agricultural Waste, Wood and Woody Biomass, Solid Waste and Others), By Application (Power Generation, Heat Generation, Transportation, and Others), and Regional Forecast, 2020–2017.

Global CCS Institute, 2019. Perspective: Bioenergy and Carbon Capture and Storage. Melbourne https://www.globalccsinstitute.com/wp-content/uploads/2019/03/BECCS-Perspective_FINAL_18-March.pdf. Budinis, S., Going Carbon-Negative—What Are the Technology Options? IEA, 20 January 2020. https://www.iea.org/commentaries/going-carbon-negative-what-are-the-technology-options.

Global Ethanol Production. Renewable Fuels Association. Archived from the original on 8 April 2008; Retrieved 29 May 2019.

Goettemoeller, J., Goettemoeller, A., 2007. Sustainable Ethanol: Biofuels, Biorefineries, Cellulosic Biomass, Flex-Fuel Vehicles, and Sustainable Farming for Energy Independence. Prairie Oak Publishing, Maryville, MO, ISBN: 978-0-9786293-0-4, pp. 56–61.

Government of Canada Statistics, 2016. (Chapter 1) www.statcan.gc.ca. Retrieved 29,3,2017.

Greentumble, 2018. Biomass Energy Advantages and Disadvantages. Greentumble.

Home Biogas Industrial Biogas in the United States of America, n.d. Biogas Basics.

IEA, 2017a. Energy Technology Perspectives 2017. International Energy Agency. Retrieved from https://webstore. iea.org/energy-technology-perspectives-2017.

IEA, 2017b. Technology Roadmap—Delivering Sustainable Bioenergy. International Energy Agency. Retrieved from https://webstore.iae.org/technology-roadmap-delivering-sustainable-boenergy.

IEA, et al., 2019. Tracking SDG 7: The Energy Progress Report 2019. International Energy Agency, Washington, DC. https://trackingsdg7.esmap.org/downloads.

IEA PVPS, 2019. Trends in Photovoltaic Applications 2019. International Energy Agency.

IEA PVPS, 2020. Snapshot of Global PV Markets 2020. International Energy Agency.

IEA Renewables 2018. International Energy Agency. https://www.iea.org/reports/renewables-2018.

IEA Renewables 2019. International Energy Agency. https://www.iea.org/reports/renewables-2019.

IEA Renewables 2020, 2020. Analysis and Forecast to 2025. International Energy Agency. https://webstore.iea.org/ download/direct/4234.

IEA Transport Biofuels.

IEA World Energy Investment 2019, 2019. Investing in Our Energy Future; Flagship Report. International Energy Agency. May 2019.

IRENA, 2015. Renewable Power Generation Costs in 2014. International Renewable Energy Agency, Abu Dhabi.

IRENA, 2020a. Renewable Power Generation Costs in 2019. International Renewable Energy Agency, Abu Dhabi.

IRENA, 2020b. Global Renewables Outlook: Energy Transformation 2050 (2020 edition). International Renewable Energy Agency, Abu Dhabi. www.irena.org/-/media/Files/IRENA/Agency/Publication/2020/Apr/IRENA_ Global_Renewables_Outlook_2020.pdf.

IRENA and CPI, 2020. Global Landscape of Renewable Energy Finance. International Renewable Energy Agency, Abu Dhabi, ISBN: 978-92-9260-237-6.

IRENA Bioenergy. https://www.irena.org/bioenergy.

IRENA Bioenergy for Power. https://www.irena.org/costs/Power-Generation-Costs/Bioenergy-for-Power.

IRENA Renewable Cost Database, n.d.

IRENA Renewable Energy Statistics 2020. IRENA.

Laan, T., Steenblik, R., Litman, T.A., 2009. Biofuels—At What Cost?: Government Support for Ethanol and Biodiesel in Canada. International Institute for Sustainable Development. Global Studies Initiative; Gibson Library Connections, Winnipeg, MB, ISBN: 978-1894784283. OCLC 435739721.

Mabee, W.E., 2007. Policy options to support biofuel production. In: Olsson, L. (Ed.), Biofuels, Advances in Biochemical Engineering/Biotechnology. vol. 108. Springer Berlin Heidelberg, ISBN: 9783540736509, pp. 329–357, https://doi.org/10.1007/10_2007_059. 17846726.

McFarland, K., 2017. Biomass Advantages and Disadvantages. Syntech Bioenergy.

Morales Pedraza, J., 2015. Electrical Energy Generation in Europe: The Current Situation and Perspectives in the Use of Renewable Energy Sources and Nuclear Power for Regional Electricity Generation. Springer International Publishing Switzerland, ISBN: 978-3-319-16082-5, https://doi.org/10.1007/978-3-319-16083-2. ISBN 978-3-319-16083-2 (eBook). Library of Congress Control Number: 2014950679.

Mukhopadhyay, K., Thomassin, P.J., 2011. Macroeconomic effects of the ethanol biofuel sector in Canada. Biomass Bioenergy 35 (7), 2822–2838. https://doi.org/10.1016/j.biombioe. 21,3,2011.

Murray, B.C., Gaik, C.S., Vegh, T., 2014. Biogas in the United States: An Assessment of Market Potential in a Carbon-Constrained Future. Nicholas Institute for Environmental Policy Solutions, Duke University, USA.

NREL Biogas Potential in the United States. National Renewable Energy Laboratory; Energy Analysis.

Reid, W.V., Mooney, H.A., Cropper, A., Capistrano, D., Carpenter, S.R., Chopra, K., Zurek, M., 2005. Ecosystems and Human Well-Being: Synthesis. Island Press, Washington, DC.

Reid, W.V., Ali, M.K., Field, C.B., 2019. The future of bioenergy. Glob. Chang. Biol. https://doi.org/10.1111/gch.14883 (Accepted October 2019).

Renewable Energy in Canada. https://db0nus869y26v.cloudfront.net/en/Renewable_energy_in_Canada.

Renewable Fuels Association, 2020. Renewable Fuels Association – Leading Trade Association for US Ethanol (ethanolrfa.org).

Renewables 2018. OECD/IEA. https://www.iea.org/renewables2018.

Renewables 2019 Global Status Report. https://www.ren21.net/wp-content/uploads/2019/05/ gsr_2019_full_report_en.pdf.

Renewables 2020 Global Status Report. https://www.ren21.net/wp-content/uploads/2019/05/gsr_2020_full_report_en.pdf.

Report on Conversion Efficiency of Biomass BASIS—Biomass Availability and Sustainability Information System. Version #2—July 2015; Supported by EU.

Rinkesh, 2020. Various Pros and Cons of Biomass Energy. Conserve Energy Future.

Rogelj, J., Shindell, D., Jiang, K., Fifita, S., Forster, P., Ginzburg, V., Vilariño, M.V., 2018. Mitigation pathways compatible with 1.5°C in the context of sustainable development. In: Masson-Delmotte, V., Zhai, P., Pörtner, H.-O., Roberts, D., Skea, J., Shukla, P.R., et al. (Eds.), Global Warming of 1.5°C. An IPCC Special Report on the Impacts of Global Warming of 1.5°C Above Pre-Industrial Levels and Related Global Greenhouse Gas Emission Pathways, in the Context of Strengthening the Global Response to the Threat of Climate Change, Sustainable Development, and Efforts to Eradicate Poverty. Intergovernmental Panel on Climate Change, Geneva, pp. 313–443.

Searchinger, T.D., Wirsenius, S., Beringer, T., Dumas, P., 2018. Assessing the efficiency of changes in land use for mitigating climate change. Nature 564 (7735), 249. https://doi.org/10.1038/s41586-018-0757-z.

Selasa, 2009. Renewable Energy.

Statistics Canada, 2016. https://www150.statcan.gc.ca/n1/daily-quotidien/181005/dq181005d-eng.htm.

Tavares Kennedy, H., 2020. COVID-19 Causes First Contraction in Biofuel Output in 20 Years. Biofuels Digest.

The Advantages and Disadvantages of Biomass Energy. Ecavo.com/Renewables.

The American Biogas Council, 2018.

Thompson, M., 2019. Siouxland energy temporarily idles production. Ethanol Producer Magazine. 16 September 2019 http://ethanolproducer.com/articles/16547/siouxland-energytemporarily-idles-production. POET LLC, Oil bailouts force POET to lower production. Ethanol Producer Magazine, 20 August 2019. http://ethanolproducer.com/articles/16466/oil-bailouts-force-poet-to-lower-production.

US Biomass Power Plants. Biomass Magazine.

US EIA Monthly Energy Review. Table 10.3, Fuel ethanol overview https://www.eia.gov/totalenergy/data/monthly.

Vis, M.W., van den Berg, D., 2010. Best Practices and Methods Handbook. Harmonization of Biomass Resource Assessments, vol. I Biomass Energy Europe.

Voegele, E., 2020. Bioenergy, biofuels account for nearly half of US renewable use. Biodiesel Magazine.

Wikipedia—Biofuel.

Winter Fuels Outlook. US Energy Information Administration (EIA). https://www.eia.gov/outlooks/steo/special/winter/2018_winter_fuels.pdf.

Wood Gas Used as Alternative Energy to Fuel Cars and Trucks. Woodgas.net. Retrieved 15 July 2010.

# 7

# The use of nuclear energy for electricity generation

## Introduction

Without a doubt, nuclear energy is a realistic option for electricity generation in the North American region. For this reason, during the 2000s, the nuclear industry expected an increase in the construction of new nuclear power reactors due to concerns about environmental contamination due to carbon dioxide emissions. However, in 2009, Petteri Tiippana, the director of STUK's nuclear power plant division, told the BBC that it was challenging to deliver a Generation III reactor project on schedule because builders were not used to working to the exacting standards required on nuclear construction sites since so few new reactors had been built in recent years (Meirion, 2009).

Besides, advanced reactor developers are at various stages of commercializing new nuclear power reactor designs and, for this reason, must take into account the market environments that will exist when their new designs are available. It is, therefore, critical to have a clear understanding of what the new market environment will look like, the cost of the new nuclear power reactors designs, the need that these costs be attractive for investors, and what performance characteristics of the new nuclear power reactors designs will create the most value for power plant owners (Ingersoll et al., 2020).

In 2018, the MIT Energy Initiative study on the future of nuclear energy[a] concluded that, together with the strong suggestion that government should financially support the development and demonstration of new advanced nuclear technologies (Generation IV) for a worldwide renaissance of the use of nuclear energy or electricity generation commence, a global standardization of regulations needs to be adopted, with a move toward serial manufacturing of standardized units. Today, it is common for each country to demand several changes to design current types of nuclear power reactors to satisfy varying national regulatory bodies, often to benefit domestic engineering supply firms.

---

[a] The Future of Nuclear Energy in a Carbon-Constrained World: An Interdisciplinary MIT Study (2018).

The MTI report mentioned above notes that the most cost-effective projects have been built using a standarized design with multiple (up to six) reactors per site. The same component suppliers and construction crews work on each unit in a continuous workflow (Geuss, 2018). According to the same MTI report, "the costs of building nuclear power plants have only increased over the recent decades, unlike the costs of constructing other forms of electricity generation, which have become cheaper. In part, that is because nuclear reactor construction requires specialized skills that are employed over the years on one project. Nuclear power plants are also only built rarely, so there is less opportunity to develop more people with specialized skills" (Geuss, 2018).

Despite the above comments, the use of nuclear energy for electricity generation, in 2019, in the North American region reached 8.97% of the total electricity generated at the world level, according to Statista database 2020, surpassed only by natural gas with 43.91% and coal with 22.15%.

## The next generation of nuclear power reactors

The majority of the nuclear power reactors today in operation in the world are from the so-called "Generation II." This generation of nuclear power reactors began to be built in the 1970s and is still operating in large commercial power plants in several countries around the world. Most countries expanding their nuclear power programs are constructing nuclear power reactors of the so-called "Generation III," which are more reliable and with several built-in safety features. Generation III reactors were developed in the 1990s by introducing several evolutionary designs that offer significant safety and economic advances. Advances to Generation III are underway, the so-called "Generation III +," resulting in several near-term deployable nuclear power plants that are right now under development and are being considered for deployment in several countries during the coming years. New nuclear power reactors to be built between now and 2030 will likely be chosen using this type of reactor.

However, it is important to mention that there is no clear definition of what constitutes a Generation III reactor design accepted by all countries and the nuclear industry, apart from being designed in the last 15 years. However, the main characteristic features quoted by the nuclear industry are the following:

- A standardized design to expedite licensing, reduce capital cost and construction time;
- A simpler and more rugged design, making them easier to operate and less vulnerable to operational changes;
- Higher availability and longer operating life, typically 60 years;
- Reduced possibility of core melt accidents;
- Minimal effect on the environment in case of a nuclear incident or accident;
- Higher burn-up to reduce fuel use and the amount of nuclear waste;
- Burnable absorbers ('poisons') to extend fuel life (Advanced Nuclear Power Reactors, 2020).

These characteristics are very imprecise and do not define very well what a Generation III reactor is. This type of reactor is evolved from existing designs of PWR, BWR, and CANDU. Until there is much more experience with the use of these technologies, any figures on power generation costs from these designs should be treated with the utmost caution. The experience with the construction cost of the EPR in France and Finland is a clear example of what has been said before.

However, the future belongs to the fourth generation of nuclear power reactors (Generation IV). This new generation of nuclear power reactors is a revolutionary design with innovative fuel cycle technologies. Why is a new generation of nuclear power reactors needed? The answer is the following: Generation IV initiative "is the recognition that maintaining global nuclear capacity at its current level of roughly 400 GWe[b] will be insufficient to reduce and stabilize carbon dioxide emissions in the longer term against a background of increasing energy demand, even if a substantial contribution from renewables is realized" (Hesketh et al., 2004).

In other words, the current nuclear power capacity already installed is insufficient to reduce the $CO_2$ emissions to the atmosphere in the long-term and to satisfy a foresee increase in the energy demand all over the world and the impossibility of a significant increase in the use of renewable energy for electricity production during the coming years. For this reason, the international community needs a new type of nuclear power reactor, which could deliver the highest power capacity in a manner that would be regarded as long-term sustainable for at least a group of countries.

According to the document entitled "A Technology Roadmap for Generation IV Nuclear Energy Systems (2002)", it is expected that Generation IV reactors may be available for commercial application around 2030–2040. Generation IV systems will respond to the following main sustainability criteria and future market conditions:

- **Sustainability:** Generation IV systems will provide sustainable energy generation that meets clean air objectives and promotes long-term energy availability. This system will have effective fuel utilization for worldwide energy production and will minimize and manage their nuclear waste with the purpose of reducing the long-term stewardship burden, thereby improving protection for the public health and the environment;
- **Economic competitiveness:** Generation IV systems will have a clear life-cycle cost advantage over other energy sources and will have a level of financial risk compared to other energy sources;
- **Safety and reliability:** Generation IV systems operations will excel in safety and reliability, will have a very low likelihood and degree of reactor core damage, and will eliminate the need for off-site emergency response;
- **Proliferation resistance and physical protection:** It is expected that Generation IV systems will be very unattractive and the least desirable route for diversion or theft of weapons-usable materials to produce a nuclear weapon. It is also foreseen that this type of nuclear power reactor will provide increased physical protection against terrorism activities (A Technology Roadmap for Generation IV Nuclear Energy Systems, 2002).

The following are the designs of Generation IV systems already selected based on the set of criteria that have been established:

- Sodium cooled fast reactor (SFR);
- Very high-temperature gas reactor (VHTR);

[b] According to PRIS—Power Reactor Information System (2021), the total net nuclear power capacity installed is 391.7 GWe.

- Supercritical water-cooled reactor (SCWR).
- Lead cooled fast reactor (LFR);
- Gas-cooled fast reactor (GFR);
- Molten salt reactor (MSR).

The above six Generation IV system designs are very different among them. These designs have different characteristics and important challenges that need to be solved in the ongoing research and development programs carried out by several states. The purpose is to have all designs available in the market as soon as possible.

It is expected that some of the Generation IV reactor designs mentioned above could be available in the market after 2030. Other Generation IV reactor designs may need significant additional research and development work to be considered ready to be used for electricity generation. The design of the different Generation IV systems is shown in Fig. 7.1–7.6.

## Main steps for the introduction or expansion of a nuclear power program

A reliable and adequate supply of energy, particularly in the form of electricity, is a crucial element for any country's economic and social development. Based on this undisputed fact, providing safe, reliable energy in an economically acceptable way is an essential political, economic, and social requirement by any country. From an economic and social perspective, a country cannot grow if the energy system, particularly the electricity generating system, cannot provide the energy it needs.

However, it is important to note that the use of nuclear energy for electricity generation is limited to a small number of countries in the world. In 2021, only 32 states, or 15.6% of the 193 United Nations Member States, operate nuclear power plants. Five of them, the USA, China, France, Russia, and the Republic of Korea, produced, in 2018, a total of 70.2% of the

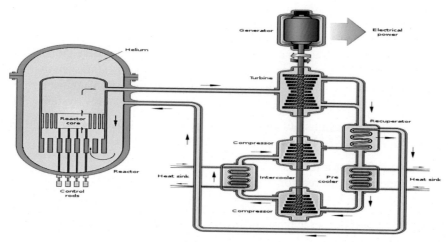

**FIG. 7.1**   Gas-cooled fast reactor system (GFR). *Source: Wikipedia.*

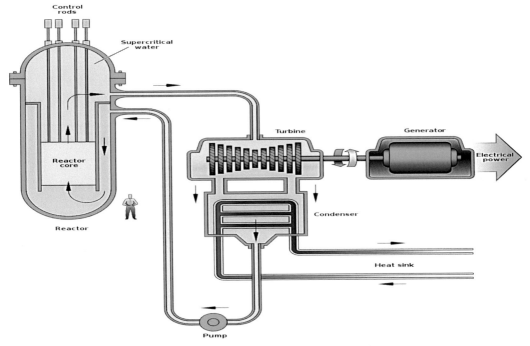

**FIG. 7.2**    Lead-cooled fast reactor system (LFR). *Source: Wikipedia.*

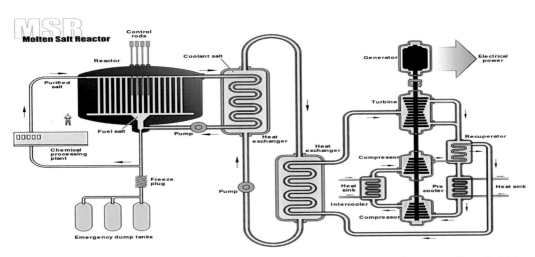

**FIG. 7.3**    Molten salt reactor system (MSR). *Source: US Department of Energy Nuclear Energy Research Advisory Committee.*

nuclear electricity in the world. Simultaneously, 40.5% of the world's nuclear power reactors are located in Western, Eastern, and Central Europe and count, in 2019, for 40.4% of the world's nuclear electricity production. In the case of the North American region, the number of nuclear power reactors in operation in April 2021 was 113, representing 25.5% of the world

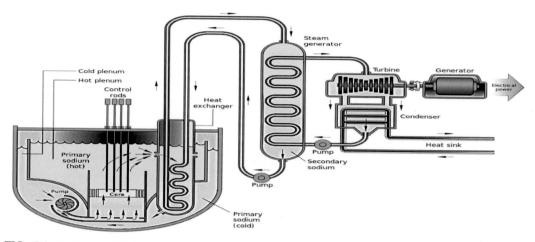

**FIG. 7.4**  Sodium-cooled fast reactor system (SFR). *Source: By Sfr.gifderivative work: Beao - Sfr.gif, Public Domain, https://commons.wikimedia.org/w/index.php?curid=8120627.*

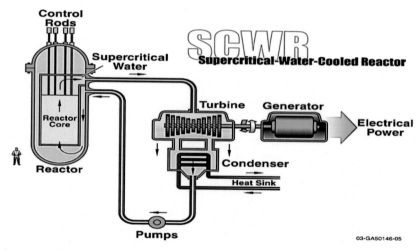

**FIG. 7.5**  Supercritical-water-cooled reactor system (SCWR). *Source: US DoE.*

total and with a net electrical capacity that represents 27.9% (110,107 MW) of the world total electricity capacity installed.

When a country decides how to expand its electricity generating system, the government and the private and public energy industry would have to carry out comprehensive assessments of all the energy options available in the country in order to have the best possible energy mix.

The reasons for choosing a specific energy option will differ from country to country, depending on local and regional energy resources available, technological capabilities, availability of finance and qualified personnel, environmental considerations, and

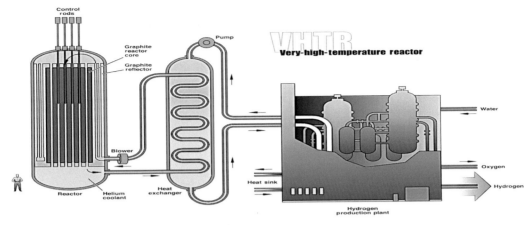

**FIG. 7.6** Very-high-temperature reactor system (VTHR). *Source: US DoE.*

the country's overall energy policy (Choosing the Nuclear Power Option: Factors to be Considered, 1998).

Governments may have different reasons for considering introducing a nuclear power program to meet their national energy needs. These reasons could be the following:

- A lack of available indigenous energy resources such as oil, natural gas, or coal, and of appropriate renewable energy sources infrastructure such as wind farms, solar parks, hydroelectric power plants, bioenergy plants, geothermal power plants, among others in order to satisfy the projected future electricity demand;
- The possibility to use conventional energy sources available in the country for exports purposes instead to use for electricity generation at the country level;
- To increase energy-intensive;
- The desire to reduce the dependency of the country upon imported energy, mainly if the imported energy is coming from unstable political regions;
- To increase the stability of the national electrical grid;
- The need to increase the diversity of the country's energy mix and the reduction of the $CO_2$ emission to the atmosphere.

It is important to note that if a country is considering using nuclear energy for electricity generation, then it should have a well-defined energy policy and strategy in place and a plan for implementing the policy and strategy adopted. The plan mentioned above should be prepared with the objective of achieving an overall energy optimization and high efficiency not only to secure the supply of different energy sources, including the import of energy to meet energy needs, but also environmental considerations, safety issues, the cost associated with the implementation of the plan, financing of the program, the management of nuclear waste, decommissioning and dismantling strategy, energy conservation, and efficiency improvements, among others.

According to the IAEA report (Choosing the Nuclear Power Option: Factors to be Considered, 1998), when a country is studying the introduction of a nuclear power program for the first time, the following relevant criteria should be considered:

- Nuclear power should be considered as a realistic energy option only when it is technically feasible, and when it would be part of an economically viable long-term energy and electricity supply expansion strategy, after considering all energy alternatives and relevant factors;
- A nuclear power program should only be launched when it has a definite likelihood of being successful, mainly when it is the first nuclear power plant to be constructed; i.e., it can be executed within the planned schedule and predicted financial limits and can be operated safely and reliably once in service;
- A nuclear power project should be finally committed only based on comprehensive planning and after specific steps have been taken to meet all necessary supporting infrastructure requirements, including assurance of financing (Choosing the Nuclear Power Option: Factors to be Considered, 1998).

Other issues that must be considered in the framework of the introduction for the first time of a nuclear power program are the following, according to the document entitled "Things to Consider Before Investing in Nuclear Energy for Electricity (n.d.)":

- The economic competitiveness of the use of nuclear energy for electricity generation in comparison with other energy sources available in the country that can be used for the same purpose;
- The adoption of all necessary laws and regulations to ensure the successful introduction of a nuclear power program;
- The safety aspects related to the licensing of the nuclear power plant and the development of a safety culture, according to the safety culture concept promoted by the IAEA;
- The size of the electrical grids[c];
- Proliferation considerations;
- Environmental impact[d];
- The cost involved in the construction of a nuclear power plant;
- The need to have well-trained and experienced personnel, and how this training should be provided, and by whom;

[c] It is important to take into account that normally no single nuclear power reactor should account for more than 10% of the installed capacity of the entire electricity network of the country. According to IAEA's recommendations, countries expanding or thinking to introduce a nuclear power program are advised to consider the capacity of their electrical grids as part of their planning process, particularly because the size of the electrical grid determine the size and type of the nuclear power reactor to be deployed. Specific issues that should be considered in the early phases of the introduction for the first time of a nuclear power program include grid capacity and future growth, historical stability and reliability, and the potential for local and regional interconnections. Assessment of the current electrical grid and plans for improving it should, therefore, be developed in a manner to be consistent with plans regarding the use of nuclear energy for electricity generation.

- The technological capability to assimilate an advanced and demanding technology;
- The safe management of nuclear waste, particularly high-level nuclear waste;
- The need to gain public acceptance;
- The international and regional cooperation in the field of nuclear technology.

For the first time in a given country of a nuclear power program, the introduction requires establishing a basic infrastructure to deal with all aspects of a nuclear power project, including the construction of its first nuclear power reactor. The creation of the appropriate infrastructure, particularly the establishment of an independent nuclear regulatory authority, should start immediately after the adoption by the government or the private or public nuclear industry of the decision to introduce a nuclear power program and should be in place well in advance of the initiation of the construction work of the first nuclear power reactor.

According to the IAEA report Basic Infrastructure for a Nuclear Power Project (2006), the different stages of the development of the basic infrastructure include:

- The development of nuclear power policy and strategy, and its formal adoption by the government;
- The confirmation of the feasibility of implementing a nuclear power project from the economic and technological point of view;
- The establishment of an independent nuclear regulatory authority;
- The adoption of the necessary laws and regulations to ensure the safe use of nuclear energy for electricity generation;
- The establishment of the physical components of the infrastructure needed to implement a nuclear power program;
- The development, contracting, and financing of the first nuclear power project;
- The construction of the first nuclear power reactor according to the established safety, quality, and economic requirements;
- Safe, secure, and efficient operation of the first nuclear power plant (Basic Infrastructure for a Nuclear Power Project, 2006).

Government and utility funding is required to establish major components of the basic nuclear infrastructure. These may include, according to the IAEA report Basic Infrastructure for a Nuclear Power Project (2006), the following:

- The adoption of a national legal framework and the ratification of international agreements related to the peaceful uses of nuclear energy;
- The establishment of a nuclear safety and environmental regulatory authority;
- The construction of all necessary physical facilities with the aim of supporting a nuclear power program;
- The adoption of finance/economics measures to support the introduction of a nuclear power program;
- The identification of all necessary human resources, education, and training with the purpose of supporting the introduction of a nuclear power program;

[d] Nuclear power plants require approximately one $km^2$ of land per typical nuclear power reactor.

- The identification of operational practices and processes to assure safety and performance throughout the life of the nuclear power plant;
- Public information and acceptance (Basic Infrastructure for a Nuclear Power Project, 2006).

The introduction of a nuclear power program by any country should be viewed within a medium to long-term electricity supply strategy adopted by the government to develop its energy sector and diversify its energy mix. Different tasks should be implemented with the purpose of ensuring the successful introduction of a nuclear power program. According to the IAEA report Considerations to Launch a Nuclear Power Programme (2007), these tasks should:

- Develop a comprehensive nuclear legal framework covering all aspects of the peaceful uses of nuclear energy, i.e., safety, security, safeguards, and liability, in addition to the commercial issues related to the use of nuclear material;
- Establish and maintain an effective regulatory system;
- Develop the human resources for the state organizations and also for the operating organizations required to adequately supervise and implement the introduction of the nuclear power program adopted;
- Ensure adequate financial support for the construction, sustained safe operation, and decommissioning of the nuclear power reactor, as well as for the long-term management of nuclear materials;
- Communicate openly and transparently with the public and the neighboring states about the considerations behind introducing a nuclear power program (Considerations to Launch a Nuclear Power Programme, 2007).

A nuclear power program should be implemented with the purpose of producing stability in power generation and electricity price, to achieve an important impact in the development of the domestic industry, and to reduce the adverse effects on the environment and population as a result of the use of different conventional energy sources available in the country for electricity generation.

The potential economic benefits of introducing a nuclear power program in a given country are reflected in maintaining the long-term stability of electricity prices. An important consideration in many developing countries has been the influence of a national nuclear power program in increasing the country's technological level and enhancing the domestic industry's global competitiveness (Choosing the Nuclear Power Option: Factors to be Considered, 1998).

One of the main elements that need to be considered to ensure the successful introduction of a nuclear energy program in a given country, particularly in a developing country, is the national industry level of participation within the program. The involvement of national industry in the implementation of a nuclear power program could be materialized in one or more of the following manners, according to IAEA report Choosing the Nuclear Power Option: Factors to be Considered (1998) and Morales Pedraza (2011):

- Providing local labor and supplying some construction materials that could be used for non-specialized purposes on-site, especially civil engineering works associated with the construction of the nuclear power plant;

- Local contractors could take full or partial responsibility for the civil engineering work related to the construction of the nuclear power plant, including some design work assigned by the primary constructor;
- Locally manufactured components from existing national factories could be used for non-critical parts of the nuclear power plant;
- Local manufacturers could extend their regular product line to incorporate nuclear designs and standards, possibly under licensing arrangements with foreign suppliers;
- Factories could manufacture heavy and specialized specific nuclear components under licensing agreements with foreign suppliers.

However, such undertakings' economic viability would have to be assessed carefully in view of the future domestic market and the availability of such equipment internationally with a better price and quality (Choosing the Nuclear Power Option: Factors to be Considered, 1998).

Well-designed, constructed, and operated nuclear power plants have proved to be reliable, economic under certain conditions, environmentally acceptable, and a safe source of electrical energy all over the world but in a limited manner. The economic competitiveness of nuclear power arises from cost reductions in construction and plant operations. However, the level of initial capital investment is still too high in some cases for several countries, particularly for most developing countries. For this reason, further cost reduction can be achieved, for example, in costs associated with waste management and decommissioning. Construction costs per kW for nuclear power plants have fallen considerably due to standardized design, shorter construction times in some types of nuclear power reactors, and more efficient generating technologies. Nevertheless, this could not be true for the new generation of nuclear power reactors available yet in the market. Recent new-build experience has demonstrated, from one side, that new nuclear power plants can be built on time and budget. Still, from another side, the construction of new design reactors could cause significant delays, as is happening today with the construction of the EPR in Finland and France.

One of the most critical decisions that need to be adopted by the competent national authority in a given country before introducing a nuclear power program is the type of nuclear power reactor to be selected. Two main groups of nuclear power reactors are available in the market: a) the PWRs, including the PHWRs, and b) the BWRs. Most of the nuclear power reactors available in the market in April 2021 are of the PWRs type. A total of 303 PWRs were operating in April 2021, with 287,988 MW net capacity. In the case of BWRs, 63 BWRs were operating in April 2021, with a net capacity of 64,122 MW. Finally, 498 PHWRs were operating in April 2021, with a net capacity of 24,505 MW (PRIS—Power Reactor Information System, 2021).

The type mentioned above of reactors is generally available in sizes of about 1000 MW or higher electrical output. However, smaller reactors of 600–700 MW output are also available. Suppose a smaller unit is required due to the capacity of the national electrical grid network. In that case, the available technology is limited to satisfy this condition, although some reactors of 200–400 MW output are being operated and developed in some countries. Several nuclear power designs are being developed for future applications, although a significant challenge is to achieve an industrial design at a smaller size and the highest safety operation. It is important to stress that the high-temperature gas-cooled reactor with a net capacity between 160 and 270 MW and several small water-cooled reactors is being developed, reaching design approval over the next years (Considerations to Launch a Nuclear Power Programme, 2007).

The decommissioning of nuclear power plants is a costly and complex activity that should begin after the nuclear power plant closure. Decommissioning a nuclear power plant means removing all regulatory controls that apply to a nuclear site while securing the public's long-term safety and the environment. Underlying this, other practical objectives are to be achieved, including releasing valuable assets such as sites and buildings for unrestricted alternative use, recycling and reuse of materials, and the restoration of environmental attractiveness. In all cases, the primary objective is to achieve a sensible endpoint in technical, social, and financial terms that properly protect workers, the public, and the environment. In summary, to comply with the basic principles of sustainable development.

In other words, the term "decommissioning" covers all of the administrative and technical actions associated with the shutdown of a nuclear power plant. These actions begin when a nuclear power plant or facility is shut down and end with its removal from the site (termed dismantling). These actions may involve some or all of the following activities:

- Dismantling of equipment;
- Decontamination of structures and components;
- Remediation of the contaminated ground;
- Disposal of the resulting wastes.

Three main decommissioning strategies can be applied for the dismantling of a nuclear power facility. These are the following:

- Immediate dismantling;
- Deferred dismantling, also called "safe enclosure";
- Entombment.

In the first case, a nuclear power plant or facility is dismantled right after removing nuclear materials and waste from the facility. In the second case, after removing nuclear materials and waste, the facility is kept in a safe enclosure state between 30 and 100 years, followed by dismantling. In the third case, a facility is encapsulated on-site and kept isolated until the radionuclides decayed to levels that allow a release from nuclear regulatory control. According to the document entitled "Selecting Strategies for the Decommissioning of Nuclear Facilities: A Status Report" (NEA-OECD, 2006), the present trend favors immediate dismantling.

The selection of the correct decommissioning strategy depends on several factors that can be grouped, according to the report mentioned above, into the following three categories:

- Policy and socio-economic factors;
- Technological and operational factors;
- Long-term uncertainties (NEA-OECD, 2006).

Policy and socio-economic factors are dominated by national and local situations, which vary from country to country. Countries with important nuclear power programs tend to immediately dismantle nuclear power plants and other facilities to use the site to construct new facilities in a possible short period. Decommissioning costs associated with a nuclear power plant or facility include, among other elements, the following components:

- Dismantling the nuclear power plant or facility;
- Nuclear waste treatment;
- Disposal of all types of radioactive waste;
- Security of the site;
- Site cleanup;
- Project management.

Undoubtedly, dismantling and disposal represent a major share of a nuclear power plant's decommissioning total cost, each accounting for approximately 30% of this cost. The average cost estimates are between US$320 and US$420 per kW for most nuclear power reactor types. In general, GCR is more expensive to decommission than water-cooled reactors because they must dispose of large graphite quantities. Union Fenosa has recently estimated the cost of dismantling the older 160 MWe nuclear power reactor at Zorita nuclear power plant in Spain at €850 per kW, and the German plant's dismantling Obrigheim was estimated at €1400/kW.

National energy policy may influence the decommissioning strategy to be selected either directly or indirectly. If a national decommissioning policy is reflected in the national legislation, then the direct influence is exerted through the legal framework. The extent of this influence will depend on the degree to which laws are either prescriptive or enabling. Policies and regulations related to a nuclear power program vary from country to country and affect some or all of the following issues:

- Public and occupational health and safety;
- Environmental protection;
- The definition of end-state, waste management, reuse and recycling of nuclear materials;
- Arrangements for the release of nuclear materials from regulatory control and matters concerning regional development.

However, a national energy policy may influence the decommissioning strategy indirectly. In this case, influence may be by way of national energy policies that are not explicitly concerned with the process of decommissioning a nuclear power plant or facility but may be linked to it by way of broader issues. These may include matters such as the future use of nuclear power for electricity generation and other purposes, economic and societal issues associated with the effects of shutting down major industrial facilities, safety issues, broad financial issues concerned with costs, the use of available funds, and the timing of their deployment. Although perhaps not associated with national energy policy, as such, the prospects for the continued availability of qualified and well-trained staff may also have such an influence, according to the document entitled "Selecting Strategies for the Decommissioning of Nuclear Facilities: A Status Report" (NEA-OECD, 2006).

Although decommissioning technology is already available, technological and operational factors will influence the pursuit of strategic choices. For example, long-term uncertainties are important during the decommissioning strategy selection process to be chosen and later implemented.

Policies and legal/regulatory frameworks are subject to change. But the direction of change is uncertain, although regulatory standards have tended to become more stringent with time.

It is important to single out that assessing the factors mentioned above is a challenge, mainly when long periods are involved, and because most of these factors are not quantitative and need subjective assessment, not always very objectives.

Techniques for decontaminating and dismantling nuclear power plants and other nuclear facilities are already available for their safe use. Many nuclear facilities, including several nuclear power plants in different countries, have already been successfully decommissioned and dismantled using these techniques. Some sites have already been returned to a condition suitable for unrestricted reuse. It is recognized that funding for dismantling and decommissioning nuclear facilities needs to be made during its operating lifetime. The challenges are to ensure that dismantling and decommissioning costs are calculated correctly, and that sufficient funds will be available when required. Waste management costs are a substantial element of the overall costs of dismantling and decommissioning activities associated with a nuclear power plant or facility. They may dominate in some cases, depending on how residual spent fuel management costs are assigned. Hence, it is essential that waste quantities are minimized and that the costs of waste treatment, storage, and disposal are individually identified and assigned (Morales Pedraza, 2012).

Without a doubt, early planning is an essential element in any decommissioning project due to its complexity. The shift from operations to decommissioning requires a well-defined work program, similar to the engineering industry's methodologies. For a successful outcome, decommissioning must be treated as an engineering project with modern project management. A dedicated decommissioning organization is also required. This new 'mindset' often poses difficulties. The nature of the forward is radically different, requiring both new technical skills and the need to control and manage budgets proactively, to achieve cost and time targets. Such changes create tensions as the order of priorities change. The decommissioning phase can lead to the loss of experienced and younger staff as they may face redundancy or significant changes in their jobs (Morales Pedraza, 2012).

Decommissioning and dismantling nuclear power plants and other nuclear facilities are the nuclear power plant operator's responsibility and must be conducted under a competent national authorities' license. The critical points in decommissioning and dismantling of nuclear power plants and other nuclear facilities are, among others, the following:

- The purpose of decommissioning and dismantling activities is to allow the removal of all regulatory controls apply to a nuclear site;
- There is no unique or preferable approach to decommissioning and dismantling of nuclear facilities;
- Techniques for decommissioning and dismantling are available, and experience is being fed back to plant design and decommissioning plans;
- Many nuclear facilities have been successfully decommissioned and dismantled in Germany, Belgium, France, the USA, Russia, and the UK, among others;
- Current institutional arrangements for decommissioning and dismantling are sufficient for today's needs;
- Existing systems for the protection and safety of workers, the public, and the environment are satisfactory to carry out successfully decommissioning and dismantling of nuclear facilities, particularly nuclear power plants;

- Arrangements are in place for the funding of decommissioning and dismantling of nuclear facilities, but evaluation of costs requires further attention by competent national authorities;
- Local communities are increasingly demanding a more active involvement in planning, decommissioning, and dismantling nuclear facilities, particularly nuclear power plants (Morales Pedraza, 2012).

Another important element for the successful introduction of a nuclear power program is the existence of a well-prepared and trained workforce in several areas of the peaceful uses of nuclear energy, particularly for electricity production and nuclear safety. In the last two decades, particularly after the Chernobyl and Fukushima Daichi nuclear accidents, several countries, particularly in Europe and North America, have had an increasing lack of interest in developing and using nuclear energy for electricity generation. In addition to the retirement of the specialized workforce working in the nuclear sector and its lack of perspectives, the lack of interest significantly reduced the number of well-trained professionals and skilled workforce available in several countries' nuclear sectors. The situation is even more serious regarding the new workforce, which will eventually substitute the currently specialized workforce working in the nuclear energy sector. The lack of a specialized trained workforce available in some countries can be overcome by adopting different initiatives. The purpose of these initiatives is to increase the number of professional and skilled workforce available in the nuclear sector in the coming years to ensure the energy sector's development. In several of these countries, has been an increase in the recruitment and training of a new workforce in the nuclear sector in the last years (Morales Pedraza, 2012).

Companies have been promoted to build new nuclear power plants supported by different government institutions and agencies, including the IAEA. However, it is essential to single out that companies in charge of constructing new nuclear power plants require specialized personnel in the nuclear energy sector. To avoid that the lack of a specialized workforce in the nuclear sector has a negative effect on developing this sector in the future, the UK and the Russian Federation recently decided to establish universities and institutes specifically dedicated to preparing professionals and technicians in nuclear sciences and technologies. The goal is to satisfy the foresee demand for a specialized workforce in their respective nuclear sectors in the future.

Simultaneously, the IAEA has also been promoting the preparation of new professionals and technicians in sciences and nuclear technologies in its member states to satisfy the foresee demand for a specialized workforce in their respective nuclear sectors in the future. If the situation of lack of specialized workforce in the energy sector is not reverse shortly, then the nuclear sector will face an insufficient replacement of specialized workforce duly prepared to assume the responsibility to operate not only the current number of nuclear power reactors already constructed but also the new units to be built in the future.

The consideration of three crucial elements could facilitate introducing a nuclear power program in any given country. These elements are the following:

- The elaboration and adoption of national energy policy and strategy supporting the use of nuclear energy for electricity generation;

- The elaboration and approval of regional energy policy and strategy supporting the use of nuclear energy for electricity generation and enhancing the cooperation in the implementation of the strategy adopted at that level;
- The correct planning for the introduction of a nuclear power program.

## Economic optimization in the use of different energy sources available in the United States for electricity generation

Before the competent national authorities decide to build new power plants, all necessary measures must be taken to achieve economic optimization using all other energy sources available in the country for electricity generation. That is particularly important if the decision adopted is to build a nuclear power plant due to the high capital investment involved to carry such an undertaking. The cost of a nuclear power plant depends on its output, the construction company's experience, resources available, and the type of contract signed, among others. Usually, the construction cost of a nuclear power plant of 1000 MW could be around US$1.5–US$2.5 billion or a little more. If there is a delay in the construction process for any reason, this figure could go up significantly. The recent experience in constructing some nuclear power plants worldwide demonstrates how the initial cost associated with constructing these power plants can increase due to the delay in this activity. For this reason, the current cost of the construction of the first EPR in Finland is now around US$8.5 billion, which is much more than the initial construction costs of US$3.2 billion (Wikipedia Nuclear Power in Finland, 2020).

Regarding the economic optimization of the energy electricity supply, it is assumed that energy and electricity supply systems should be economically optimized using different models, including computer models.[e] The planning period should be between 10 and 15 years to make the optimization meaningful. In the USA, this period could be 15 years. The objectives of security and diversity of energy supplies could be factored in through appropriate assumptions on future fossil fuel prices produced domestically or imported. In addition to considering future prices of fossil fuels, energy policy decisions have to be made, considering other costs, such as environmental and health costs and infrastructure development costs (Choosing the Nuclear Power Option: Factors to be Considered, 1998). In the specific case of nuclear energy use for electricity generation, the decommissioning and management of spent nuclear fuel costs should also be included.

## Stability of the national electrical grid

The national electrical grid's stability is an essential element that needs to be examined very carefully by the competent national authorities in charge of the energy sector. It is a crucial prerequisite for the successful introduction of a nuclear power program in any country.

[e] The IAEA's Secretariat has a computer model to carry this type of studies called "Wien Automatic Planning System" (WASP-IAEA, n.d.) available to all of its member states.

Necessary measures to minimize any frequency and voltage fluctuations during the regular operation of nuclear power plants must be identified, adopted, and implemented (Choosing the Nuclear Power Option: Factors to be Considered, 1998). The adoption of these measures should create the necessary conditions to introduce a nuclear power program successfully. At the same time, the government should ensure that the electrical grid's capacity is adequate for the nuclear power reactor's size to be built. According to the IAEA and other experts' opinions, the electrical grid's capacity should be between 7 and 10 times the nuclear power reactor's capacity to be constructed. The adoption of specific measures to ensure the successful introduction of a nuclear power program may improve, in some particular cases, the quality of the country's electricity supply system.

Finally, it is essential to single out the following. Besides, to ensure that the electrical grid will provide reliable off-site power to nuclear power plants connected to the grid, other important factors should also be considered before this connection is carried out. If a nuclear power plant is too large for the electrical grid's size, then the plant operators may face several problems. These problems are, among others, the following:

- Off-peak electricity demand might be too low for a large nuclear power plant to be operated in baseload mode, i.e., at constant full power;
- There must be enough reserve generating capacity in the electrical grid to ensure grid stability during the nuclear power plant planned outages for refueling and maintenance;
- Any unexpected sudden disconnect of the nuclear power plant from an otherwise stable electrical grid could trigger a severe imbalance between power generation and consumption, causing an immediate reduction in grid frequency and voltage. That could even cascade into the electrical grid's collapse if additional power sources are not connected to the grid in time.

It is important to single out that grid interconnectivity and redundancies in transmission paths and generating sources are key elements in maintaining high-performance electrical grids' reliability and stability. However, operational disturbances can still occur even in well-maintained electrical grids. Similarly, even a nuclear power plant running in baseload steady-state conditions can encounter unexpected operating circumstances that may cause a complete shutdown in the plant's electrical generation. When relatively large nuclear power plants are connected to the electrical grid, abnormalities can lead to the electrical grid's shutdown or collapse. The technical issues associated with the interface between nuclear power plants and the electrical grid includes:

- The magnitude and frequency of load rejections and the loss of load to nuclear power plants;
- Grid transients causing degraded voltage and frequency in the power supply of vital safety and operational systems of nuclear power plants;
- A complete loss of off-site power to a nuclear power plant due to electrical grid disturbances;
- A nuclear power plant unit trip causes an electrical grid disturbance resulting in severe degradation of the grid voltage and frequency or even the electrical power grid collapse.

## Security of electricity supply

A secure supply of energy is a crucial element for the economic and social development of any country. Thus, if a country plans to introduce an important economic and social development plan, it must guarantee the electricity supply to ensure its implementation using all available energy sources. It is important to be aware that the security of electricity supply and the implementation of an effective industrial development plan are two indispensable development components of any country's economic development and social policy.

## Environmental issues

The international community is well aware that fossil fuel use for a different purpose, particularly for electricity generation, is one of the leading causes of climate change. For this reason, any energy policy and strategy adopted for the development of the energy sector by any government should include an assessment of the impact of this policy and strategy on the environment and population. There is no doubt that fossil-fueled power plants are sources of pollution and climate change. For this reason, they have become an essential factor in all considerations regarding environmental protection, not only at the national level but also at the regional and international levels as well. It is becoming increasingly unlikely that a country can avoid considering environmental issues when setting national energy policy and electricity generation strategy in the future (Choosing the Nuclear Power Option: Factors to be Considered, 1998).

Transboundary effects of pollution, such as acid rain and the emission of $CO_2$ to the atmosphere, are increasing the concern of many countries about the use of fossil fuels for electricity generation and the need to use nuclear energy for the same purpose. For this reason, the consideration of environmental issues is an indispensable component of any regional energy policy and strategy of Canada and the USA that supports the adoption of an increasing role of nuclear power within their energy mix in the future.

## Electrical grid integration policy and strategy

Adopting a regional electrical grid integration policy and strategy within the North American region could directly benefit Canada and the USA by increasing the security of electricity supply and improving electricity supply and operation reliability. It is essential to single out that adopting a regional electrical grid integration policy and strategy within the North American region will demand a cooperative planning effort to expand electricity generation in Canada and the USA. Without a doubt, the planning process of such integration policy and strategy could be facilitated as a result of the broader experience in the elaboration of such policy and strategy available in the region.

A regional electrical grid integration policy and strategy may also permit the use of large units for electricity generation and an increase in the capacity of the nuclear power plants now in operation in Canada and the USA to a size larger than any national electrical grid could accept. However, the adoption of a regional electrical grid integration policy is not an easy task.

It requires, among others, the approval of national energy policies and strategies that creates the necessary political conditions to promote and supports such regional integration and the use of nuclear energy as part of the national energy mix of the countries involved.

## Nuclear safety

Nuclear safety is one of the most important components of any national and regional policy associated with the use of nuclear energy for electricity generation. Close cooperation at the regional level in nuclear safety matters should be an indispensable component of the nuclear power policy and strategy to be adopted by Canada and the USA in the future. This cooperation can give additional assurances about the safety of the nuclear power plant in operation or that are going to enter into service in the future by providing immediate access to information about nuclear incidents and accidents that could occur in the plant affecting the other country, and for the coordination of emergency plans. It can also provide the necessary assistance in case of a nuclear accident or incident in a nuclear power plant, particularly the use of available medical treatment facilities to those affected, among other indispensable assistance.

## Sharing of nuclear power plant services

If more than one country in a specific region has nuclear power plants in operation, which is the case in the North American region, then there are clear advantages in trying to share plant services, such as plant maintenance, repair, and spare parts, where this is a feasible option. However, many difficulties and problems impede plant service sharing due to different political, economic, and technological considerations. In the North American region, Canada and the USA use different types of nuclear power reactors. In Canada's case, the only type of nuclear power reactor used for electricity generation is the so-called "CANDU reactor" (PHWR type). In the USA, mainly two types of nuclear power reactors are used (PWR and BWR). In April 2021, 94 units were in operation in the country: 62 PWR and 32 BWR with a total capacity of 96,553 MW, generating 809,409 GWh or 19.7% of the total electricity generated in the country in that year. A total of 19 units with a capacity of 13,554 MW were operating in the country in April 2021, generating 95,469 GWh in 2019, or 14.9% of the total electricity generated in the country in that year. Considering that the types of nuclear power reactors used by Canada and the USA are different, the adoption of a regional sharing of plant services cooperation policy and strategy in this specific area is very difficult to achieve.

## Importance of a nuclear energy program in the North American region

The energy crisis the world is now facing, the current level of the world's oil reserves, and the possibilities of its extinction in the coming decades could increase oil prices significantly compared to those registered in 2007 and 2008. The possibility of another oil price increase above US$120 per barrel in the future, and the negative impact in the environment and population due to the release to the atmosphere of a significant amount of $CO_2$ as a result of an increase in the consumption of fossil fuel in the USA and Canada, among other relevant factors, are somehow changing the negative perception of the public in these two countries regarding the future use of nuclear energy for electricity generation.

According to public information disclosed during the last US presidential campaign, the USA should plan the construction of around 45 new nuclear power reactors by 2030 and approximately 55 more units after that year to satisfy the country's energy demand in the coming decades. Undoubtedly, that is an important nuclear power program to be developed by the USA in the future if the new US administration approves it. However, until April 2021, only two nuclear power are under construction in the USA and none in Canada.

Why is the construction of new nuclear power reactors so crucial for both countries, particularly for the USA? The reason is the following: North America had an aging base of nuclear and fossil fuel power plants for electricity generation. In the specific case of nuclear power plants, the last unit to enter into commercial operation in the USA was TVA's Watts Bar Unit 1 in June 1996, more than 24 years ago. The last successful order for a US commercial nuclear power plant was presented in 1973; this means 47 years ago.

The IAEA has foreseen that nuclear energy participation in the world's energy balance will drop from 16% today to between 8% and 10% in 2030 if no decision is adopted by the EU, Canada, and the USA to build more nuclear power reactors for electricity generation during the coming years.

The USA is a pioneer of nuclear power development. The USA's annual nuclear electricity generation has more than tripled since 1980, reaching 809,409 GWh in 2019. The participation of the different energy sources in the US electricity production is the following:

- 27% from a coal-fired power plant (1146 TWh);
- 19.7% from nuclear power plants (809,409 GWh) (see Table 7.1);
- 35% from gas-fired power plants (1468 TWh);
- 7% from hydropower plants (292 TWh);
- 6.6% from wind energy (275 TWh);
- 1.6% from solar power (67 TWh);
- 1.5% from biomass (63 TWh);
- 1.4% from geothermal and other energy sources (US EIA database and PRIS—Power Reactor Information System, 2021).

In 2019, nuclear energy generated in the USA reached 809,409 GWh. It achieved an average above 90% in the load capacity factor, which is very high within the nuclear energy sector. US annual electricity demand is projected to increase from 4178 TWh in 2018 to 5000 TWh in

TABLE 7.1    The production of electricity using nuclear energy in North America.

| Country | Number of nuclear power reactors connected to the grid April 2021 | Nuclear electricity generation (net GWh) 2019 | Nuclear percentage of total electricity supply (%) 2019 |
|---|---|---|---|
| United States of America | 94 | 809,409 | 19.7 |
| Canada | 19 | 95,469 | 14.9 |
| Total | 113 | 904,478 | – |

*Source: April 2021. PRIS—Power Reactor Information System. www.iaea.org/pris. © IAEA.*

2030, increasing 19.7% in 12 years. From 2006 to 2030, the USA is expected to add 12.7 GW capacity at newly-built nuclear power plants and 3.7 GW from up rates of existing plants, offset in part by the retirement of 4.4 GW capacity at older nuclear power plants. Almost all the US nuclear-generating capacity comes from reactors built between 1967 and 1990. Between 1977 and 2013, there had been no new construction of nuclear power reactors.

For many years, electricity generation by gas-fired power plants was considered more economically attractive than the use of nuclear energy for the same purpose. Besides, the construction schedules of new nuclear power plants during the 1970s and 1980s had frequently been extended by the opposition to the use of nuclear energy for electricity generation within the American people following the Three Mile Island accident in 1979. After that accident, only one PWR type of nuclear power reactor – Watts Bar 2 – started up in 2016 following the Tennessee Valley Authority's (TVA's) decision in 2007 to complete the construction of unit 2 (World Nuclear Association-Nuclear Power in the USA, 2020).

In Canada, nuclear energy produced, in 2019, a total of 95,469 GWh of electricity or 14.9% of the total (see Table 7.1). In 2019, Canada generated a total of 660.4 TWh of electricity, 57.8% of which from hydropower plants (382 TWh), 8.3% from coal (54.6 TWh), 10.5% from gas (69.3 TWh), 1% from oil and other energy sources (4.8 TWh), and 7.5% from wind, solar and other renewables (49.3 TWh) (Nuclear Power in Canada, 2020; PRIS—Power Reactor Information System, 2021; BP Statistical Review of World Energy 2020, 2020).

For most of this coming decade, the construction of new nuclear power reactors in Canada is uncertain based on the most recent electricity market outlooks. According to the Canadian authorities, the return to service of the remaining laid-up nuclear units and the completion of gas-fired power units already under construction should satisfy the foresee energy demand in the coming years. While market prospects for the construction of new nuclear power reactor sales in the near-to medium-term are not too promising, the refurbishment of existing units holds more promise. At least in the medium-term, it would avoid the replacement of nuclear-generating capacity with fossil-fueled power plants increasing the $CO_2$ emission to the atmosphere.

## Shut down of nuclear power reactors in the North American region

Until April 2021, in North America, the number of nuclear power reactors that have been shut down reached 45 units. Canada shut down six nuclear power reactors between 1957 and April 2021, reducing its net nuclear power capacity in 2143 MW. In the USA's case, the number of nuclear power reactors shut down between 1957 and April 2021 reached 39 units, reducing its net nuclear power capacity in 18,141 MW (PRIS—Power Reactor Information System, 2021).

## The need for nuclear energy for electricity generation in the North American region

What is the current situation concerning nuclear energy for electricity generation and the North American region's energy infrastructure? The region consumed, in 2018, a total of 152.9 million tons of oil equivalent for electricity generation and produced, in 2019, a total of 5093.7 TWh (BP Statistical Review of World Energy 2020, 2020). The electricity generation using nuclear power plants in the region reached 952.4 TWh in 2019, representing around 19% of the total electricity generated in that year (BP Statistical Review of World Energy 2020, 2020).

The average global energy consumption per capita increased by 1.8% in 2018, a growth significantly higher than the historical average (0.3%) registered during the period 2007–17. North America is the region with the highest energy consumption per capita, followed by CIS and the Middle East. Africa remains the region with the lowest average energy consumption. South and Central America and Europe were the only regions where average energy consumption per capita decreased in 2018 (BP Statistical Review of World Energy 2019, 2019).

In North America, specifically in the USA, existing electricity power plants continued to pollute the atmosphere with greenhouse gases significantly due to the use of fossil fuel, particularly coal, as a fuel for electricity generation. The USA is the second world polluting country after China. However, there is a significant difference between the USA and Canada on this important subject. What is the reason for such a difference? The main reason is the following: the USA's and Canada's governments' reactions concerning the atmosphere's contamination due to fossil fuel consumption are different. The Canadian government became increasingly concerned about the emission of a large quantity of $CO_2$ into the atmosphere and, for this reason, signed the Kyoto Protocol and the Paris agreement on climate change committing the energy sector to reduce the emission level of this gas to the atmosphere. The primary energy consumption by fuel in 2019 in Canada was the following:

- Oil: 107.8 million tons of oil equivalent (31.7% of the total);
- Natural gas: 97.9 million tons of oil equivalent (30.5% of the total);
- Coal: 16.3 million tons of oil equivalent (3.9% of the total);
- Nuclear energy: 22.6 million tons of oil equivalent (6.3% of the total);
- Renewables: 98 million tons of oil equivalent (27.7 of the total) (BP Statistical Review of World Energy 2020, 2020 and Author's own calculations).

The electricity generation by fuel in Canada in 2019 was the following:

- Oil: 4.1 TWh (0.6% of the total);
- Natural gas: 69.3 TWh (10.5% of the total);
- Coal: 54.6 TWh (8.3% of the total);
- Nuclear energy: 100.5 TWh (15.2% of the total)[f];
- Renewables and others: 432 TWh (65.4% of the total) (BP Statistical Review of World Energy 2020, 2020).

On the contrary, and even though the USA is the second most polluted country in the world, the USA administration refused to sign the Kyoto Protocol and had said many times that it has no intention to sign it in the future, and refused to ratify the Paris agreement on climate change. The primary energy consumption by fuel in 2019 in the USA was the following:

- Oil: 922.7 million tons of oil equivalent (39.1% of the total);
- Natural gas: 680.5 million tons of oil equivalent (32.2% of the total);
- Coal: 371.2 million tons of oil equivalent (12% of the total);

[f] According to PRIS—Power Reactor Information System (2021), the electricity generated by nuclear power plants in the Canada in 2019 was 95,469 GWh or 14.9% of the total.

- Nuclear energy: 192.2 million tons of oil equivalent (8% of the total);
- Renewables: 165.8 million tons of oil equivalent (8.7% of the total) (BP Statistical Review of World Energy 2020, 2020 and Author's own calculations).

The electricity generation by fuel in the USA in 2019 was the following:

- Oil: 20 TWh (0.5% of the total);
- Natural gas: 1700.9 TWh (35.4% of the total);
- Coal: 1053.5 TWh (27.9% of the total);
- Nuclear energy: 852 TWh (19.1% of the total)[g];
- Renewables and others: 775 TWh (17% of the total) (BP Statistical Review of World Energy 2020, 2020).

Based on the above data, the following can be stated: in the North American region, the main energy sources for electricity generation in 2019 were the following:

- Fossil fuels with 2902.4 TWh of electricity generated using this type of fuel (57.3% of the regional total[h]);
- Renewable energy sources with 1206.8 TWh (23.8% of the regional total);
- Nuclear energy with 952.4 TWh (18.9% of the regional total).[i]

It is expected that this situation regarding the use of primary energy sources for electricity generation will not change during the coming years.

## Limiting factors in raising the use of nuclear energy for electricity generation in the North American region

The main limiting factor in raising the use of nuclear energy for electricity generation in North America are the following:

- The energy policy adopted by the governments of the USA and Canada regarding the use of nuclear energy for electricity generation;
- The small number of nuclear power reactors under construction in the region (two units in the USA only);
- The lack of a plan to build new units during the coming decades;
- The increasing number of nuclear power reactors reaches the point where choices must be made between extensions of their useful life or decommissioning, particularly in the USA.

However, other limiting factors need to be considered as well. These additional limiting factors are the following:

- In the USA's case, one of the main limiting factors to be considered is the restricted incentives and measures to encourage the private energy industry's participation in

[g] According to PRIS—Power Reactor Information System (2021), the electricity generated by nuclear power plants in the USA in 2019 was 809,409 GWh or 19.7% of the total.

[h] The regional total was 5061.7 TWh.

[i] According to PRIS—Power Reactor Information System (2021), the electricity generated by nuclear power plants in the North American region in 2019 was 904,878 GWh or 19% of the total.

constructing new nuclear power plants. "The hard truth about nuclear energy is that it requires strong government backing. Sure, a new reactor can have great engineers and be backed by a large and reputable company, but to succeed in the global electricity generation sector, it must have a stamp of approval from a credible regulator" (Bryce, 2020);

- The opposition of the majority of the regional banks to provide to the private energy industry the necessary funds for the construction of new nuclear power reactors, particularly now due to the present economic and financial crisis that the world is facing and the lack of credit for fresh money to support substantial investment in risk project such as the construction of a nuclear power plant;
- The difficulties facing designers to improve the current designs of nuclear power reactors significantly to meet the increasingly stringent safety requirements adopted by different governments to avoid a new nuclear accident;
- The problem causing the final disposal of high-level nuclear waste. In the particular case of the USA, the definitive approval of the national nuclear waste repository at Yucca Mountain, Nevada, could have a positive impact in increasing the participation of nuclear energy in the energy balance of the country in the future;
- The high construction cost and the time needed for constructing the current type of nuclear power reactors, despite all measures already adopted to reduce the construction time and the time required to receive the approval, by the regulatory authorities, of the operation license required for the beginning of a nuclear power plant's operation.

It is easy to verify that several nuclear power plants in North America are operating successfully and efficiently, which is one of the reasons why nuclear energy should continue to be a generally competitive, profitable source of electricity. In the USA's case, the average load factor is above 90%, which is considered a very high load factor within the whole energy sector. However, the decision to begin the construction of new nuclear power plants should be adopted after considering the following elements:

- The USA and Canada are rich in other energy sources, making very competitive the use of any of them in comparison with nuclear energy;
- The overall current and future electricity demand, how fast it is growing, and the level of the energy reserves of the different energy sources available in the region that can be used for electricity generation;
- The market structure and investment environment, particularly the government's position to support the construction of new nuclear power plants.

In general, banks were somehow reluctant to finance the construction of new nuclear power plants in the region, particularly in the USA. Only companies with their own capital resources are ready to go ahead with constructing new nuclear power plants. For this reason, the Energy Policy Act (EPA) (2005) introduced a production tax credit (PTC)[j] of 1.8 cents per kWh of electricity produced by all new nuclear power plants. However, it is important to

---

[j] Tax reductions by subject area are the following: US$4.3 billion for nuclear power, US$2.8 billion for fossil fuel production, US$2.7 billion to extend the renewable electricity production credit, US$1.6 billion in tax incentives for investments in "clean coal" facilities, US$1.3 billion for energy conservation and efficiency, US$1.3 billion for alternative fuel vehicles and fuels (bioethanol, biomethane, liquified natural gas, propane), and US$500 million Clean Renewable Energy Bonds (CREBS) for government agencies for renewable energy projects.

single out that the production tax credit of 2.1 cents for kWh is available only for the first 6000 MWe of new nuclear capacity constructed and lasts only for the first eight years of operation (Wikipedia Nuclear Energy Policy of the United States, 2020).

Under the terms of the EPA 2005, to qualify for the PTC, a nuclear power plant must be in service on or before 31 December 2020, and the maximum value of the nuclear PTC is US$6 billion over eight years (or US$750 million per year). In February 2018, "an extension to the PTC was passed by the US Senate and Congress. The extension allows reactors entering service after 31 December 2020 to qualify for the tax credits and allows the US Energy Secretary to allocate credit for up to 6000 MWe of new nuclear capacity, which enters service after 1 January 2021. The PTC cannot be claimed until assets begin generating electricity and is not inflation-adjusted" (US Nuclear Power Policy, 2020). It is seen as an essential component for completing US nuclear power plants already under construction and first-of-a-kind small modular reactor (SMR)[k] construction (World Nuclear Association-Nuclear Power in the USA, 2020).

Independently of the decisions to be adopted regarding the future development of the nuclear energy sector within the North American region, there is an understanding among all factors involved that most of the existing nuclear power plants should continue operating in the USA. That should be done with the maximum security levels and the appropriate management of the high-level nuclear waste generated during nuclear power plant operation to protect the environment.

## The nuclear power program in Canada

Canada has one of the most diversified energy mixes in the world. The Canadian energy mix includes hydroelectricity, natural gas, oil, coal, nuclear energy, and renewables, and produced 3% of the total energy at the world level, after China with 17%, the USA with 14%, Russia with 10%, Saudi Arabia with 5%, and India with 4% (Energy and the Economy, 2019).

Canada already has one of the cleanest electricity systems in the world, with more than 80% of electricity generated in the country in 2018 from sources that do not produce greenhouse gas emissions, such as renewables (66.1% of the total) and nuclear power (15.3% of the total).[l] At the same time, the country "has been a world leader in nuclear energy since the development of Canada's own CANDU reactor technology by Atomic Energy of Canada Ltd. (AECL) in 1952" (Country Nuclear Power Profiles-Canada, 2018).

Without a doubt, nuclear energy is a relevant part of Canada's current clean energy mix and will continue to play a crucial role in achieving its low carbon future commitments. That includes not only the contributions of nuclear energy "to Canada's clean energy mix but continued investments in Canada's nuclear science and technology capabilities and exploration of the potential applications of new nuclear technologies, including small modular reactors. Nuclear energy is the second-largest source of non-emitting electricity in Canada, with 19 commercial nuclear power reactors located in Ontario and Brunswick. These nuclear power

[k] For more information on SMRs see Morales Pedraza (2015, 2017).

[l] In 2019, in Canada the nuclear electricity share was 14.9% of the total (PRIS—Power Reactor Information System, 2021).

plants produced in 2019 approximately 15% of Canada's electricity supply" (Country Nuclear Power Profiles-Canada, 2018), and with a total net capacity of 13.55 MW (gross capacity of 14,269 MW) (PRIS—Power Reactor Information System, 2021).

The Canadian nuclear power plants generated, in 2019, around 15% of Canada's electricity supply (see Fig. 7.7). That includes "approximately 60% of the electricity supply in the province of Ontario and 33% in New Brunswick. During the period 2016–18, the Unit Capability Factor of the Canadian nuclear power reactors in operation in the country reached 80.6%, occupying the place 18th according to this factor's level at the world level" (Country Nuclear Power Profiles-Canada, 2018; PRIS—Power Reactor Information System, 2021).

Based on the data included in Table 7.2, the following can be stated:

- In 2022, two units with a total capacity of 1084 MW are planned to be closed or their operating licenses extended;
- In 2024, the number of units to be closed or their operating licenses extended are four with a total capacity of 2160 MW;
- In 2025, four units with a total capacity of 3736 MW should be closed or their operating licenses extended;
- In 2035, two units with a total capacity of 1660 MW should be closed or their operating licenses extended;
- In 2036, two units with a total capacity of 1660 MW should be closed or their operating licenses extended;
- In 2037, one unit with a 705 MW capacity should be closed or their operating license extended.

During the period 2022–37, a total of 15 units or 79% of the total number of nuclear power reactors in operation in the country should be closed, or their operating licenses extended.

It is important to single out that provincial governments and power companies have decided to extend the operating life of several nuclear power reactors currently in operation to meet current and future electricity needs. "Refurbishing CANDU units consists of such steps as replacing fuel channels and steam generators and upgrading ancillary systems to current standards. While refurbishing usually takes less time and is less costly than building a new

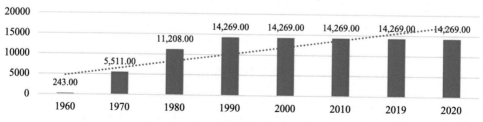

**FIG. 7.7**  Evolution of the nuclear power reactors capacity in Canada during the period 1960–2020 (MW). *Source: April 2021. PRIS—Power Reactor Information System. www.iaea.org/pris. © IAEA.*

**TABLE 7.2** Nuclear power reactors in operation in Canada (April 2021).

| Name | Type | Status | Location | Planned close or license to | Gross Electrical Capacity [MW] | First Grid Connection |
|------|------|--------|----------|------------------------------|-------------------------------|-----------------------|
| BRUCE-1 | PHWR | Operational | TIVERTON | 2035 | 830 | 1977-01-14 |
| BRUCE-2 | PHWR | Operational | TIVERTON | 2035 | 830 | 1976-09-04 |
| BRUCE-3 | PHWR | Operational | TIVERTON | 2036 | 830 | 1977-12-12 |
| BRUCE-4 | PHWR | Operational | TIVERTON | 2036 | 830 | 1978-12-21 |
| BRUCE-5 | PHWR | Operational | TIVERTON | – | 872 | 1984-12-02 |
| BRUCE-6 | PHWR | Operational | TIVERTON | – | 891 | 1984-06-26 |
| BRUCE-7 | PHWR | Operational | TIVERTON | – | 872 | 1986-02-22 |
| BRUCE – 8 | PHWR | Operational | TIVERTON | – | 872 | 1987-03-09 |
| DARLINGTON-1 | PHWR | Operational | TOWN OF NEWCASTLE | 2025 | 934 | 1990-12-19 |
| DARLINGTON – 2 | PHWR | Operational | TOWN OF NEWCASTLE | 2025 | 934 | 1990-01-15 |
| DARLINGTON-3 | PHWR | Operational | TOWN OF NEWCASTLE | 2025 | 934 | 1992-12-07 |
| DARLINGTON-4 | PHWR | Operational | TOWN OF NEWCASTLE | 2025 | 934 | 1993-04-17 |
| DOUGLAS POINT | PHWR | Permanent Shutdown | TIVERTON | – | 218 | 1967-01-07 |
| GENTILLY-1 | HWLWR | Permanent Shutdown | BECANCOUR | – | 266 | 1971-04-05 |
| GENTILLY-2 | PHWR | Permanent Shutdown | BECANCOUR | – | 675 | 1982-12-04 |
| PICKERING-1 | PHWR | Operational | PICKERING | 2022 | 542 | 1971-04-04 |
| PICKERING-2 | PHWR | Permanent Shutdown | PICKERING | – | 542 | 1971-10-06 |
| PICKERING-3 | PHWR | Permanent Shutdown | PICKERING | – | 542 | 1972-05-03 |
| PICKERING-4 | PHWR | Operational | PICKERING | 2022 | 542 | 1973-05-21 |
| PICKERING-5 | PHWR | Operational | PICKERING | 2024 | 540 | 1982-12-19 |
| PICKERING-6 | PHWR | Operational | PICKERING | 2024 | 540 | 1983-11-08 |
| PICKERING-7 | PHWR | Operational | PICKERING | 2014 | 540 | 1984-11-17 |
| PICKERING-8 | PHWR | Operational | PICKERING | 2024 | 540 | 1986-01-21 |
| POINT LEPREAU | PHWR | Operational | SAINT JOHN | 2037 | 705 | 1982-09-11 |
| ROLPHTON NPD | PHWR | Permanent Shutdown | ROLPHTON | – | 25 | 1962-06-04 |

*Source: April 2021. PRIS—Power Reactor Information System. www.iaea.org/pris. © IAEA, and 2020. Nuclear Power in Canada. World Nuclear Association.*

plant, there have been several cost overruns that, in some cases, have made it almost as expensive as new construction"[m] (Nuclear Power in Canada, 2020).

A summary of all nuclear power reactors currently in operation in Canada is shown in Table 7.3.

According to IAEA sources, two units were connected to the grid in the 60s, nine units in the 70s, ten units in the 80s, and four units in the 90s. Out of these 25 units connected to the electrical grid during the period 1960–90, a total of six units have already been shut down; two units in the 60s, three units in the 70s, and one unit in the 80s.

The evolution of the nuclear power reactor's capacity in Canada during the period 1960–2019 is shown in Fig. 7.7.

Based on the data included in Fig. 7.7, the following can be stated: Canada's nuclear power gross capacity increased 58.7-fold during the period 1960–2020. However, it is important to note that no new nuclear capacity was added or withdrawn until April 2021. It is expected that in the future, the country's nuclear capacity will be lower than in 2020 due to the closure of some units, the absence of new nuclear power reactors under construction in the country in April 2021, and the absence of plans for the construction of new units over the next few years.

Summing up, the following can be stated:

- In 2018, Canada's energy sector directly employed more than 269,000 people and indirectly supported over 550,500 jobs. "The Canadian Nuclear Association estimates that the Canadian nuclear industry employs approximately 30,000 people, and creates another 30,000 jobs indirectly through contracting. The industry generates revenues of approximately US$6.6 billion and contributes US$1.5 billion in federal and provincial taxes" (Nuclear Power in Canada, 2020);
- Canada's energy sector accounts for over 11% of the nominal Gross Domestic Product (GDP);
- Government revenues from the energy sector reached US$14.1 billion in 2017;
- Canada traded energy with 165 countries in 2018;
- More than US$799 million was spent on energy research, development, and deployment by the Canadian government in 2017–18 (Energy and the Economy, 2019).

Canada is the sixth-largest energy producer, the fifth largest net energy exporter, and the eighth largest energy consumer at the world level. Canada is also the world's second-largest

TABLE 7.3    Number of nuclear power reactors in operation in Canada and the electricity generated in 2018.

| Country | Number of nuclear power reactors in April 2021 | Gross electrical capacity installed in April 2021 (MW) | Electricity generated in 2019 (GWh) | Nuclear share in 2019 (%) |
|---------|---------|---------|---------|---------|
| Canada | 19 | 14,269 | 95,469 | 14.9 |

*Source: April 2021. PRIS—Power Reactor Information System. www.iaea.org/pris. © IAEA.*

[m] In Canada, the refurbishment of the Darlington nuclear power plant began with work on the first reactor in 2016 and is expected to be completed by 2026. Unit 2 refurbishment remains on budget and schedule with completion of the works expected in time. In August 2018, the Canadian Nuclear Safety Commission (CNSC) approved Ontario's plan to keep the Pickering nuclear power plant in operation until the end of 2024, four years longer than previously planned (Nuclear Energy Data, 2019).

uranium producer, with over 20% of total uranium world production coming from mines in Saskatchewan. Eighty-five percent of Canada's uranium production is exported for its use in nuclear power plants worldwide.

The Canadian primary energy production, including uranium, reached, in 2017, a total of 29,642 petajoules. The breakdown by fuel is the following: (see Fig. 7.8):

- Crude oil 32%;
- Uranium 29%;
- Natural gas 24%;
- Hydropower 5%;
- Coal 4%;
- Other renewables 3%;
- Natural gas liquids 3% (Energy and the Economy, 2019).

The Canadian energy sector enjoys a strong presence in all primary energy commodities and strong electricity and energy efficiency industries. Electricity accounts for 15% of domestic energy requirements (Country Nuclear Power Profiles-Canada, 2003, 2018).

All nuclear power reactors operating in Canada are of the PHWR type.[n] The electricity generation using nuclear energy reached, in 2019, a total of 95,469 GWh, representing, in that year, 14.9% of total Canada's electricity production (640,391 GWh). In Ontario, nuclear power plants generated today, approximately 60% of the electricity produced by the country and in New Brunswick, a total of 36.1% (Fig. 7.9).[o] The unit capability factor of all nuclear power

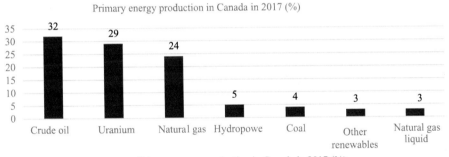

Primary energy production in Canada in 2017 (%)

FIG. 7.8 Primary energy production in Canada including uranium. *Source: 2019. Energy and the Economy. Natural Resources of Canada.*

[n] However, in February 2018, Natural Resources Canada (NRCan) initiated a process bringing together provincial and territorial governments, utilities, industry and other interested stakeholders to develop a roadmap for the potential development of SMRs in Canada. In April 2018, the Canadian Nuclear laboratories (CNL) initiated an Invitation for Demonstration, inviting further discussions with SMR vendors interested in building a demonstration unit at a CNL-managed site. The CNSC has been approached by a number of SMR vendors to initiate an optional preliminary step before the licensing process, called a "Vendor Design Review" (VDR). The VDR is completed at a vendor's request and expense to assess their understanding of Canada's regulatory requirements and the acceptability of a proposed design (Nuclear Energy Data, 2019).

[o] The Bruce Nuclear Power Plant is the world's largest operating nuclear facility with eight units and a capacity of 6827 MW.

reactors in operation in Canada in 2018 was 79.7%, which was a little lower than the level reached during the period 2016–18. Canada shut down permanently six nuclear power reactors. These nuclear power reactors are the following:

- Douglas Point, in1967, with a gross capacity of 218 MW;
- Gentilly 1, in 1971, with a gross capacity of 266 MW;
- Gentilly-2, in 1982, with a gross capacity of 675 MW;
- Pickering-2, in 1971, with a gross capacity of 542 MW;
- Pickering-3, in 1972, with a gross capacity of 542 MW;
- Rolphton NPD, in 1962, with a gross capacity of 25 MW (PRIS—Power Reactor Information System, 2021).

## The uranium resources in Canada

The nuclear energy sector is an important part of the Canadian economy converting the country into the world's sixth-largest electricity producer using this type of energy source. Canada needed to have enough uranium to fuel its nuclear power plants' operation to achieve this position worldwide. Besides this need, Canada was researching the possible use of uranium for military purposes. Due to this specific need, in earnest in 1942, Canada began uranium exploration in response to a demand for civil and military purposes. "The strategic nature of such material resulted in a ban on prospecting and mining of all radioactive materials across Canada" (Uranium in Canada, 2020).

FIG. 7.9    Bruce nuclear power plant. *Source: Courtesy By Chuck Szmurlo - Own work, CC BY 2.5, https://commons. wikimedia.org/w/index.php?curid=1841973.*

[P] In 2017, Kazakhstan produced 39% of the world uranium production and exported 43% of the total, while Canada produced 22% of the world total and exported 21% of the total (Energy Fact Book 2019-20 Canada, 2019).

Canada was the world's largest producer of uranium until 2009, when Kazakhstan overtook it.[P] The uranium production comes mainly from the McArthur River and Cigar Lake uranium mines located in the northern Saskatchewan province. Both mines are the largest and highest grade in the world.

Canada's uranium production during the period 2010–19 is shown in Fig. 7.10. Although uranium production in the country was "relatively constant over the last few years, its share of world production dropped from about 20% to 15% before recovering to about 22% in 2016, worth about US$2 billion." Over 85% of uranium production is exported to several countries (World Nuclear Association, 2021).

Without a doubt, with known uranium resources of 606,600 tons of $U_3O_8$ (514,400 tU), as well as much continuing exploration, Canada has a significant role in meeting future uranium world demand. Canada is a country rich in uranium resources and has a long history of exploration, mining, and electricity generation using nuclear power. By 2019, more uranium had been mined in Canada than any other country – 539,773 tU, about one-fifth of the world total (World Nuclear Association, 2021).

According to Fig. 7.10, Canada's uranium production decreased by 29.3% during the period 2010–19, falling from 11,540 tons in 2010 to 8165 tons in 2019. The mentioned period can be divided into three phases. During the first phase, covering the period 2010–14, the uranium production decreased by 6.7%, falling from 11,540 tons in 2010 to 10,771 tons in 2014. During the second phase, covering the period 2014–16, the uranium production increased by 53.6%, rising from 10,771 tons in 2014 to 16,541 tons in 2016, the highest level reached during the whole period under consideration. During the third phase, covering the period 2016–19, the uranium production decreased once again by 50.6%, falling from 16,541 tons in 2016 to 8165 tons in 2019, the lowest level reached during the period under consideration.

The reduction of uranium production since 2016 is because the majority of Canada's uranium resources are in high-grade deposits, some one hundred times the world average, and the weakness of the uranium price. "Many of these deposits have difficult mining conditions, which require ground freezing" (World Nuclear Association, 2021).

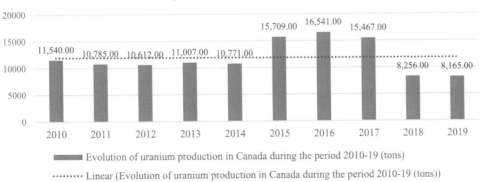

**FIG. 7.10** Evolution of uranium production in Canada during the period 2010–19 (tons $U_3O_8$). *Source: World Nuclear Association, 2021. Uranium in Canada.*

"All operating uranium mines and mills in Canada are located in northern Saskatchewan. Orano Canada, and Cameco Corporation are the licensees of the active mining and milling facilities" (Uranium Mines and Mills, 2020). These facilities are the following:

- Cigar Lake mine;
- Key Lake mill;
- McArthur River mine;
- McClean Lake mill;
- Rabbit Lake mine and mill.

It is important to single out that despite uncertainty about the use of nuclear energy for electricity generation in Canada in the future, "new uranium mine and mill projects are currently being proposed in Saskatchewan, Quebec, and Nunavut" (Uranium Mines and Mills, 2020).

The production target of McArthur River mining and Key Lake mill was to reach 9908 tons per year by 2018, "but in November 2017, Cameco announced the temporary suspension of production from the McArthur River mining and Key Lake milling operations from January 2018." Due to persistent uranium price weakness, the suspension was expected to last for ten months, but in July 2018, Cameco announced that the mine was closed for an undefined period. "Cameco earlier said that it would take 18-24 months for the mine to ramp up to full production." This situation explains why uranium production fell 50% in 2018 concerning 2017 (World Nuclear Association, 2021).

The location of the uranium operating mines, the proposed mine, the former mines now closed, the uranium fuel cycle, and the sites of the nuclear power plants operating in Canada are shown in Fig. 7.11.

## The nuclear energy policy in Canada

Canada's energy policy supports using various types of energy sources to ensure a secure and sustainable energy supply. The following are the three major areas of active federal energy policy development:

- Conventional and renewable energies;
- Nuclear energy;
- Environment.

In the specific case of nuclear energy, the Canadian government supports using this energy source for electricity generation. It considers it an important component of a diversified energy mix, which effectively reduces $CO_2$ emissions to the atmosphere. "Constitutionally, nuclear energy falls within the jurisdiction of the federal government" (Nuclear Energy, 2017). The government's roles include carrying out research and development activities and adopting all necessary regulations related to all nuclear materials and Canadian activities. The Canadian government places the highest priority on health, safety, security, and the environment concerning Canada's nuclear activities. For this reason, the Canadian government "has established a comprehensive legislation framework that focuses on protecting health, safety, security, and the environment. It consists of the following:

**FIG. 7.11** Location of the uranium mines. *Source: World Nuclear Association, 2021. Uranium in Canada.*

- Nuclear Safety and Control Act (Regulation);
- Nuclear Energy Act (Nuclear Research and Development);
- Nuclear Fuel Waste Act (Nuclear Waste);
- Nuclear Liability and Compensation Act (Liability)" (Nuclear Energy, 2017).

While the Canadian federal government has important responsibilities relating to nuclear energy, the decision to invest in electric generation rests with the province's authorities. In coordination and cooperation with the relevant provincial energy organizations and power utilities, it is up to the provinces to determine whether new nuclear power plants should be built. It is important to single out that Canada's government considers nuclear energy an essential component of a diversified energy mix. For this reason, it has taken all necessary measures to ensure nuclear power's long-term development as a sustainable energy source in meeting current and future energy requirements.

Without a doubt, "when properly managed, nuclear energy can contribute effectively and significantly to sustainable development objectives. The Canadian nuclear energy program is a crucial component of Canada's economy and energy mix" (Nuclear Energy, 2017).

Who is in charge of promoting the use of nuclear energy for electricity generation in Canada? The Atomic Energy of Canada Limited (AECL) is the federal government agency in charge of promoting the use of nuclear energy for electricity generation in the country. AECL has both a public and commercial mandate and has overall responsibility for Canada's nuclear research and development programs and the Canadian reactor design (CANDU), engineering and marketing program.

According to the Canadian Constitution, it is important to single out that electricity generation, transmission, and distribution fall under the provinces' jurisdiction. The provincial governments, along with the Yukon and Northwest Territories governments, have full control of their natural resources and are responsible for most regulation and energy sector development within their territories. This authority includes electricity policy and planning.

Corresponding to the current text of the Constitution, three territories do not share these constitutional rights. Still, jurisdiction over electrical energy has been devolved under the Yukon Act, the Northwest Territories Act, and the Nunavut Act. Therefore, in principle, all provinces and territories have the authority to govern their electricity systems and are free to decide their electricity supply sources and the design of their electricity markets. The federal government's role "is restricted to nuclear energy policy and regulation, the regulation of international transmission lines and electricity exports, and the regulation of interprovincial transmission lines that are designated by the Governor in Council. Both levels of governments are involved in electricity research" (Country Nuclear Power Profiles-Canada, 2018).

Each Canadian province and territory has its own electricity policy and regulatory framework based on its Constitutional rights. Except for Yukon, Newfoundland, Labrador, and Northwest Territories, all Canadian provinces have adopted a regulated monopoly model for electricity transmission and distribution. Some Canadian provinces and territories have also adopted the same electricity production model, while others have opened electricity generation to the competition. In many Canadian provinces and territories, electricity is mainly supplied by a vertically integrated electric utility. "Although some of these utilities are privately owned, most are corporations owned by the provincial and territorial governments" (Country Nuclear Power Profiles-Canada, 2018).

Since the 1990s, the structure of the Canadian electricity industry has undergone significant change. Most provinces and territories have moved from the traditional model of provincially regulated and vertically integrated monopolies within the energy sector toward a more competitive system. Under the new energy system, the private energy sector is playing an increasing role. A bid-based model exists between local distribution companies and both large and small generators in Alberta and Ontario. At the same time, "in other provinces and territories, independent power producers are able to sell power only to the major utility that provides most of the generation, transmission, and distribution services" (Country Nuclear Power Profiles-Canada, 2018).

# CANDU technology and the nuclear industry in Canada

Canada has an indigenous nuclear power industry established around the CANDU (CANada Deuterium Uranium) nuclear technology. Without a doubt, the Canadian nuclear industry is a significant contributor to the Canadian economy in terms of GDP, government revenue, and employment. Between 170 and 200 companies supply products and services to the country's utilities and broader nuclear industry. While the nuclear industry is concentrated in four Canadian provinces, Ontario, Saskatchewan, New Brunswick, and Quebec, nuclear research and development have been undertaken in nine provinces and 37 institutions across the country. The most recent data indicated that the nuclear industry in Canada provides over 30,000 direct jobs. It is important to single out that "the Canadian companies are important global suppliers of medical isotopes, reaching approximately 75% of the world's supply of Cobalt-60 used to sterilize 45% of the world's single-use medical supplies" (The Canadian Nuclear Industry and Its Economic Contributions, 2016).

One of the main characteristics of the Canadian nuclear industry is it encompasses the entire nuclear energy fuel cycle, including:

- Research and development;
- Uranium mining;
- Fuel fabrication to nuclear power reactor construction as well as servicing of CANDU reactors in Canada and abroad;
- Nuclear power plant operation;
- Nuclear waste management and decommissioning.

Canada is one of seven countries that use CANDU nuclear technology in its nuclear power plants (19 units). The other states where CANDU technology has been exported are Argentina (one unit), China (two units), India (two units), Pakistan (one unit), Romania (two units), and South Korea (four units) for a total of 12 units. In India, are operating 13 CANDU derivate units constructed using local design.

Which are the main characteristics of CANDU technology? The CANDU is a nuclear pressurized heavy water reactor (PHWR) that uses heavy water (deuterium oxide) as a moderator and coolant and natural uranium for fuel. The majority of nuclear power reactors in use in the world are light water reactors (LWR), which use normal water as the moderator and coolant and enriched uranium as a fuel.

It is important to stress that the CANDU technology continues to evolve to enable the use of alternative fuels in nuclear power reactors in the future. Work is underway in Chinese CANDU nuclear power reactors to demonstrate that they can recycle used fuel from other types of nuclear power reactors, reducing the volume of nuclear waste produced by a nuclear power plant in Canada (Energy Fact Book 2019-20 Canada, 2019).

There are two major types of CANDU reactors. One is the original design of around 500 MWe intended to be used in multi-reactor installations in large nuclear power plants. The other is the rationalized CANDU-6 in the 600 MWe class designed to be used in single stand-alone units or small multi-unit plants. CANDU-6 units were built in Quebec and New Brunswick and exported to Pakistan, Argentina, South Korea, Romania, and China. A single example of a non-CANDU-6 design was sold to India. "The multi-unit design was used only in Ontario, Canada, and grew in size and power as more units were installed in the province, reaching around 880 MWe in the units installed at the Darlington nuclear power plant. An effort to rationalize the larger units in a fashion similar to CANDU-6 led to the CANDU-9" (Wikipedia CANDU Reactor, 2020).

By the early 2000s, sales prospects for the original CANDU designs declined due to the introduction of newer designs from other foreign nuclear companies. The response from AECL to this unique situation was to cancel the CANDU-9 development and moving to the Advanced CANDU reactor (ACR) design. However, ACR failed to find any buyers, and its last potential sale was for an extension at Darlington nuclear power plant, but this was canceled in 2009. For this reason, all efforts to sell the ACR to foreign countries failed too. In October 2011, the Canadian Federal Government licensed the CANDU design to CANDU Energy (a wholly-owned subsidiary of SNC-Lavalin), which also acquired the former reactor development and marketing division of AECL at that time. CANDU Energy offers support services for existing sites and completes formerly stalled installations in Romania and Argentina through a China National Nuclear Corporation partnership. SNC Lavalin is pursuing new CANDU-6 reactor sales in Argentina (Atucha 3), China, and the UK (Wikipedia CANDU Reactor, 2020).

Private sector firms, which undertake the manufacturing of CANDU components and the engineering and project management work for nuclear power reactors, both inside and outside of Canada, act as subcontractors to AECL (Country Nuclear Power Profiles-Canada, 2003). The basic structure of the CANDU reactors is shown in Fig. 7.12.

The basic operation of the CANDU design is similar to the operation of other nuclear power reactors. However, the CANDU design differs from most other nuclear power reactor

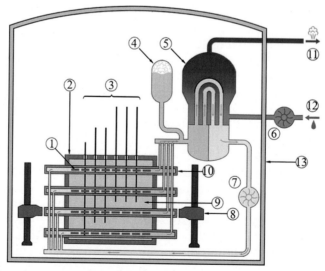

**FIG. 7.12**  Pressurized heavy water reactor (CANDU). Note: Schematic diagram of a CANDU reactor. 1-Fuel bundle; 2-Calandria (reactor core); 3-Adjuster rods; 4-Heavy-water pressure reservoir; 5-Steam generator; 6-Light-water pump; 7-Heavy-water pump; 8-Fueling machines; 9-Heavy-water moderator; 10-Pressure tube; 11-Steam going to the steam turbine; 12-Cold water returning from the turbine; 13-Containment building made of reinforced concrete. *Source: https://commons.wikimedia.org/wiki/File:CANDU_Reactor_Schematic.svg.*

designs in the fissile core and the primary cooling loop details. Natural uranium consists of a mix of uranium-238 with small amounts of uranium-235 and trace amounts of other isotopes. Fission in these elements releases high-energy neutrons, causing other uranium-235 atoms in the fuel to undergo fission. That process is much more effective when the neutron energies are much lower than the reactions release naturally. Most nuclear power reactors in operation in many countries use neutron moderators to lower the neutrons' power or "thermalize" them, making the reaction more efficient. During this moderation process, the energy lost by the neutrons heats the moderator and is extracted for power (Wikipedia CANDU Reactor, 2020).

It is important to single out that commercial nuclear power reactor designs use normal water as the moderator. Water absorbs some of the neutrons enough that it is not possible to keep the reaction going in natural uranium. CANDU replaces this light water with heavy water. Heavy water's extra neutron decreases its ability to absorb excess neutrons, resulting in a better neutron economy. That allows CANDU reactors to run on unenriched natural uranium, or uranium mixed with various other materials such as plutonium and thorium. That was an important goal of the CANDU design; by operating on natural uranium, the cost of enrichment is removed. That also presents an advantage in nuclear proliferation terms because there is no need to have enrichment facilities, which might also be used to produce a nuclear weapon (Wikipedia CANDU Reactor, 2020).

There are now 34 CANDU nuclear power reactors in operation in seven countries all over the world. Nineteen units are operating in Canada, and 12 units are operating in six countries. A total of 13 CANDU derivate reactors are also operating in India. The unit capability factor up to 2018 of this type of reactor was 77.4%, slightly lower than the total for that year (77.7%). The construction time for a CANDU-6 nuclear power reactor is between 4 and 5 years.

## The evolution of the nuclear energy sector in Canada

One of the main characteristics of electricity generation in Canada is that this responsibility falls within the provinces' jurisdiction, which at the same time owns the natural resources located within the provinces. It is important to single out that "given the diversity of provincial and territorial electricity markets and electricity resources, the use of various energy sources for baseload, intermediate, and peak load depends on the province or territory. However, in most cases, hydro, nuclear, and coal are used for baseload, while natural gas, petroleum, and hydro are used in intermediate and peak situations" (Country Nuclear Power Profiles-Canada, 2018).

As a result of the specific structure of the energy jurisdiction within the country, the Canadian electricity industry is not only organized along provincial lines but, in most provinces, the electricity industry is highly integrated, with the bulk of the power generation, transmission, and distribution provided by a few leading companies.

Canada ranked fifth, in 2019, according to the level of electricity produced by its nuclear power plants. That year, the electricity generated by these power plants reached 100.5 TWh,[q] below the USA with 852 TWh, China with 348.7 TWh, Russia with 209 TWh, and South Korea

---

[q] According to PRIS—Power Reactor Information System (2021), the electricity generated by the Canadian nuclear power plants in 2019 reached 95,469 GWh (95.5 TWh).

with 146 TWh. Nuclear energy in Canada was, in 2019, the second energy source providing electricity after hydropower plants.

The evolution of the electricity generation using nuclear energy in Canada during the period 2014–19 is shown in Fig. 7.13.

Based on the data included in Fig. 7.13, the following can be stated: electricity generation by the Canadian nuclear power plants decreased by 5.5% during 2014–16, falling from 101.21 TWh in 2014 to 95.65 TWh in 2016. After that year, the electricity generation using nuclear energy increased by 5%, rising from 95.69 TWh in 2016 to 100.5 TWh in 2019. It is important to single out that Canada's electricity generation during the period 2017–19 was almost the same. It is expected that Canada will continue using nuclear power for electricity generation as one of the main energy sources available in the country, at least during the coming years. However, the role of nuclear energy in the Canadian energy mix will decrease in the near future due to the following reasons:

- The shut down of some nuclear power reactors during the coming years;
- The lack of plans for the construction of new nuclear power reactors.

## The Canadian nuclear safety commission

The Canadian Nuclear Safety Commission (CNSC) regulates all stages of the useful life of each nuclear power reactor operating in Canada, from the environmental assessment required before plant construction to the decommissioning of the facility once operations are ended. In order words, the CNSC's mission is to regulate the use of nuclear energy and materials within the country. The goals to be achieved are the following:

- To protects the health of the Canadian citizens;
- The safe operation of all nuclear power reactors, with the lowest negative impact on the environment and population;

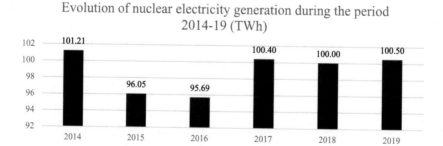

Evolution of nuclear electricity generation during the period 2014-19 (TWh)

■ Evolution of nuclear electricity generation during the period 2014-19 (TWh)

FIG. 7.13    Evolution of the electricity generation using nuclear energy in Canada during the period 2014–19. *Source: Statistics Canada. Electric Power, Annual Generation by Class Producer 2019, 69th ed. Table 25-10-0019-01 Electricity from fuels, annual generation by electric utility thermal plants. doi: 10.25318/2510001901-eng and 2020. BP Statistical Review of World Energy 2020.*

- To implement Canada's international commitments on the peaceful use of nuclear energy;
- To disseminates objective scientific, technical, and regulatory information to the public (Nuclear Energy, 2017).

Under the Nuclear Safety and Control Act (2000) and the Canadian Environmental Assessment Act (2012), the CNSC's mandate involves four major areas:

- Regulation of the development, production, and use of nuclear energy in Canada, to protect the health of the Canadian citizens, safety, the environment, and population;
- Regulation of the production, possession, use, and transport of nuclear materials, as well as the production, possession, and use of prescribed equipment and information;
- Implementation of measures related to international control of the development, production, transport, and use of nuclear energy and nuclear materials, including measures related to the non-proliferation of nuclear weapons and nuclear explosive devices;
- Dissemination of scientific, technical, and regulatory information concerning the activities of the CNSC and the effects on the environment, on the health and safety of the population, of the development, production, possession, transport, and use of nuclear energy and nuclear substances (Country Nuclear Power Profiles-Canada, 2018).

CNSC licensing of nuclear power plants "is comprehensive and covers 14 separate topics referred to as safety and control areas, such as radiation protection, emergency preparedness, environmental protection, and equipment fitness for service" (Canadian Nuclear Safety Commission-Nuclear Power Plant, 2020).

The CNSC assesses license applications to ensure that safety measures are technically and scientifically sound, that all requirements for the safe operation of a nuclear power plant are met, and that the appropriate safety systems are in place to protect the environment and population.

The Canadian licensing process offers significant opportunities for public participation, including Commission hearings and community meetings open to the public. After a license is issued, the CNSC consistently evaluates compliance with all government regulations for the nuclear sector. In addition to having a team of onsite inspectors, CNSC staff with specific technical expertise regularly visit nuclear power plants to verify that operators meet the regulatory requirements and license conditions. The CNSC carefully reviews any items of non-compliance and follows up to ensure all items are immediately corrected. The CNSC published a regulatory oversight report for Canadian nuclear power plants annually. In 2018, for the first time, the CNSC assessed the safety performance not only of Canada's nuclear power plants but also of adjacent waste management facilities located in each nuclear power plant site (Canadian Nuclear Safety Commission-Nuclear Power Plant, 2020):

The CNSC also collaborates with international partners, including the IAEA and its foreign counterparts, to ensure the highest safety level in all nuclear power plants operating in the country. Every three years, the CNSC publishes a comprehensive report as part of the Convention on Nuclear Safety, following up on its commitment to maintaining a high level of safety at all nuclear power plants in operation in Canada (Canadian Nuclear Safety Commission-Nuclear Power Plant, 2020).

## Decommissioning activities in Canada

The CNSC regulates the entire lifecycle of nuclear power plants operating in Canada. The decision to stop operating a nuclear power reactor is taken solely by its operator but under the country's energy regulation on the subject. The CNSC's role is to ensure that decommissioning activities are carried out in accordance with its own regulatory requirements. The purpose is to ensure the protection of the workers, the public, and the environment and, at the same time, implement Canada's international commitments on the peaceful use of nuclear energy during the implementation of all decommissioning activities. It is important to single out that plans related to the decommissioning of nuclear power plants take, on average, 50 years to implement completely. As a requirement for obtaining a license to operate a nuclear power plant, the operator must submit a decommissioning plan that delineates how the operator plans to manage the plant's dismantling at the end of its useful life. The financial guarantee, which is based on this plan, is also used by the CNSC to evaluate how the operator will ensure the financing of decommissioning activities.

Three prototypes of nuclear power reactors constructed and owned by Atomic Energy of Canada Limited were shut down and prepared for decommissioning in the 1980s (Rolphton and Kincardine, Ontario, and Bécancour). "Today, the CNSC continues to license these facilities as waste management facilities, which continue to store radioactive waste on-site safely. Canada's Nuclear Legacy Liabilities Program provides funding to proceed with the decommissioning activities. In the future, all radioactive waste will be removed from those locations, and decommissioning will eliminate any unreasonable risks to health, safety, security, and the environment. At that point, those facilities will be issued abandonment licenses to release them from CNSC regulatory control" (Canadian Nuclear Safety Commission-Nuclear Power Plant, 2020).

Until April 2021, six nuclear power reactors were shut down, and four were shut down and decommissioned. These four units are the following:

- Gentilly 1 shut down in 1977;
- Douglas Point shut down in 1984;
- Rolphton NPD shut down in 1987;
- Gentilly 2, a more modern CANDU 6 type, shut down in 2012.

The first three units are expected to be demolished in about 30 years. "Gentilly 1 was a steam-generating heavy water reactor with vertical pressure tubes, light water coolant, and heavy water moderation. It was not a successful nuclear power reactor and had only 180 full-power days in six years of operation. The other two were prototype CANDU designs" (Nuclear Power in Canada, 2020). In the case of the fourth unit, Gentilly 2, "a decommissioning license was issued for 2016 to 2026. The reactor's main part will be closed up and left for 40 years to allow radioactivity to decay before the nuclear power plant's demolition. All 27,000 fuel bundles are expected to be in dry storage by 2020." (Nuclear Power in Canada, 2020). The decommissioning cost has been estimated at C$1.8 billion over 50 years.

It is important to single out that the "Gentilly-2 decommissioning plan includes the dismantling of all systems and structures, the remediation of any radioactive contamination or harmful or potentially dangerous substances to bring the values under the prescribed limits, and the long-term management of radioactive waste, including spent fuel. As the decommissioning strategy adopted by Hydro-Québec has a 40 year dormancy period, the dismantling will be gradually phased in" (Country Nuclear Power Profiles-Canada, 2018).

## Canada trade within the nuclear sector

Canada is one of the leading suppliers of nuclear technology in the nuclear sector, particularly CANDU technology and uranium material, but this role has been reduced significantly in recent years. The countries where CANDU technology has been exported to six countries.

Although uranium production in the country was "relatively constant over the last few years, its share of world production dropped from about 20% to 15% before recovering to about 22% in 2016, worth about US$2 billion" (World Nuclear Association, 2021). In 2019, around 76% of Canadian uranium production was exported for nuclear power use worldwide. In 2019, Canadian uranium production represented 12.9% of the world's uranium production (63,273 tons $U_3O_8$) and, since 2009, is the second at that level below Kazakhstan (41%). In 2019, only Cigar Lake produced uranium in Canada (8165 tons $U_3O_8$). The uranium exported to Asia represented, in 2018, 42% of the total, followed by North America and Latin America with 42%, and Europe with 17%. The USA imported from Canada in 2018 around 24% of the total uranium used in the country to ensure the operation of its nuclear power plants.

## The use of nuclear energy for electricity generation in Canada and its impact on the environment

Canada was one of the first countries to develop a civilian nuclear power program after the end of World War II. Since then, nuclear power plants have been producing commercial electricity in Canada. Four active nuclear power plants are in operation in Canada, with 19 operating nuclear power reactors. All of them are CANDU reactors. Three of these nuclear power plants are located in Ontario and one in New Brunswick. Quebec is the only other province to have used nuclear power for electricity generation in the past, but in December 2012, Quebec's Gentilly-2 nuclear facility was permanently shut down.

Canada's electricity system is diverse and focuses on clean energy electricity generation, with about 80% of electricity supply coming from non-greenhouse gases emitting sources.

The generation mix in Canada in 2018 and 2019 is shown in Fig. 7.14.

Based on the data included in Fig. 7.14, the following can be stated: Canada has one of the cleanest electricity generation systems worldwide, with more than 80% of electricity supply coming from non-greenhouse gases emitting sources. In 2019, Canada's hydroelectric resources represented a large part of this supply, around 58% of the total electricity supply. About 20% of the electricity supply comes from fossil fuel sources, with coal (8.3%) and natural gas (10.5%). The electricity generation by fuel in Canada in 2019 was the following:

- Oil: 4.1 TWh (0.6% of the total);
- Natural gas: 69.3 TWh (10.5% of the total);
- Coal: 54.6 TWh (8.3% of the total);

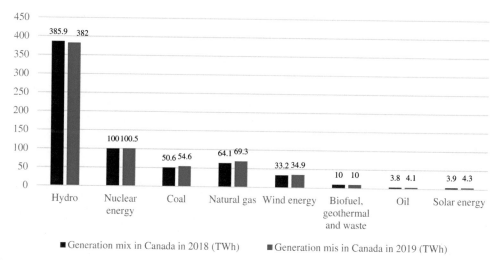

**FIG. 7.14** Generation mix in Canada in 2018 and 2019. *Source: 2020. Nuclear Power in Canada. World Nuclear Association, and 2020. BP Statistical Review of World Energy 2020, 69th ed. and Author own work.*

- Nuclear energy: 100.5 TWh (15.2% of the total);
- Renewables and others: 432 TWh (65.4% of the total).

The generation mix in Canada (see Fig. 7.15) has shifted from 2005 to 2017 as technology changed and investments were made in renewables energy. In total, electricity generation in Canada during the period 2005–17 increased by 50 TWh. Coal and oil generation decreased, while hydro, natural gas, non-hydro renewables, and nuclear energy generation increased. The nuclear generation increased by 8 TWh, or 9%, due to refurbishments at existing nuclear power plants.

According to Fig. 7.15, fossil fuel sources have lost their share in electricity generation during the period 2005–17, except natural gas. Its participation has increased in the period

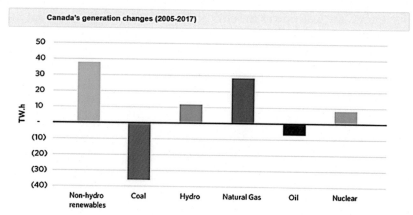

**FIG. 7.15** Canada's generation changes during the period 2005–17. *Source: Nuclear Energy in Canada: Energy Market Assessment, published in August 2018.*

under consideration. All of the non-greenhouse sources have enlarged their participation in electricity generation.

The electricity generation mix in Canada varies significantly from province to province. Several factors are determined by the provincial electricity mix, including access to natural resources, historical infrastructure decisions, and policy initiatives implemented by provincial governments. So the provinces and territories have the authority to govern their electricity system and are free to decide its electricity supply sources and the design of its electricity markets.

As a result of this situation, Newfoundland and Labrador, British Columbia, Manitoba, Yukon, and Quebec rely primarily on hydroelectricity (with shares exceeding 90% of provincial supply). Hydro and nuclear power represent the major sources of electricity in Ontario. On the other hand, coal and natural gas dominate in Alberta, Saskatchewan, and Nova Scotia. In Brunswick, nuclear, hydro, natural gas, and coal each play dominant roles. Northwest Territories and Nunavut rely primarily on diesel, while wind and diesel play relatively equal roles on Prince Edward Island (Country Nuclear Power Profiles-Canada, 2018).

It is well known that nuclear power reactors do not produce carbon dioxide emissions or air pollution during their operation. It already displaces over 80 million tons of greenhouse gas emissions annually in Canada, equivalent to taking 15 million cars off the road every day. However, mining and refining uranium ore and making reactor fuel all require large amounts of energy in the form of fossil fuel. Nuclear power plants also need large amounts of metal and concrete for their construction, which requires large amounts of energy to manufacture them. Because fossil fuels are used for mining, and refining uranium ore or fossil fuels are used during the construction of a nuclear power plant, the emissions from burning those fuels could be associated with the electricity that nuclear power plants generate.

According to Pembita's study (Winfield et al., 2006), environmental impacts of the use of nuclear energy in Canada is measured by each of the four major stages of nuclear energy production:

- Uranium mining and milling;
- Uranium refining;
- Conversion and fuel fabrication;
- Nuclear power plant operation and waste fuel management.

According to the study mentioned above, the most significant short-term environmental impact of using nuclear power to produce electricity in Canada is the generation of several extremely large waste streams containing a wide range of radioactive and hazardous contaminants. An estimated 575,000 tons of tailings are produced per year, of which 90,000–100,000 tons can be attributed to uranium production. Due to uranium refining, conversion, and production, nearly 1000 tonnes of solid wastes and 9000 m$^3$ of liquid wastes are produced each year. Approximately 85,000 waste fuel bundles are generated by the Canadian nuclear power reactor operation each year (Winfield et al., 2006).

The second major area of environmental impacts is water-related, according to the same study mentioned above. Discharges to surface waters from uranium mines and mills in Canada include uranium, molybdenum, arsenic, nickel, selenium, and ammonia. Concentrations of major ions (potassium, magnesium, among others) ranging from 10 to 200 times the levels of un-impacted groundwater have been found in areas near uranium mine and tailings management facilities.

Without a doubt, nuclear power generation is a major consumer of water in a nuclear power plant. For this reason, the nuclear power plant site should be located close to a water source. Uranium mining operations involve extensive dewatering, in the range of at least 16–17 billion liters per year, with the implication of impacts on groundwater and surface water storage and flows.

Storing, transporting, and managing toxic nuclear waste is the major environmental concern in Canada relative to the use of nuclear power for electricity generation. The Nuclear Waste Management Organization (NWMO) is responsible for designing and implementing Canada's safe, long-term nuclear fuel management plan. When spent nuclear fuel is removed from a nuclear power reactor, they are placed in a water-filled pool where their heat and radioactivity decrease. After seven to ten years, spent nuclear fuel is placed in dry storage containers, silos, or vaults (with a minimum design life of 50 years).

In summary, the Canadian government reported ten nuclear accidents since 1952 involving nuclear power plants in operation and other nuclear facilities with different negative impacts on the environment. Some of these nuclear accidents are, according to Wikipedia Nuclear Power in Canada (2020), the following:

- Chalk River: on December 12, 1952, the world's first major nuclear reactor accident happened at Chalk River Laboratories, 180 km northwest of Ottawa. A power excursion and partial loss of coolant led to severe damages to the NRX reactor core, resulting in fission products being released through the reactor stack and 4.5 tonnes of contaminated water collecting in the building's basement (Jedicke, 2016);
- 24 May 1958, a fuel rod caught fire and ruptured as it was removed from the NRU reactor, leading to the building's complete contamination. As in 1952, the military was called in to aid, and approximately 679 people were employed in the clean-up (A Closer Look at Canada's Nuclear Plants, 2012);
- Pinawa: in November 1978, a loss of coolant accident affected the experimental WR-1 reactor at Whitshell Laboratories in Pinawa, Manitoba. 2739 L of coolant oil (terphenyl isomer) leaked, most of it into the Winnipeg River, and three fuel elements broke with some fission products being released. The repair took several weeks for workers to complete (Taylor, 2011);
- Pickering: on August 1, 1983, pressure tubes — which hold fuel rods — ruptured due to hydriding at the Pickering nuclear power reactor No. 2. Some coolant escaped but was recovered before it left the nuclear power plant, and there was no release of radioactive material from the containment building. All four reactors were re-tubed with new materials (Zr-2.5% Nb) over ten years (A Closer Look at Canada's Nuclear Plants, 2012). On August 2, 1992, a heavy water leak at Pickering nuclear power reactor No. 1 heat exchanger released 2.3 petabecquerels (PBq) of radioactive tritium into Lake Ontario, resulting in increased levels of tritium in drinking water along the lake shoreline.[r]

On December 10, 1994, a pipe break at Pickering nuclear power reactor No. 2 resulted in a major loss of coolant accident and a spill of 185 tonnes of heavy water. The Emergency Core Cooling System had to be used to prevent a core meltdown (Canada's Nuclear Reactors: How

---

[r] See Ontario's Nuclear Generating Facilities: A History and Estimate of Unit Lifetimes and Refurbishment Costs (2017).

Much Safety Is Enough?, 2001) called "the most serious nuclear accident in Canada" by The Standing Senate Committee on Energy, the Environment and Natural Resources in 2001.

On January 12, 2020, a nuclear incident alert was sent out to all residents of Ontario. The alert stated that "an incident was reported at Pickering Nuclear Generating Station" and that "people near the Pickering Nuclear Generating Station DO NOT need to take any protective actions at this time." It was later revealed that in a statement by MPP Sylvia Jones, "the cause of the alert was revealed to be an error during a 'routine training exercise' being conducted by the Provincial Emergency Operations Center (PEOC)"[s];

- In 2009, Darlington contained more than 200,000 L of water containing trace amounts of tritium and hydrazine spilled into Lake Ontario after workers accidentally filled the wrong tank with tritiated water. However, the level of the isotope in the lake was not enough to pose harm to residents (A Closer Look at Canada's Nuclear Plants, 2012);
- Point Lepreau: on December 13, 2011, a radioactive spill happened at New Brunswick's Point Lepreau nuclear power plant during refurbishment. Up to six liters of heavy water splashed to the floor, forcing an evacuation of the reactor building and halt operations. Then, on December 14, NB Power issued a news release, admitting there had been another type of spill three weeks earlier (MacKinnon, 2012).

## The public opinion on the use of nuclear energy for electricity generation in Canada

According to a Canadian Nuclear Association (CAN) study carried out by ABACUS DATA, one in three Canadians sees climate change as one of the country's main issues. Some 78% of the surveyed said they are very worried about the negative impact of climate change on future generations. A random sample of 1500 Canadian adults was conducted for CAN. Some 86% of the population believes that the Canadian government should invest in clean technology to address climate change. Around 55% surveyed said they support the use of nuclear energy to generate electricity in Canada, knowing that nuclear power reactors contribute about 17% of Canada's energy needs and have the potential to increase supply thoroughly[t] (World Nuclear News, 2020).

The problem with using this type of energy source is the deep lack of awareness of the Canadian people on nuclear energy works for electricity production. One in three Canadians indicated their understanding of how nuclear energy works for electricity production comes from different sources such as films, TV, or fictional books. Knowledge is critical to make the right informed decision around Canada's energy mix to fight climate change.

Another survey released by Friends of the Earth with over 2000 Canadians surveyed said that more than six in ten (62%) believe that Ontario, Saskatchewan, and New Brunswick governments are on the wrong path in dealing with climate change, 31% said they are on the right path, and 7% were unsure. The results are showing in Fig. 7.16.

[s] See OPG Statement about Pickering Nuclear (Media release > OPG statement about Pickering Nuclear | OPG) (Ontario Power Generation, 2020).

[t] See Climate Change, Meeting Canada's Climate Targets and the Future of Energy (World Nuclear News, 2020).

Opinion if Canadian government are on right or wrong path
about climate change (%)

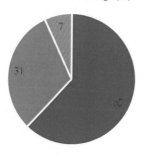

■ Wrong path   ■ Right path   ■ Unsure

**FIG. 7.16** Opinion if the Canadian government is on the right or wrong path about climate change. *Source: Friends of the Earth, 2020. Small Modular Nuclear Reactors: Climate Change.*

The majority of all Canadian provinces are against using nuclear energy for electricity generation. One exception is the Quebec province. Quebec province authorities are changing their opinion about using nuclear energy to meet its electricity needs in order to diminish the amount of greenhouse gases released to the environment. They are a long way from reaching the majority in favor of nuclear energy, but they move in that direction. The tendency in the Canadian opinion about using nuclear energy for electricity production during the period 2011–20 is shown in Table 7.4.

Based on the data included in Table 7.4, the following can be stated: most Canadian people do not support using nuclear energy to satisfy their electricity needs. One in three Canadians (31%) sees nuclear energy as a good option to fight climate change and curb greenhouse gas emissions. In 2011, 38% of Canadians supported nuclear energy for electricity production; in 2020, 31% are in favor; this decreases by 7%. It is expected that this trend will continue without a change in the near future. The decrease of the Canadian people in supporting using nuclear

**TABLE 7.4**   The tendency from 2011 to 2020 in the Canadian opinion about using nuclear energy for electricity production.

| Segments | 2011 | 2012 | Δ | 2020 | Δ |
| --- | --- | --- | --- | --- | --- |
| Alberta | 46% | 38% | − 8% | 36% | − 2% |
| Atlantic | 31% | 40% | 9% | 31% | − 9% |
| BC | 33% | 33% | 0% | 28% | − 5% |
| Ontario | 53% | 54% | 1% | 35% | − 19% |
| Man/Sask. | 37% | 44% | 7% | 37% | − 7% |
| Quebec | 17% | 12% | − 5% | 21% | 9% |
| National | 38% | 37% | − 1% | 31% | − 6% |

*Source: Friends of the Earth, 2020. Small Modular Nuclear Reactors: Climate Change and Innovative Research Group, 2012. 2012 Public Opinion Research: National Nuclear Attitude Survey.*

energy for electricity generation is due to fear of a new nuclear disaster like the one that occurred in the Fukushima Daiichi nuclear plant in 2011 in Japan. The fear has increased further due to the emergency alert about an incident at the Pickering nuclear power plant reported on January 12, 2020. Other reasons are the high initial cost and the delay in constructing a nuclear power plant. It is a fact that the new generation of nuclear power reactors, like Generation IV or SMR, excel in safety and reliability and are more economically competitive than previous types of nuclear power reactors.

According to the Friends of the Earth (2020), the concerns about nuclear power reactors and the use of nuclear energy for electricity generation are the following:

- Spills that contaminate drinking water: 82%;
- Neighborhood safety and security risks close to nuclear power plants: 77%;
- Overall cost to build a nuclear power reactor: 75%;
- Cost of managing nuclear waste: 73%;
- The security risk to transport nuclear fuel and its waste: 70%;
- Spending large amounts of money for SMRs to be operating in five or ten years may delay adopting immediately available renewable energy technologies: 68%;
- Lack of a long-term disposal and storage plan: 61%;
- Risks from the long-term disposal of contaminated waste: 59%;
- Radiation exposure to workers: 58%.

Based on the factors described above, it is important to mention that the Canadian people are very concerned with using nuclear energy for electricity generation because of the risk of nuclear accidents with negative consequences for the environment and population. Around 82% of the Canadian people are worried about the spills that contaminate drinking water. They would rather prefer using other clean energy sources for this specific purpose, like renewable energy. If nuclear energy wants to be included in the future Canadian energy mix, then it is necessary to change the opinion of the Canadian people has now of using nuclear energy for electricity generation. Perhaps the campaign favoring the use of nuclear energy for electricity generation promoted by the Quebec authorities could be a useful tool in other province authorities' hands to achieve the same goal.

## The future of nuclear energy in Canada

Without a doubt, "Canada already has one of the cleanest energy mixes in the world, with more than 80% of electricity generated from hydro, wind, solar, and nuclear power" (Canadian Nuclear Association, 2017).

In the past decade, cutbacks, the character of the government of Canada's nuclear assets, the privatization of government-owned nuclear companies, as well as the absence of a clear, long-term strategy for the development of the use of nuclear energy for electricity generation in Canada indicated a lack of real support by the government for the Canadian nuclear industry (Hanebach and McDonald, 2017).

This lack of interest and support, and its impact, echoed throughout Canada's nuclear supply chain. Based on several factors, the tide is turning, particularly the Chalk River Laboratories' planned activities with respect to SMR and advanced nuclear power reactors.

Besides, two notable developments were reported in Canada's nuclear situation. "The first one is the 2015 Ontario decision to approve the lifetime extension of four nuclear units at the Darlington nuclear power plant and six units at the Bruce nuclear power plant. The life extension of four units at the Darlington nuclear power plant will ensure that emissions-free nuclear continues to be Ontario's single largest power source. That action will continue to boost economic activity across Ontario, create jobs, and secure a clean supply of reliable electricity for the future," stated Ontario Minister of Energy Glenn Thibeault (Canadian Nuclear Association, 2017). This C$26 billion 15-year program is one of the largest nuclear energy projects in North America.

The second development relates to international leadership regarding SMRs. In 2018, Natural Resources Canada issued its SMR Roadmap, a plan for nuclear technology development based on this type of reactor. In December 2019, New Brunswick and Saskatchewan provinces agreed to collaborate with Ontario to advance the development and deployment of SMRs to address climate change, regional energy demand, economic development, and research and innovation opportunities. Along with this, the CNSC has a pre-licensing vendor design review process to assess nuclear power plant designs based on the vendor's reactor technology – for about ten SMRs with a wide range of capacities up to 300 MWe. "Also, Canadian Nuclear Laboratories (CNL) invited expressions of interest resulting in almost 20 proposals for siting an SMR at a CNL-managed site. CNL aims to have a new SMR at its Chalk River site by 2026" (Nuclear Power in Canada, 2020).

There have been proposals to build several nuclear power reactors to begin operation in the next decade. However, these proposals have been deferred or have failed. Two units (units 5 and 6) are planned to be built in Darlington nuclear power plant, but this plan has also been deferred. Another one was proposed for Point Lepreau nuclear power plant, but the plan lapsed, and one or possibly four SMRs were proposed in Peace River nuclear power plant, but the plan also lapsed. The total capacity of the new nuclear power reactors planned is 9 GWe.

Finally, the future of nuclear energy in Canada will depend on how the government will implement the following recommendations adopted by the Standing Committee on Natural Resources Report entitled "The Nuclear Sector at a Crossroads: Fostering Innovation and Energy Security for Canada and the World (2017)." These recommendations are the following:

- The government should work with the nuclear industry, indigenous communities, provincial/territorial governments, and international partners to ensure that Canada's nuclear sector continues to advance its rigorous regulatory and safety practices;
- The government should continue its support for Canadian nuclear research, development, and innovation;
- The government should continue to support the development and commercialization of Canadian nuclear technologies in the country and abroad;
- The government should continue to work with the nuclear industry, indigenous governments and communities, provincial/territorial governments, as well as international partners to promote and advance Canadian leadership in nuclear power generation technologies at home and abroad;
- The government should work in collaboration with the nuclear industry, the academic community, indigenous governments and communities, and provincial/territorial governments to sustain and improve Canadian expertise in the nuclear sector;

- The government should continue to support the development of SMRs, recognizing the potential of this type of nuclear power reactors to provide clean and reliable power to remote and northern communities and open new areas to economically valuable resource development;
- Along with academia and innovators, the nuclear industry should establish a nuclear innovation council with representatives from the federal and provincial governments to leverage non-power applications (e.g., for health care, agriculture, manufacturing, etc.) for public benefit.

The Canadian government endorsed the above set of recommendations. It stated the following: the nuclear industry is "an important part of Canada's clean energy and climate change initiatives, and beyond energy, the nuclear sector contributes to a wide range of other scientific and economic activities, such as medicine, human health, and safety, material testing, food safety, even space exploration" (Government of Canada Response to Recommendations in the Standing Committee on Natural Resources' Report, n.d.).

## The nuclear power program in the United States

The USA is a pioneer of nuclear power development. With 94 nuclear power reactors in operation in April 2021, the USA has the highest number of nuclear power reactors in operation in the world. In 2019, it accounted for 19.7% of total electricity generation in the country (PRIS—Power Reactor Information System, 2021).

The evolution of nuclear power plant capacities in the USA during the period 1970–2020 is shown in Fig. 7.17.

According to Fig. 7.17, the USA's nuclear capacity increased 15.4-fold during the period 1970–2020, rising from 6252 MW in 1970 to 96,553 MW in 2020. However, after 1990, the USA's nuclear capacity in operation was almost the same, with a small decrease in 2000, 2019, and 2020 below 100,000 MW of capacity. It is expected that the USA's nuclear capacity will continue to decrease during the coming years, as a result of the closure of additional nuclear power

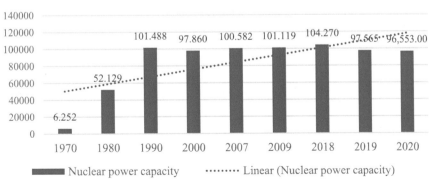

**FIG. 7.17** Evolution of nuclear power plant capacities in the USA during the period 1970–2020. *Source: April 2021. PRIS—Power Reactor Information System. www.iaea.org/pris. © IAEA, 2020. Nuclear Power Reactors in the World. Reference data series No. 2. IAEA, Vienna. 2019 edition. ISBN 978-92-0-102719-1 and Author own calculation.*

reactors for two specific reasons: the end of its useful life, and because it is too expensive to generate electricity using nuclear energy in comparison with the use of other energy sources.

According to Fig. 7.18, the following can be stated: suppose that "two cases implement fees of US$15 per ton of $CO_2$ and US$25 per ton of $CO_2$ (in 2017 dollars) starting in 2020, increasing by 5% in each subsequent year in real dollar terms. In both cases, much of the existing nuclear power reactors remain competitive, and additional nuclear power plants are constructed, so that capacity in 2050 is higher than current levels. With a US$15 per ton $CO_2$ fee, nuclear capacity increases to 106 GW in 2050; at US$25 per ton, nuclear capacity increases to 145 GW in 2050" (Scott, 2018). In the reference case, nuclear generation capacity is expected to decrease from around 100 GW in 2018 to a little higher than 75 GW in 2050.

Based on the data included in Fig. 7.19, the following can be stated: the number of nuclear power reactors in operation in the USA increased 5.5-fold during the period 1970–2020, rising from 17 units in 1970 to 94 units in April 2020. However, since 2011, the number of nuclear power reactors operating in the country has decreased by 15.4%, falling from 111 units in 1990 to 94 units in April 2021. There were 58 commercially operating nuclear power plants with 94 nuclear reactors in 29 USA states in 2020. Of these 58 nuclear power plants, 32 have two units,

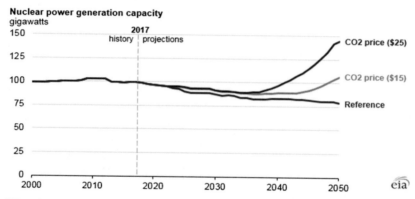

FIG. 7.18   US nuclear power generation capacity (GW). *Source: 2018. Annual Energy Outlook 2018, USA EIA. https:// www.eia.gov/about/copyrights_reuse.php. February 6, 2018.*

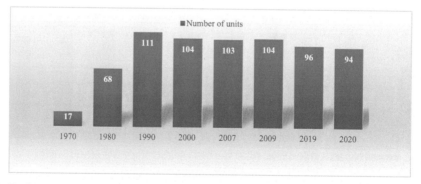

FIG. 7.19   Evolution of nuclear power reactors in the USA during the period 1970–2020. *Source: 2008 Nuclear Power Plants in the World, eighth ed. CEA and April 2021. PRIS—Power Reactor Information System. www.iaea.org/pris.* © IAEA.

and three have three units.[u] The Palo Verde nuclear power plant in Arizona is the largest nuclear power plant in the USA. It has three nuclear power reactors with combined net summer electricity generating capacity of 3937 MW (Gross electrical capacity of 4242 MW).

On the other hand, the R.E. Ginna Nuclear Power Plant in New York is the USA's smallest nuclear power plant. It has one nuclear power reactor with a net electricity generating capacity of 560 MW (Gross electrical capacity of 608 MW).

It is important to single out that after 2000, the number of units operating in the country was almost the same, except for the decrease registered in 2019 and 2020. It is expected that the number of nuclear power reactors in operation in the USA will continue to decrease during the coming years due to the closure of new units[v] and the lack of plans for the construction of new units.

The USA nuclear power industry grew to its present size following construction programs initiated during the 1960s and 1970s when the use of nuclear energy was anticipated to be a low-cost source of electricity that could satisfy the increasing electricity demand expected in several countries. Increases in nuclear-generating capacity during 1969–96 made nuclear energy the second-largest electricity generation in the USA, following coal. As a result of the USA's energy policy after the Three Miles Island nuclear accident occurred in 1979,[w] nuclear energy electricity generation dropped. In 2018, it was the third source of electricity in the country after natural gas and coal. It is crucial to indicate that the use of renewable energy for electricity generation in the country is increasing very fast. It is likely to surpass the use of nuclear energy for electricity generation during the coming decades. Wind and solar energies are the two renewable energy sources that will increase their capacities during the coming years. Costs for renewables, particularly in the case of wind and solar power, have continued to decline in the last years due to the experience gained with more builds. However, how long these high-cost reduction rates can be sustained is highly uncertain (Annual Energy Outlook 2020 With Projections to 2050, 2020).

Better utilization of generating nuclear capacity, the average load factor of nuclear power plants is above 80%, has permitted nuclear power plants to maintain this relative position despite stopping the construction of new nuclear power reactors and several units' shutdown after the Three Miles Island nuclear accident. However, it is important to note that several

[u] The Indian Point Energy Center in New York has two nuclear power reactors that the US EIA counts as two separate nuclear power plants.

[v] In 2020, two new nuclear power reactors are under construction in the Georgia (Vogtle units 3 and 4), in the USA. Both units are expected to begin operation in 2021. Since 1973, no energy generation company in the USA has been willing to order the construction of a new nuclear power plant. Only preliminary steps toward purchasing and constructing a new nuclear power plants have been carried out by some of the generating companies until today in the absence of a promise of huge federal subsidies to finance the construction of this type of power plants. The reason why only two new nuclear power reactors are under construction is not because of public opposition to the use of nuclear energy for electricity generation; not for the conditions to have a licensed geologic repository for the disposal of spent nuclear fuel; and certainly not because of the proliferation risks associated with commercial nuclear power plants. Rather, it is because new commercial nuclear power plants were uneconomical in the USA until 2008 in comparison with other available energy sources. Now the situation has change due to the increase in the oil price in 2008, the raise in the price of natural gas and clean coal and the negative impact in the climate as a consequence of the increase emission of $CO_2$ to the atmosphere due to the use of fossil fuels for generating electricity.

[w] To know more details about this accident read the paper Morales Pedraza (2013).

nuclear power reactors permanently shut down during the 1950–90s were small or prototype units with a capacity below 100 MW. The last unit of this group was shut down in 1968.

The improved operation of nuclear power plants has helped drive down the cost of nuclear-generated electricity. According to the World Nuclear Association, average nuclear generation costs have fallen from US$42 per MWh in 2012 to US$33 per MWh in 2018; this means a decrease of 21.5% during the period under consideration. Despite a near halt in the new construction of new nuclear power plants in the USA for more than 30 years, US reliance on nuclear power has grown. In 1980, nuclear power plants produced 251 billion kWh, accounting for 11% of the country's electricity generation. In 2008, that output had risen to 809 billion kWh and nearly 20% of electricity, providing more than 30% of the electricity generated from nuclear power worldwide. Much of the increase came from the 47 nuclear power reactors, all approved for construction before 1977, that came online in the late 1970s and 1980s, more than doubling US nuclear generation capacity. The US nuclear industry has also achieved notable gains in power plant utilization through improved refueling, maintenance, and safety systems at existing nuclear power plants. Average generating cost in 2012 was US$42 per MW; in 2014 decreased to US$36.27 per MWh (US$44.14 at single-unit sites and US$33.76 at multi-unit sites), including fuel and capital, and the average operating cost was US$21 per MWh, and to US$33 per MWh in 2018 (World Nuclear Association-Nuclear Power in the USA, 2020).

It is important to stress that although there are in the USA the largest number of operating nuclear power reactors in the world in a single country, the number of canceled projects, 138 units, is even larger.

According to Fig. 7.20, it is easy to confirm that most USA nuclear power plants are located in the East and Central part of the country, with very few units located in the Western part of the country. In Table 7.5, the list of all nuclear power reactors in operation in April 2021 in the USA is shown.

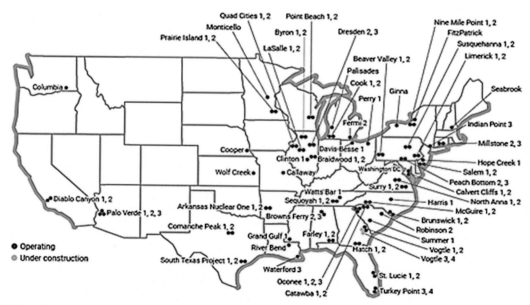

FIG. 7.20   Reactors operating in the USA. *Source: World Nuclear Association, 2020. Nuclear Power in the USA. https://www.world-nuclear.org/information-library/country-profiles/countries-t-z/usa-nuclear-power.aspx.*

**TABLE 7.5** Nuclear power reactors in operation and shut down in the USA.

| | | | | Nuclear power reactors in the USA | | |
|---|---|---|---|---|---|---|
| Name | Type | Status | Location | Reference unit Power [MW] | Gross electrical capacity [MW] | First grid connection |
| ANO-1 | PWR | Operational | POPE | 836 | 903 | 1974-08-17 |
| ANO-2 | PWR | Operational | POPE | 988 | 1065 | 1978-12-26 |
| BEAVER VALLEY- 1 | PWR | Operational | SHIPPINGPORT | 908 | 959 | 1976-06-14 |
| BEAVER VALLEY-2 | PWR | Operational | SHIPPINGPORT | 905 | 958 | 1987-08-17 |
| BIG ROCK POINT | BWR | Permanent Shut down | CHARLEVOIX | 67 | 71 | 1962-12-08 |
| BONUS | BWR | Permanent Shut down | RINCON | 17 | 18 | 1964-08-14 |
| BRAIDWOOD-1 | PWR | Operational | BRAIDWOOD | 1194 | 1270 | 1987-07-12 |
| BRAIDWOOD-2 | PWR | Operational | BRAIDWOOD | 1160 | 1230 | 1988-05-25 |
| BROWNS FERRY-1 | BWR | Operational | DECATUR | 1101 | 1155 | 1973-10-15 |
| BROWNS FERRY-2 | BWR | Operational | DECATUR | 1104 | 1155 | 1974-08-28 |
| BROWNS FERRY-3 | BWR | Operational | DECATUR | 1105 | 1155 | 1976-09-12 |
| BRUNSWICK-1 | BWR | Operational | SOUTHPORT | 938 | 990 | 1976-12-04 |
| BRUNSWICK-2 | BWR | Operational | SOUTHPORT | 932 | 960 | 1975-04-29 |
| BYRON-1 | PWR | Operational | BYRON | 1164 | 1242 | 1985-03-01 |
| BYRON-2 | PWR | Operational | BYRON | 1136 | 1210 | 1987-02-06 |
| CALLAWAY-1 | PWR | Operational | FULTON | 1215 | 1275 | 1984-10-24 |
| CALVERT CLIFFS-1 | PWR | Operational | LUSBY | 863 | 918 | 1975-01-03 |
| CALVERT CLIFFS-2 | PWR | Operational | LUSBY | 855 | 911 | 1976-12-07 |
| CATAWBA-1 | PWR | Operational | YORK COUNTY | 1160 | 1188 | 1985-01-22 |
| CATAWBA-2 | PWR | Operational | YORK COUNTY | 1150 | 1188 | 1986-05-18 |
| CLINTON-1 | BWR | Operational | HART TOWNSHIP | 1062 | 1098 | 1987-04-24 |
| COLUMBIA | BWR | Operational | BENTON | 1131 | 1190 | 1984-05-27 |
| COMANCHE PEAK-1 | PWR | Operational | GLEN ROSE | 1218 | 1259 | 1990-04-24 |
| COMANCHE PEAK-2 | PWR | Operational | GLEN ROSE | 1207 | 1250 | 1993-04-09 |

*Continued*

**TABLE 7.5** Nuclear power reactors in operation and shut down in the USA—cont'd

| | | Nuclear power reactors in the USA | | | |
|---|---|---|---|---|---|
| Name | Type | Status | Location | Reference unit Power [MW] | Gross electrical capacity [MW] | First grid connection |
| COOK-1 | PWR | Operational | BRIDGMAN | 1030 | 1131 | 1975-02-10 |
| COOK-2 | PWR | Operational | BRIDGMAN | 1168 | 1231 | 1978-03-22 |
| COOPER | BWR | Operational | BROWNVILLE | 769 | 801 | 1974-05-10 |
| CRYSTAL RIVER-3 | PWR | Permanent Shut down | | 860 | 890 | 1977-01-30 |
| CVTR | PHWR | Permanent Shut down | PARR | 17 | 19 | 1963-12-18 |
| DAVIS BESSE-1 | PWR | Operational | OTTAWA | 894 | 925 | 1977-08-28 |
| DIABLO CANYON-1 | PWR | Operational | AVILA BEACH | 1138 | 1197 | 1984-11-11 |
| DIABLO CANYON-2 | PWR | Operational | AVILA BEACH | 1118 | 1197 | 1985-10-20 |
| DRESDEN-1 | BWR | Permanent Shut down | MORRIS | 197 | 207 | 1960-04-15 |
| DRESDEN-2 | BWR | Operational | MORRIS | 894 | 950 | 1970-04-13 |
| DRESDEN-3 | BWR | Operational | MORRIS | 879 | 935 | 1971-07-22 |
| DUANE ARNOLD-1 | BWR | Operational | PALO | 601 | 624 | 1974-05-19 |
| ELK RIVER | BWR | Permanent Shut down | | 22 | 24 | 1963-08-24 |
| FARLEY-1 | PWR | Operational | DOTHAN | 874 | 918 | 1977-08-18 |
| FARLEY-2 | PWR | Operational | DOTHAN | 883 | 928 | 1981-05-25 |
| FERMI-1 | FBR | Permanent Shut down | LAGOONA BEACH | 61 | 65 | 1966-08-05 |
| FERMI-2 | BWR | Operational | LAGOONA BEACH | 1095 | 1198 | 1986-09-21 |
| FITZPATRICK | BWR | Operational | OSWEGO | 813 | 849 | 1975-02-01 |
| FORT CALHOUN-1 | PWR | Permanent Shut down | FORT CALHOUN | 482 | 512 | 1973-08-25 |
| FORT ST. VRAIN | HTGR | Permanent Shut down | PLATTEVILLE | 330 | 342 | 1976-12-11 |
| GE VALLECITOS | BWR | Permanent Shut down | PLEASANTON, SUNOL | 24 | 24 | 1957-10-19 |
| GINNA | PWR | Operational | ONTARIO | 560 | 608 | 1969-12-02 |

| GRAND GULF-1 | BWR | Operational | PORT GIBSON | 1401 | 1500 | 1984-10-20 |
| HADDAM NECK | PWR | Permanent Shut down | HADDAM NECK | 560 | 603 | 1967-08-07 |
| HALLAM | SGR | Permanent Shut down | LINCOLN | 75 | 84 | 1963-09-01 |
| HARRIS-1 | PWR | Operational | NEW HILL | 932 | 960 | 1987-01-19 |
| HATCH-1 | BWR | Operational | BAXLEY | 876 | 911 | 1974-11-11 |
| HATCH-2 | BWR | Operational | BAXLEY | 883 | 921 | 1978-09-22 |
| HOPE CREEK-1 | BWR | Operational | SALEM | 1172 | 1240 | 1986-08-01 |
| HUMBOLDT BAY | BWR | Permanent Shut down | EUREKA | 63 | 65 | 1963-04-18 |
| INDIAN POINT-1 | PWR | Permanent Shut down | BUCHANAN | 257 | 277 | 1962-09-16 |
| INDIAN POINT-2 | PWR | Operational | BUCHANAN | 998 | 1067 | 1973-06-26 |
| INDIAN POINT-3 | PWR | Operational | BUCHANAN | 1030 | 1085 | 1976-04-27 |
| KEWAUNEE | PWR | Permanent Shut down | CARLTON | 566 | 595 | 1974-04-08 |
| LACROSSE | BWR | Permanent Shut down | GENOA | 48 | 55 | 1968-04-26 |
| LASALLE-1 | BWR | Operational | MARSEILLES | 1137 | 1207 | 1982-09-04 |
| LASALLE-2 | BWR | Operational | MARSEILLES | 1140 | 1207 | 1984-04-20 |
| LIMERICK-1 | BWR | Operational | LIMERICK | 1099 | 1194 | 1985-04-13 |
| LIMERICK-2 | BWR | Operational | LIMERICK | 1134 | 1194 | 1989-09-01 |
| MAINE YANKEE | PWR | Permanent Shut down | WISCASSET | 860 | 900 | 1972-11-08 |
| MCGUIRE-1 | PWR | Operational | CORNELIUS | 1158 | 1215 | 1981-09-12 |
| MCGUIRE-2 | PWR | Operational | CORNELIUS | 1158 | 1215 | 1983-05-23 |
| MILLSTONE-1 | BWR | Permanent Shut down | WATERFORD | 641 | 684 | 1970-11-29 |
| MILLSTONE-2 | PWR | Operational | WATERFORD | 869 | 918 | 1975-11-09 |
| MILLSTONE-3 | PWR | Operational | WATERFORD | 1210 | 1280 | 1986-02-12 |
| MONTICELLO | BWR | Operational | MONTICELLO | 628 | 691 | 1971-03-05 |
| NINE MILE POINT-1 | BWR | Operational | SCRIBA | 613 | 642 | 1969-11-09 |
| NINE MILE POINT-2 | BWR | Operational | SCRIBA | 1277 | 1320 | 1987-08-08 |

*Continued*

**TABLE 7.5** Nuclear power reactors in operation and shut down in the USA—cont'd

| Name | Type | Status | Location | Reference unit Power [MW] | Gross electrical capacity [MW] | First grid connection |
|------|------|--------|----------|---------------------------|-------------------------------|----------------------|
| NORTH ANNA-1 | PWR | Operational | MINERAL | 948 | 990 | 1978-04-17 |
| NORTH ANNA-2 | PWR | Operational | MINERAL | 944 | 1011 | 1980-08-25 |
| OCONEE-1 | PWR | Operational | OCONEE | 847 | 891 | 1973-05-06 |
| OCONEE-2 | PWR | Operational | OCONEE | 848 | 891 | 1973-12-05 |
| OCONEE-3 | PWR | Operational | OCONEE | 859 | 900 | 1974-09-18 |
| OYSTER CREEK | BWR | Permanent Shut down | FORKED RIVER | 619 | 652 | 1969-09-23 |
| PALISADES | PWR | Operational | SOUTH HAVEN | 805 | 850 | 1971-12-31 |
| PALO VERDE-1 | PWR | Operational | WINTERSBURG | 1311 | 1414 | 1985-06-10 |
| PALO VERDE-2 | PWR | Operational | WINTERSBURG | 1314 | 1414 | 1986-05-20 |
| PALO VERDE-3 | PWR | Operational | WINTERSBURG | 1312 | 1414 | 1987-11-28 |
| PATHFINDER | BWR | Permanent Shut down | | 59 | 63 | 1966-07-25 |
| PEACH BOTTOM-1 | HTGR | Permanent Shut down | YORK COUNTY | 40 | 42 | 1967-01-27 |
| PEACH BOTTOM-2 | BWR | Operational | YORK COUNTY | 1232 | 1412 | 1974-02-18 |
| PEACH BOTTOM-3 | BWR | Operational | YORK COUNTY | 1251 | 1412 | 1974-09-01 |
| PERRY-1 | BWR | Operational | PERRY | 1240 | 1303 | 1986-12-19 |
| PILGRIM-1 | BWR | Permanent Shut down | PLYMOUTH | 677 | 711 | 1972-07-19 |
| PIQUA[a] | – | Permanent Shut down | PIQUA | 12 | 12 | 1963-07-01 |
| POINT BEACH-1 | PWR | Operational | TWO CREEKS | 591 | 640 | 1970-11-06 |
| POINT BEACH-2 | PWR | Operational | TWO CREEKS | 591 | 640 | 1972-08-02 |
| PRAIRIE ISLAND-1 | PWR | Operational | RED WING | 522 | 566 | 1973-12-04 |
| PRAIRIE ISLAND-2 | PWR | Operational | RED WING | 519 | 560 | 1974-12-21 |

| Name | Type | Status | Location | | | Date |
|---|---|---|---|---|---|---|
| QUAD CITIES-1 | BWR | Operational | CORDOVA | 908 | 940 | 1972-04-12 |
| QUAD CITIES-2 | BWR | Operational | CORDOVA | 911 | 940 | 1972-05-23 |
| RANCHO SECO-1 | PWR | Permanent Shut down | SACRAMENTO | 873 | 917 | 1974-10-13 |
| RIVER BEND-1 | BWR | Operational | ST.FRANCISVILLE | 967 | 1016 | 1985-12-03 |
| ROBINSON-2 | PWR | Operational | HARTSVILLE | 741 | 780 | 1970-09-26 |
| SALEM-1 | PWR | Operational | SALEM | 1169 | 1254 | 1976-12-25 |
| SALEM-2 | PWR | Operational | SALEM | 1158 | 1200 | 1981-06-03 |
| SAN ONOFRE-1 | PWR | Permanent Shut down | SAN CLEMENTE | 436 | 456 | 1967-07-16 |
| SAN ONOFRE-2 | PWR | Permanent Shut down | SAN CLEMENTE | 1070 | 1127 | 1982-09-20 |
| SAN ONOFRE-3 | PWR | Permanent Shut down | SAN CLEMENTE | 1080 | 1127 | 1983-09-25 |
| SAXTON | PWR | Permanent Shut down | | 3 | 3 | 1967-03-01 |
| SEABROOK-1 | PWR | Operational | SEABROOK | 1246 | 1296 | 1990-05-29 |
| SEQUOYAH-1 | PWR | Operational | DAISY | 1152 | 1221 | 1980-07-22 |
| SEQUOYAH-2 | PWR | Operational | DAISY | 1139 | 1200 | 1981-12-23 |
| SHIPPINGPORT | PWR | Permanent Shut down | | 60 | 68 | 1957-12-02 |
| SHOREHAM | BWR | Permanent Shut down | SHOREHAM | 820 | 849 | 1986-08-01 |
| SOUTH TEXAS-1 | PWR | Operational | BAY CITY | 1280 | 1354 | 1988-03-30 |
| SOUTH TEXAS-2 | PWR | Operational | BAY CITY | 1280 | 1354 | 1989-04-11 |
| ST. LUCIE-1 | PWR | Operational | FORT PIERCE | 981 | 1045 | 1976-05-07 |
| ST. LUCIE-2 | PWR | Operational | FORT PIERCE | 987 | 1050 | 1983-06-13 |
| SUMMER-1 | PWR | Operational | JENKINSVILLE | 973 | 1006 | 1982-11-16 |
| SURRY-1 | PWR | Operational | GRAVEL NECK | 838 | 890 | 1972-07-04 |
| SURRY-2 | PWR | Operational | GRAVEL NECK | 838 | 890 | 1973-03-10 |
| SUSQUEHANNA-1 | BWR | Operational | SALEM | 1257 | 1330 | 1982-11-16 |

*Continued*

**TABLE 7.5** Nuclear power reactors in operation and shut down in the USA—cont'd

| | | | | Nuclear power reactors in the USA | | | |
|---|---|---|---|---|---|---|---|
| Name | Type | Status | Location | Reference unit Power [MW] | Gross electrical capacity [MW] | First grid connection |
| SUSQUEHANNA-2 | BWR | Operational | SALEM | 1257 | 1330 | 1984-07-03 |
| THREE MILE ISLAND-1 | PWR | Permanent Shut down | DAUPHIN | 819 | 880 | 1974-06-19 |
| THREE MILE ISLAND-2 | PWR | Permanent Shut down | DAUPHIN | 880 | 959 | 1978-04-21 |
| TROJAN | PWR | Permanent Shut down | PRESCOTT | 1095 | 1155 | 1975-12-23 |
| TURKEY POINT-3 | PWR | Operational | FLORIDA CITY | 837 | 829 | 1972-11-02 |
| TURKEY POINT-4 | PWR | Operational | FLORIDA CITY | 821 | 829 | 1973-06-21 |
| VERMONT YANKEE | BWR | Permanent Shut down | VERNON | 605 | 635 | 1972-09-20 |
| VOGTLE-1 | PWR | Operational | WAYNESBORO | 1150 | 1229 | 1987-03-27 |
| VOGTLE-2 | PWR | Operational | WAYNESBORO | 1152 | 1229 | 1989-04-10 |
| VOGTLE-3 | PWR | Under Construction | WAYNESBORO | 1117 | 1250 | |
| VOGTLE-4 | PWR | Under Construction | WAYNESBORO | 1117 | 1250 | |
| WATERFORD-3 | PWR | Operational | TAFT | 1168 | 1250 | 1985-03-18 |
| WATTS BAR-1 | PWR | Operational | SPRING CITY | 1123 | 1210 | 1996-02-06 |
| WATTS BAR-2 | PWR | Operational | SPRING CITY | 1135 | 1218 | 2016-06-03 |
| WOLF CREEK | PWR | Operational | BURLINGTON | 1200 | 1285 | 1985-06-12 |
| YANKEE NPS | PWR | Permanent Shut down | ROWE | 167 | 180 | 1960-11-10 |
| ZION-1 | PWR | Permanent Shut down | ZION | 1040 | 1085 | 1973-06-28 |
| ZION-2 | PWR | Permanent Shut down | ZION | 1040 | 1085 | 1973-12-26 |

a Organic cooled and moderated nuclear reactor.
Source: April 2021. PRIS—Power Reactor Information System. www.iaea.org/pris. © IAEA.

A list of the US nuclear power reactors planned and proposed is shown in Table 7.6.

Of the above, for the first four AP1000 units, construction is well underway at Vogtle, Georgia, with about US$4 billion invested in the project before it was technically classified as "under construction." Construction was also well underway in Summer, South Carolina, but has been put on hold. In addition to sites listed in Table 7.7, US Southern Company evaluates several possible sites, including existing plants and greenfield locations, for additional AP1000 reactors (World Nuclear Association-Nuclear Power in the USA, 2020).

Finally, Table 7.8 shows the list of nuclear accidents reported by the US authorities since 1955 (Sovacool, 2009, 2010; Blade, 2004; Bel, 2000).

**TABLE 7.6**  US nuclear power reactors planned and proposed.

| Site | Technology | MWe gross | Proponent/utility | COL lodgement and issue dates | Loan guarantee; start operation |
|---|---|---|---|---|---|
| Turkey Point 6 and 7, FL | AP1000 | $2 \times 1250$ | Florida Power & Light | 30/6/09, COL April 2017 | 2027, 2028 |
| UAMPS Carbon-Free Power Project, ID | Nuscale | 50 | Western Initiative for Nuclear, Utah AMPS, Energy NW | Design certification application Jan 2017, COL application mid-2020 | 2017 loan guarantee application; 2023 |
| **Subtotal planned: 2 large and 1 small unit (2550 MWe gross)** | | | | | |
| Fermi 3, MI | ESBWR | 1600 | Detroit Edison | 18/9/08, COL issued May 2015 | No decision to proceed |
| North Anna 3, VA | ESBWR | ~ 1500 | Dominion | 20/11/07, COL issued June 2017, ESP issued | On hold from Sept 2017 |
| Clinch River, TN | Uncertain, was mPower | $2 \times 360$? up to $2 \times 800$ | TVA | ESP application May 2016, issued Dec 2019 | |
| Bellefonte 1 and 2,[g] AL | B&W PWR (partly built) | $2 \times 1263$ | Nuclear Development LLC (sale pending from Tennessee Valley Authority) | 30/10/07 for units 3 and 4,[h] but COL withdrew in 2016 | Seeking loan guarantee |
| UAMPS Carbon-Free Power Project, ID | Nuscale | $11 \times 50$ | Western Initiative for Nuclear, Utah AMPS, Energy NW | Design certification application Jan 2017, COL application mid-2020 | 2017 loan guarantee application |
| Salem 3/Hope Creek, NJ | unspecified | 1200? | PSEG Nuclear | ESP issued May 2016 | |
| **Subtotal proposed: 7 large units, 11 small (ca. 8000 MWe gross)** | | | | | |

COL, combined license.

*Source: World Nuclear Association, 2020. Nuclear Power in the USA. https://www.world-nuclear.org/information-library/country-profiles/countries-t-z/usa-nuclear-power.aspx.*

**TABLE 7.7** Other proposals, less definite, suspended, or canceled.

| Site | Technology | MWe gross | Proponent/ utility | COL lodgement and issue dates | Status |
|------|-----------|-----------|---------------------|-------------------------------|--------|
| Victoria County, TX | ESBWR | 3200 | Exelon (merchant plant) | 03/9/08 but withdrawn, ESP application 25/3/10, but withdrawn Oct 2012 | |
| Piketon (DOE site leased to USEC), OH | US EPR | 1710 | Duke Energy | | |
| Payette County, ID | APWR | 1700 | Alternate Energy Holdings Inc. (merchant plant) | Plans stalled since 2012 | |
| Fresno, Ca | US EPR | 1710 | Fresno Nuclear Energy Group | | |
| Amarillo, TX | US EPR | 2×1750 | Amarillo Power (merchant plant) | | |
| Levy Country, FL | AP1000 | 2×1250 | Duke Energy (formerly Progress Energy) | 30/07/08, COLs approved Oct 2016 and canceled April 2018 | Project canceled Aug 2017 |
| Callaway, MO | Westinghouse SMR | 5×225 | Ameren Missouri | 24/07/08 for EPR then canceled, SMR proposal suspended, COL application for EPR withdrawn | |
| Shearon Harris 2&3, NC | AP1000 | 2×1250 | Duke Energy (formerly Progress Energy) | 19/02/08, COL suspended May 2013 | |
| Grand Gulf, MS | ESBWR | 1600 | Entergy | 27/02/2008, COL application withdrawn 9/15, ESP issued | |
| Comanche Peak, TX | US-APWR | 2×1700 | Luminant (merchant plant) | 19/09/08, COL suspended 11/13 | |
| Bell Bend (near Susquehanna), PA | US EPR | 1710 | PPL/Talen (merchant plant) | 10/10/08, COL review suspended 2014, but EIS approved. COL application was withdrawn in Aug 2016 | Suspended indefinitely |
| Calvert Cliffs, MD | US EPR | 1710 | UniStar Nuclear (merchant plant) | 07/07 and 03/08, terminated in 2012, COL application withdrew in 07/15 | Refused an offered loan guarantee, needs US equity |
| Green River, UT | AP1000 | 2×1250 | Blue Castle/ Transition Power Development | | 2030 |
| River Bend, LA | ESBWR | 1600 | Entergy | 25/09/08, COL application withdrawn | |

**TABLE 7.7** Other proposals, less definite, suspended, or canceled—cont'd

| Site | Technology | MWe gross | Proponent/ utility | COL lodgement and issue dates | Status |
|------|-----------|-----------|-------------------|------------------------------|--------|
| South Texas Project,[e] TX | ABWR | 2×1356 | Toshiba, NINA, STP Nuclear (merchant plant) | COLs issued Feb 2016, but a design certification application was withdrawn | Canceled May 2018 |
| Nine Mile Point, NY | US EPR | 1710 | UniStar Nuclear (merchant plant) | 30/09/08, COL application withdrawn 2013 | |
| Stewart County, GA | AP1000 | 1250 | Georgia Power (Southern Co) | COL application deferred in 2017 | Build after 2030 |
| William States Lee III, SC | AP1000 | 2×1250 | Duke Energy | 13/12/07, COL issued Dec 2016 | Plans canceled Aug 2017 |

*Source: World Nuclear Association, 2020. Nuclear Power in the USA. https://www.world-nuclear.org/information-library/country-profiles/countries-t-z/usa-nuclear-power.aspx.*

**TABLE 7.8** List of nuclear accidents reported by the US authorities since 1955.

| | | Nuclear reactor accidents in the USA | | | |
|------|----------|-------------|------------|---------------------------|--------------|
| Date | Location | Description | Fatalities | Cost (in millions 2006 US$) | INES rating |
| November 29, 1955 | Idaho Falls, Idaho, US | Power excursion with partial core meltdown at National Reactor Testing Station's EBR-1 Experimental Breeder Reactor I | 0 | 5 | |
| July 26, 1959 | Simi Valley, California, USA | Partial core meltdown at Santa Susana Field Laboratory's Sodium Reactor Experiment | 0 | 32 | |
| January 3, 1961 | Idaho Falls, Idaho, US | Explosion with three fatalities at National Reactor Testing Station's SL-1 Stationary Low-Power Reactor Number One | 3 | 22 | 4 |
| July 24, 1964 | Charlestown, Rhode Island, USA | An error by a worker at a United Nuclear Corporation fuel facility led to an accidental criticality | 1 | – | |
| October 5, 1966 | Monroe, Michigan, USA | Sodium cooling system malfunctions at Enrico Fermi demonstration breeder reactor, caused partial core meltdown | 0 | 19 | |
| July 16, 1971 | Cordova, Illinois, USA | A live cable electrocuted an electrician at the Quad Cities Unit 1 reactor on the Mississippi River | 1 | 1 | |
| August 11, 1973 | Covert Township, Michigan, USA | Steam generator leak at the Palisades Nuclear Generating Station caused manual shut down of the pressurized water reactor | 0 | 10 | |

*Continued*

**TABLE 7.8**   List of nuclear accidents reported by the US authorities since 1955—cont'd

| | | Nuclear reactor accidents in the USA | | | |
|---|---|---|---|---|---|
| Date | Location | Description | Fatalities | Cost (in millions 2006 US$) | INES rating |
| March 22, 1975 | Athens, Alabama, USA | Fire burns for seven hours and damaged more than 1600 control cables for three nuclear reactors at Browns Ferry, disabling core cooling systems | 0 | 240 | |
| November 5, 1975 | Brownville, Nebraska, USA | The hydrogen gas explosion damaged the Cooper Nuclear Facility's auxiliary building | 0 | 13 | |
| June 10, 1977 | Waterford, Connecticut, USA | The hydrogen gas explosion damaged three buildings and forced the shut down of the Millstone-1 Boiling Water Reactor | 0 | 15 | |
| February 4, 1979 | Surry, Virginia, USA | Surry Unit 2 shut down in response to failing tube bundles in steam generators | 0 | 12 | |
| March 28, 1979 | Middletown, Pennsylvania, US | Loss of coolant and partial core meltdown, see Three Mile Island accident and Three Mile Island accident health effects | 0 | 2400 | 5 |
| November 22, 1980 | San Clemente, California, USA | Worker cleaning breaker cubicles at San Onofre Pressurized Water Reactor contacted an energized line and is electrocuted | 1 | 1 | |
| January 25, 1982 | Ontario, New York, USA | Ginna Nuclear Generating Station (then operated by Rochester Gas & Electric now by Constellation Energy Nuclear Group) experienced a steam tube rupture, releasing radioactivity into the environment. | 0 | 1 | |
| February 26, 1982 | San Clemente, California, USA | Southern California Company shuts down San Onofre Unit 1 out of concerns about earthquake | 0 | 1 | |
| March 20, 1982 | Scriba, New York, USA | Recirculation system piping failed at Nine Mile Point Unit 1, forcing a two-year shut down | 0 | 45 | |
| March 25, 1982 | Buchanan, New York, USA | Damage to steam generator tubes and main generator resulting in a shutdown of Indian Point Energy Center Unit 3 for more than a year | 0 | 56 | |
| June 18, 1982 | Seneca, South Carolina, USA | Feedwater heat extraction line failed at Oconee 2 Pressurized Water Reactor, damaging the thermal cooling system | 0 | 10 | |

**TABLE 7.8** List of nuclear accidents reported by the US authorities since 1955—cont'd

| | Nuclear reactor accidents in the USA | | | | |
|---|---|---|---|---|---|
| Date | Location | Description | Fatalities | Cost (in millions 2006 US$) | INES rating |
| February 12, 1983 | Forked River, New Jersey, USA | Oyster Creek Nuclear Generating Station failed safety inspection, forced to shut down for repairs | 0 | 32 | |
| February 26, 1983 | Fort Pierce, Florida, USA | Damaged thermal shield and core barrel support at St. Lucie Unit 1 provoked a 13-month shut down | 0 | 54 | |
| September 15, 1984 | Athens, Alabama, US | Safety violations, operator error, and design problems forced a six-year outage at Browns Ferry Unit 2 | 0 | 110 | |
| March 9, 1985 | Athens, Alabama, US | Instrumentation systems malfunction during start-up, which led to the suspension of operations at all three Browns Ferry Units | 0 | 1830 | |
| June 9, 1985 | Oak Harbor, Ohio, US | Loss of feedwater event at Davis-Besse reactor after main pumps shut down and auxiliary pumps tripped due to operator error. NRC review determined site area emergency should have been declared | 0 | – | |
| April 11, 1986 | Plymouth, Massachusetts, US | Recurring equipment problems forced an emergency shut down of Boston Edison's Pilgrim Nuclear Power Plant | 0 | 1001 | |
| December 9, 1986 | Surry, Virginia, USA | Feedwater line-burst at Surry Nuclear Power Plant kills 4 | 4 | | |
| March 31, 1987 | Delta, Pennsylvania, US | Peach Bottom units 2 and 3 shut down due to cooling malfunctions and unexplained equipment problems | 0 | 400 | |
| July 15, 1987 | Burlington, Kansas, USA | A safety inspector died from electrocution after contacting a mislabeled wire at Wolf Creek Nuclear Generating Station | 1 | 1 | |
| December 19, 1987 | Scriba, New York, US | Malfunctions forced Niagara Mohawk Power Corporation to shut down Nine Mile Point Unit 1 | 0 | 150 | |
| March 29, 1988 | Burlington, Kansas, USA | A worker at the Wolf Creek Generating Station fell through an unmarked manhole and electrocuted himself when trying to escape | 1 | 1 | |
| September 10, 1988 | Surry, Virginia, USA | The refueling cavity seal failed and destroyed the internal pipe system at Surry Unit 2, forcing a 12-month outage | 0 | 9 | |

*Continued*

**TABLE 7.8**  List of nuclear accidents reported by the US authorities since 1955—cont'd

**Nuclear reactor accidents in the USA**

| Date | Location | Description | Fatalities | Cost (in millions 2006 US$) | INES rating |
|------|----------|-------------|-----------|-----------------------------|-------------|
| March 5, 1989 | Tonopah, Arizona, USA | Atmospheric dump valves failed at Palo Verde Unit 1, leading to main transformer fire and emergency shut down | 0 | 14 | |
| March 17, 1989 | Lusby, Maryland, US | Inspections at Calvert Cliff Units 1 and 2 reveal cracks at pressurized heater sleeves, forcing extended shut downs | 0 | 120 | |
| November 17, 1991 | Scriba, New York, USA | Safety and fire problems forced the shut down of the FitzPatrick nuclear reactor for 13 months | 0 | 5 | |
| April 21, 1992 | Southport, North Carolina, USA | NRC forced the shut down of Brunswick Units 1 and 2 after emergency diesel generators fail | 0 | 2 | |
| February 3, 1993 | Bay City, Texas, USA | Auxiliary feed-water pumps failed at South Texas Project Units 1 and 2, prompting a rapid shut down of both reactors | 0 | 3 | |
| February 27, 1993 | Buchanan, New York, USA | New York Power Authority shuts down Indian Point Energy Center Unit 3 after AMSAC system fails | 0 | 2 | |
| March 2, 1993 | Soddy-Daisy, Tennessee, USA | Equipment failures and broken pipes caused the shut down of Sequoyah Unit 1 | 0 | 3 | |
| December 25, 1993 | Newport, Michigan, USA | Shut down of Fermi Unit 2 after the main turbine experienced major failure due to improper maintenance | 0 | 67 | |
| 14 January 1995 | Wiscasset, Maine, USA | Steam generator tubes unexpectedly crack at Maine Yankee nuclear reactor; shut down of the facility for a year | 0 | 62 | |
| May 16, 1995 | Salem, New Jersey, USA | Ventilation systems failed at Salem Units 1 and 2 | 0 | 34 | |
| February 20, 1996 | Waterford, Connecticut, US | Leaking valve forced the shut down of Millstone Nuclear Power Plant Units 1 and 2, multiple equipment failures found | 0 | – | |
| May 15, 1996 | Morris, Illinois, US | Plunging water levels around the nuclear fuel in the reactor's core prompt shut down at Dresden Generating Station | 0 | – | |
| September 2, 1996 | Crystal River, Florida, US | Balance-of-plant equipment malfunction forced the shut down and extensive repairs at Crystal River Unit 3 | 0 | 384 | |

**TABLE 7.8** List of nuclear accidents reported by the US authorities since 1955—cont'd

| | Nuclear reactor accidents in the USA | | | | |
|---|---|---|---|---|---|
| Date | Location | Description | Fatalities | Cost (in millions 2006 US$) | INES rating |
| September 5, 1996 | Clinton, Illinois, USA | The reactor recirculation pump failed, prompting shut down of Clinton boiling water reactor | 0 | 38 | |
| September 20, 1996 | Seneca, Illinois, USA | The service water system failed and resulted in the closure of LaSalle Units 1 and 2 for more than two years | 0 | 71 | |
| September 9, 1997 | Bridgman, Michigan, USA | Ice condenser containment systems failed at Cook Units 1 and 2 | 0 | 11 | |
| May 25, 1999 | Waterford, Connecticut, USA | The steam leak in the feed-water heater caused the manual shut down and damaged the control board annunciator at the Millstone Nuclear Power Plant | 0 | 7 | |
| September 29, 1999 | Lower Alloways Creek Township, New Jersey, USA | A major Freon leak at Hope Creek Nuclear Generating Station caused the ventilation train chiller to trip, releasing toxic gas and damaging the cooling system | 0 | 2 | |
| February 15, 2000 | Buchanan, New York, USA | NRC Alert issued after steam tube rupture Indian Point Unit 2 | 0 | 2 | |
| February 16, 2002 | Oak Harbor, Ohio, US | Severe boric acid corrosion of reactor head forced 24-month outage of Davis-Besse reactor | 0 | 605 | 3 |
| January 15, 2003 | Bridgman, Michigan, USA | A fault in the main transformer at the Donald C. Cook Nuclear Generating Station caused a fire that damaged the main generator and backup turbines | 0 | 10 | |
| June 16, 2005 | Braidwood, Illinois, USA | Exelon's Braidwood nuclear station leaks tritium and contaminated local water supplies | 0 | 41 | |
| August 4, 2005 | Buchanan, New York, USA | Entergy's Indian Point Nuclear Plant leaks tritium and strontium into underground lakes from 1974 to 2005 | – | 30 | |
| March 6, 2006 | Erwin, Tennessee, USA | Nuclear Fuel Services plant spilled 35 L of highly enriched uranium, provoking a 7-month shut down | 0 | 98 | |
| February 1, 2010 | Vernon, Vermont, US | Deteriorating underground pipes from the Vermont Yankee Nuclear Power Plant leak radioactive tritium into groundwater supplies | 0 | 700 | |

*Continued*

TABLE 7.8　List of nuclear accidents reported by the US authorities since 1955—cont'd

| | | Nuclear reactor accidents in the USA | | | |
|---|---|---|---|---|---|
| Date | Location | Description | Fatalities | Cost (in millions 2006 US$) | INES rating |
| July 15, 2011 | Morris, Illinois, US | A chemical leak of sodium hypochlorite restricted access to a vital area that houses plant cooling water pumps at Dresden Generating Station | 0 | – | |
| January 30, 2012 | Byron, Illinois, US | Unusual Incident reported at Byron Nuclear Generating Station. Partial loss of offsite power led to a loss of nearly all power and safety functions until operators manually disconnected the grid from the plant. This exposed an electrical design flaw present in nearly every US nuclear reactor. | 0 | Undetermined | |
| March 31, 2013 | Russellville, Arkansas, US | One worker was killed and two others injured when part of a generator fell as it was being moved at the Arkansas Nuclear One. | 1 | – | |
| September 2009 | Crystal River, Florida, USA | When cutting into the Crystal River 3 Nuclear Power Plant containment building to create a large opening to replace the Steam generator (nuclear power), the structure was severely cracked, resulting in the facility's permanent closure. | 0 | 1000 + | |
| July 2016 | Bridgman, Michigan, US | Heavy steam leak into the turbine building of D.C. Cook Nuclear Station | 0 | – | |
| July 2018 | Genoa, Wisconsin, US | La Crosse Boiling Water Reactor Deconstruction leak into Mississippi River | 0 | – | |

*Source: Wikipedia Nuclear Reactors Accidents in the United States Until 2018.*

## The United States nuclear energy policy

The USA energy policy regarding the use of nuclear energy for electricity generation is implemented through the Department of Energy (DoE) and involves three important areas. These areas are the following:

• Research on new nuclear power reactor technologies;
  Reinitiating nuclear power plant construction;
• Radioactive waste management.

One of the USA nuclear energy policy's main elements is developing new standardized light-water nuclear power reactors. That is the type of reactor more popular in the country, with new safety features that make them more secure than the current models available in the market. The purpose is to introduce the new types of light-water nuclear power reactors to the market as soon as possible. The DoE's Advanced Light Water Reactor (ALWR) program in the 1980s sought to create a standardized light-water reactor at the earliest possible time. This program's implementation helped secure the Nuclear Regulatory Commission (NRC) certification for General Electric's ABWR and the combustion engineering's System 80 + advanced pressurized water reactor. The NRC gave final design approval to the ABWR and System 80 + during the summer of 1994. Programs initiated during the mid-1990s co-funded smaller (600 MWe) light-water reactors incorporating passive safety features. Westinghouse's AP-600 received design approval in 1998 (Nuclear Power in the USA, 2008).

Another central element of the US nuclear energy policy is a research program of a new type of nuclear power reactor generation, the so-called "Generation IV," with new safety features, reducing nuclear waste and proliferation risk, among other important innovations. The objective of this program is to produce a Generation IV reactor by 2040.

New bodies were established to provide technical and scientific advice to DoE on nuclear energy matters based on the current US nuclear energy policy. Two of these bodies were the Nuclear Energy Research Advisory Committee (NERAC) and the International Nuclear Energy Research Initiative (I-NERI). I-NERI was established to serve as a key mechanism to set up bilateral agreements for international collaboration in developing Generation IV systems. NERAC was established on October 1, 1998, to provide DoE and the Office of Nuclear Energy, Science, and Technology with independent advice on science and technical issues related to the DoE's nuclear energy program. NERAC reviews the nuclear energy program elements and provides advice and recommendations on long-range plans, priorities, and strategies. NERAC also provides advice on national policy and scientific aspects of nuclear energy research as requested by the Secretary of Energy or the Director of Nuclear Energy. The DoE created a Nuclear Energy Research Initiative (NERI) to address the technical and scientific issues affecting the future use of nuclear energy for electricity generation in the USA. NERI is expected to help preserve the nuclear science and engineering infrastructure within US universities, laboratories, and industry, advance nuclear energy technology within the country and maintain a competitive position worldwide (Nuclear Power in the USA, 2008).

Another vital program adopted by DoE is the Nuclear Energy Plant Optimizer (NEPO) program. This new program aimed to improve performance in the existing nuclear power plants' operation in the USA. The primary areas of focus for the NEPO program include plant aging and optimization of electrical production. NEPO is also a public-private research and development partnership with equal or higher matching funds coming from the nuclear industry (Nuclear Power in the USA, 2008).

In education and research, the DoE adopted the Nuclear Engineering Education Research (NEER) program to support universities and professionals carrying out fundamental research in engineering programs or research reactors. "DoE's role in workforce development has primarily focused upon the support of undergraduates, graduate students, and postdoctoral researchers through research and development awards at universities and at the DoE national laboratories. Today, that role also includes supporting educational and training programs to promote science and energy literacy. DoE funding has enabled tens of thousands of scientists, engineers, and technicians to tackle scientific questions of the day in physics, chemistry, biology, and other basic science and technology areas, impacting energy, environment, and national security challenges" (Workforce Development for Teachers and Scientists, 2020).

The DoE's Office of Nuclear Energy (NE) is the government's office responsible for advancing nuclear power as a resource capable of making significant contributions in meeting the USA's energy supply, environmental, and energy security needs. The NE seeks to resolve technical, cost, safety, security, and regulatory issues through research, development, and demonstration. By focusing on developing advanced nuclear technologies, NE supports the US administration's goals of providing domestic sources of secure energy, reducing greenhouse gases, and enhancing national security.

Summing up, the following can be stated. In the USA, the private sector plays a vital role in producing civilian nuclear power, greater than in any other nation, while the US administration is heavily involved in safety and environmental regulations, research and development funding, and setting national energy goals. Thus, since the late 1990s, US administration policy and funding decisions have encouraged developing a higher civilian nuclear capacity to increase nuclear energy's role within the country's energy mix. The Energy Policy Act (EPA) (2005) included the following incentives to increase the participation of nuclear power in the energy mix of the country:

- Production tax credit (PTC) of 1.8 cents per kWh for the first 6000 MWe of new nuclear capacity in the first eight years of operation. Under the initial terms of the EPA, to qualify for the nuclear PTC, a nuclear power plant must be in service on or before 31 December 2020, and the maximum value of the nuclear PTC is US$6 billion over eight years (or US$750 million per year). However, in February 2018, the PTC's extension was passed by the US Senate and Congress. The extension allows nuclear power reactors entering service after 31 December 2020 to qualify for the tax credits and allows the US Energy Secretary to allocate credit for up to 6000 MWe of new nuclear capacity, which enters service after 1 January 2021. The PTC cannot be claimed until assets begin generating electricity and are not inflation-adjusted;
- Federal risk insurance of US$2 billion to cover regulatory delays in the full-power operation of the first six advanced new nuclear power reactors;
- Rationalized tax on decommissioning funds (some reduced);
- Federal loan guarantees for advanced nuclear power reactors or other emission-free technologies up to 80% of the project cost;
- Extension for 20 years of the Price–Anderson Nuclear Industries Indemnity Act (1957) for nuclear liability protection;
- Support for advanced nuclear technology (US Nuclear Power Policy, 2020).

In mid-2008, the DoE invited applications for loan guarantees to support the construction of advanced nuclear power reactors up to US$18.5 billion and uranium enrichment plants up to US$4 billion. Loan guarantees are to encourage the commercial use of new or significantly improved energy technologies and "will enable project developers to bridge the financing gap between pilot and demonstration projects to full commercially viable projects that employ new or significantly improved energy technologies." (US Nuclear Power Policy, 2020). They are a form of support that allows companies to finance debt at reduced rates.

Without a doubt, the commitment to use nuclear power for electricity generation as part of the USA's long-term energy strategy continues, but there has been a reduction in some nuclear power programs due to greater emphasis on the use of renewable energy sources and the economics of gas-fired power for electricity generation (US Nuclear Power Policy, 2020).

The US administration remains more involved in commercial nuclear power plants than in any other industry in the USA, but this is not enough to encourage new private or public investment to construct new nuclear power plants in the country.

One of the main limitations of building new nuclear power reactors in the USA is the lengthy and costly detailed requirements for the construction and operation of all nuclear power reactors and uranium conversion, including enrichment, fuel fabrication, mining, and milling facilities. The review process preceding the construction of new nuclear power reactors can take between three and five years in a process that could cost millions of dollars. The US administration is the main funding source for advanced nuclear power reactor and fuel cycle research through its own national research laboratories and projects at university and industry facilities. It also promises to provide incentives for building new nuclear power plants through loan guarantees and tax credits, although owners have to raise their own capital. US domestic energy policy is also closely linked to foreign, trade, and defense policy on such matters as mitigating climate change and nuclear non-proliferation of nuclear weapons (US Nuclear Power Policy, 2020).

## The evolution of the nuclear power sector in the United States

Westinghouse designed the first fully commercial PWR type with a gross electrical capacity of 180 MWe in the Yankee Rowe nuclear power plant. This unit was connected to the electric grid in November 1960, started its commercial operation in July 1961, and was shut down in October 1991. Meanwhile, the Argonne National Laboratory developed another type of nuclear power reactor, the BWR type. The first commercial nuclear power reactor of this type, Dresden 1, with a gross electrical capacity of 207 MWe designed by General Electric, was connected to the grid in April 1960 and started operation in the same year. The Dresden 1 nuclear power reactor was shut down in October 1978. By the end of the 1960s, several orders were placed on constructing PWR and BWR types of more than 1000 MWe capacity.

From the first nuclear power reactor constructed in 1957 until 2020, 131 nuclear power reactors were built and entered the USA's commercial operation. During the 1950s, eight nuclear power reactors started their construction; two were connected to the electrical grid and entered commercial operation. During the 1960s, 62 nuclear power reactors started their construction, and 19 units were connected to the electrical grid and began commercial service. During the 1970s, 61 nuclear power reactors started their construction, 59 nuclear power reactors were connected to the electrical grid and entered commercial operation. During the 1980s, no nuclear power reactors were constructed, but 47 units were connected to the electrical grid and entered commercial operations. During the 1990s, no nuclear power reactors were built, but four units were connected to the electrical grid and entered commercial operations.

As can be seen from the above figures, the peak in the construction of nuclear power reactors in the USA was achieved in the 1960s. A total of 62 units initiating their construction in that decade. The peak of nuclear power reactors connected to the USA's grid was reached in the 1970s, with 59 nuclear power reactors connected to the electrical grid. The peak of nuclear power reactors entering into commercial operation was reached in the 1970s with 59 nuclear power reactors in this situation.

The PWRs and BWRs type of nuclear power reactors remain practically the only types of reactors built commercially in the USA until today. In 2020, there were 62 PWRs and 32 BWRs in operation in the country, with a total capacity of 96.55 GW. Thirty states within the USA have at least one commercial nuclear power reactor in operation.

The number of nuclear power reactors in operation in the USA in April 2021 and the electricity generated in 2019 are shown in Table 7.9. The US's nuclear power program's development suffered a significant setback after Three Mile Island nuclear accident occurred in 1979 (see Fig. 7.21), even though no one was injured or exposed to harmful radiation due to the accident. Many orders and projects were canceled or suspended, and the nuclear construction industry went into the doldrums (see Table 7.6). Since 1979, only two nuclear power reactors have been under construction in the country, and it is expected that both units could be connected to the electrical grid in 2021.

The last nuclear power reactor to be constructed in the USA was Watts Bar 1, which began its construction in 1996. The construction license on an additional four units (Watts Bar 2, Bellefonte 1 and 2, and WNP 1) was recently extended. However, there are no active construction activities on these sites. In October 2007, TVA announced that it had chosen the Bechtel group to complete the two-thirds built Watts Bar 2 reactor for US$2.5 billion. Construction started in 1972 but was frozen in 1985 and abandoned in 1994. It is important to note that Watts Bar 1 was one of the most expensive nuclear power reactors ever built in the USA within its nuclear power program. The completion of the mentioned unit took 23 years. The total amount of nuclear power reactors planned to be constructed in the USA is indicated in Table 7.10. Eight of them still are under consideration by the different suppliers.

Fig. 7.22 shows the nuclear-generated electricity in the USA in 2019 in comparison with other countries. The USA occupied place number 1 at the world level in accordance with the amount of electricity generation produced by nuclear power plants in 2019.

The electricity generation by nuclear power plants in the USA during the period 1960–2019 is shown in Fig. 7.23.

According to Fig. 7.23, the following can be stated: the electricity generation by nuclear power plants in the USA grew continuously during the period 1960–2019, a total of 809-fold, rising from one billion kWh in 1960 to 809.4 billion kWh in 2019. However, it is important to single out that since 2010 the electricity generation using nuclear energy in the USA has stagnated, remaining at approximately 807 billion kWh during the last eight years until 2019. It is expected that nuclear energy participation in the country's energy mix will decrease during the coming decades due to several nuclear power reactors' closure.

The relationship between the USA's nuclear power plant's electricity generation and generation capacity during the period 1957–2019 is shown in Fig. 7.24.

**TABLE 7.9** Number of nuclear power reactors and the generation of electricity in the USA (April 2021).

| | Number of nuclear units connected to the grid | Nuclear electricity generation (net TWh) (2019) | Nuclear percentage of total electricity supply (%) (2019) |
|---|---|---|---|
| United States of America (USA) | 94 | 809.4 | 19.7 |

*Source: April 2021. PRIS—Power Reactor Information System. www.iaea.org/pris. © IAEA.*

FIG. 7.21  Three Mile Island nuclear power plant near Middletown, Pennsylvania. *Source: DoE. http://ma.mbe.doe. gov/me70/history/photos.htm. Copyright status: Identified on DoE page as "DoE photo", i.e. not copyrighted. Public Domain. https://commons.wikimedia.org/w/index.php?curid=1111348.*

According to Fig. 7.24, the following can be stated: since 2007, nuclear power plants' capacity and electricity generation in the USA have stagnated. According to the Annual Energy Outlook 2020 Reference case, the total US nuclear power generating capacity is expected to decrease by 25.6%, falling from 98 GW in 2019 to 79 GW in 2050. Like EIA's projections for coal retirements, most of the projected nuclear power retirements are expected to occur between 2020 and 2025. In the Reference case, nearly 24 GW of nuclear capacity retires; 7.5 GW

TABLE 7.10  Nuclear power reactors planned for construction in the USA as known on 3.12.2018.

| Country | Reactor | Type | Model | Capacity MW (net) | Operator | Expected construction start |
|---------|---------|------|-------|-------------------|----------|-----------------------------|
| USA | US-5033 FERMI-3 | BWR | ESBWR | 1520 | – | – |
| USA | US-5017 NORTH ANNA-3 | PWR | US-APWR | 1500 | – | – |
| USA | US-5012 SOUTH TEXAS-3 | BWR | ABWR | 1350 | – | – |
| USA | US-5013 SOUTH TEXAS-4 | BWR | ABWR | 1350 | – | – |
| USA | US-5040 TURKEY POINT-6 | PWR | AP-1000 | 1117 | – | – |
| USA | US-5041 TURKEY POINT-7 | PWR | AP-1000 | 1117 | – | – |
| USA | US-5018 WILLIAM STATES LEE III-1 | PWR | AP-1000 | 1117 | – | – |
| USA | US-5019 WILLIAM STATES LEE III-2 | PWR | AP-1000 | 1117 | – | – |
| TOTAL | 8 | | | 10,188 | | |

*Source: 2020. Nuclear Power Reactors in the World. Reference data series No. 2. IAEA, Vienna. 2019 edition. ISBN 978-92-0-102719-1 and April 2021. PRIS—Power Reactor Information System. www.iaea.org/pris. © IAEA.*

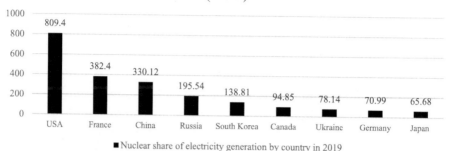

**FIG. 7.22** Nuclear share of electricity generation in the USA in comparison with other countries in 2019. *Source: April 2021. PRIS—Power Reactor Information System. www.iaea.org/pris. © IAEA, and Author own calculation.*

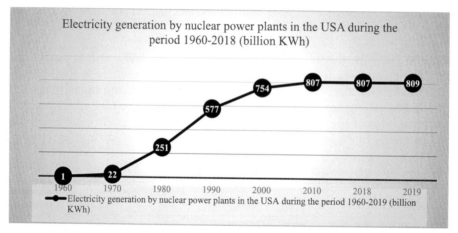

**FIG. 7.23** Electricity generation by nuclear power plants in the USA during the period 1960–2019. *Source: 2020. Nuclear Power Reactors in the World. Reference data series No. 2. IAEA, Vienna. 2019 edition. ISBN 978-92-0-102719-1, April 2021. PRIS—Power Reactor Information System. www.iaea.org/pris. © IAEA, and Author own work.*

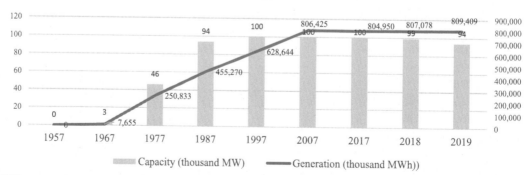

**FIG. 7.24** Relationship between electricity generation and generation capacity of the nuclear power plants in the USA during the period 1957–2019. *Source: 2019. Nuclear Explained: US Nuclear Industry. US EIA, and April 2021. PRIS—Power Reactor Information System. www.iaea.org/pris. © IAEA.*

has already been announced. EIA's Reference case projects another 16 GW will retire through 2050 in response to competitive market conditions (Fig. 7.25).

## Extension of the lifetime of the current nuclear power reactors in operation in the United States

The average age of US commercial nuclear power reactors that were operational as of October 31, 2019, was about 38.5 years. The oldest operating reactor is Nine Mile Point 1 in New York, which entered commercial service in December 1969. The newest reactor to enter service is Tennessee's Watts Bar Unit 2, which began operation in June 2016. The next-youngest operating reactor is Watts Bar Unit 1, Tennessee, which entered service in May 1996.

The aging base of nuclear power plants for electricity generation and the increasing need for additional power in the USA, along with improved economic and safety performance of the majority of the nuclear power reactors currently in operation in the country, has pushed the US administration to accept the possibility of a life extension of almost all US nuclear power reactors, particularly those that were built during the 1960s and 1970s. Simultaneously, the US administration promotes the construction of new nuclear power reactors in the coming years to satisfy a foresee increase in energy demand. However, only two new units are under construction until October 2020.

The US administration's position led many licensees to renew their operating licenses for an additional 20 years beyond their initial 40-year limits. According to Slater-Thompson (2014a), a total of 23 nuclear power reactors have extended their operating licenses since 2000; this means 24% of the total nuclear power reactors in operation in the country in that year. In 2019, applications to extend the licenses of 51 additional units were presented and approved by the NRC; this means 53.1% of the total nuclear power reactors currently in operation. In 2019, NRC reviewed license renewal applications for an additional 17 units. It is expected that other operators of nuclear power plants will eventually apply for operating license renewals of seven additional operating nuclear power reactors in the USA in the coming years.

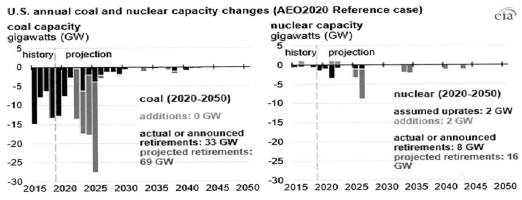

FIG. 7.25   US annual coal and nuclear capacity changes. *Source: 2020. US Energy Information Administration, Preliminary Monthly Electric Generator Inventory (AEO 2020). EIA and 2020. Annual Energy Outlook 2020 With Projections to 2050. US Energy Information Administration.*

Summing up, the following can be stated: in 2014, on the total number of nuclear power reactors in operation in the country (101 units), NRC already approved an extension of the useful life of 74 units; 11 units have not submitted to NRC an extension request, and 16 units have submitted a license extension request to the NCR (see Fig. 7.26). However, it is important to single out that in 2020, 96 units were in operation in the USA. For this reason, a total of five units were shut down between 2014 and 20, most of them within the group of nuclear power reactors that are not submitted as an extension of their useful life.

The US NRC has resumed, in 2014, issuing license renewals for nuclear power plants. On October 20, the NRC renewed the operating licenses for Limerick Generating Station Units 1 and 2, located northwest of Philadelphia, extending their license expiration dates by 20 years to 2044 and 2049. With this action, the NRC has granted license renewals providing a 20-year extension to 74 of the 96 operating nuclear power reactors in the USA; this means for 77.1% of the total nuclear power reactors in operation in the country in 2019. It is important to single out that NRC has the authority to issue initial operating licenses for commercial nuclear power plants for 40 years. The nuclear power plant owner is the authority responsible for applying for operating license renewal. NRC renews operating licenses for 20 years. Renewing an operating license is contingent on several factors, including safe management and disposal of the nuclear power plant's nuclear waste. "The NRC must determine that it has reasonable confidence that spent nuclear fuel can and will, in due course, be disposed of safely. That is known as a waste of confidence. Waste confidence enables the NRC to license new reactors or renew their operating licenses without examining the effects of extended waste storage for each site pending ultimate disposal" (Slater-Thompson, 2014b).

In parallel to the extension of an important number of nuclear power reactors' operational licenses, certain power companies show interest in constructing new nuclear power plants. Three utilities, Diminion Resources, Exelon, and Entergy, have applied during 2004 the so-called "Early Site Permits (ESP)." These applications will allow the utilities mentioned above to initiate nuclear power site clearances prior to commitments to build a nuclear power plant. Several other firms have indicated that they might be interested in requesting the ESP process initiation in the near future. Recently, the Detroit Edison company had selected the ESBWR reactor from General Electric-Hitachi for the possible construction of a new nuclear power

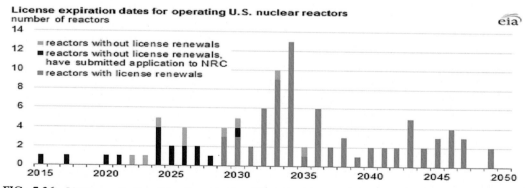

**FIG. 7.26** License expiration dates for operating US nuclear power reactors. *Source: US Energy Information Administration, on US Nuclear Regulatory Commission.*

reactor with a capacity of 1520 MWe to be built in Michigan at the site of the Fermi nuclear power plant. The license request was presented at the end of 2008.

## Reduction of the construction time of nuclear power plants in the United States

Nuclear suppliers and utilities have made progress in recent years to reduce the costs of building new nuclear power reactors by reducing construction time. The NRC has also streamlined its licensing process to construct future nuclear power reactors with the same purpose. These changes aim to shorten construction lead times and improve the economics of new nuclear power reactor technology. The US administration's goal is that these and other similar actions might encourage the power industry to restart new nuclear power plants' construction before the end of the current decade.

However, it is important to stress the following: investments in commercial nuclear power plants will only be forthcoming if investors are sure that the construction time will be reduced significantly, the cost of producing electricity using nuclear energy will be lower than the costs associated with the use of alternative electric generation technologies, and the negative impact on the environment and population will be minimum. The renewed interest in the use of nuclear power for electricity generation has resulted due:

- To higher prices for oil and natural gas registered in 2007 and 2008;
- The decrease in oil proved reserves worldwide;
- A significant improvement of the operation of existing nuclear power reactors;
- An increase in nuclear safety;
- Uncertainty about future restrictions on coal emissions;
- An expected increase in oil and natural gas prices during the coming years.

Until the recent price volatility reported worldwide since the 1980s, low fuel costs had allowed gas-fired power plants to dominate the market for new electricity generation capacity. Relatively stable costs and low air emissions of contaminating gases associated with nuclear power plants' operation may now appear more attractive, particularly combined with a substantial tax credit for nuclear generation and other incentives provided by the Energy Policy Act (EPA) (2005). However, high operating costs have played a major role in the recent retirement decisions of a group of nuclear power reactors in the USA. At least five currently operating nuclear plants have requested state-level price support to continue operating. In two sensitivity cases, assumed operating costs were raised or lowered by 20%. It is important to single out that significant uncertainties about nuclear power plant construction costs remain in the USA, along with doubts about the progress on nuclear waste disposal and concerns about public opposition to using nuclear energy for electricity generation, in addition to the opposition of many banks to provide resources for the financing of the construction of a nuclear power plant. All those problems helped cause the prolonged cessation of US nuclear power reactor orders until today. They will need to be addressed before financing new multibillion-dollar nuclear power plants is likely to be obtained (Parker and Holt, 2007).

The main problems faced by the US nuclear industry related to the construction of new nuclear power reactors could be summarized in the following manner: economic difficulties in construction and opposition to them, which led to increased construction time and subsequently increased construction costs. Many utilities went bankrupt over nuclear power projects. The estimated cost of building a nuclear power plant rose from less than US$400 million in the 1970s to around US$4000 million by the 1990s, while construction times doubled from the 1970s to the 1980s. These facts led the US business magazine *Forbes* in 1985 to describe the industry as "the largest managerial disaster in US business history, involving US$100 billion in wasted investments and cost overruns, exceeded in magnitude only by the Vietnam War and the then Savings and Loan crisis" (Schneider and Froggatt, 2007).

## Management of high-level nuclear waste in the United States

One of the main problems that are limiting the construction of new nuclear power reactors in the USA is the management of nuclear waste, particularly the final disposal of high-level nuclear waste.

Spent nuclear fuel produced by nuclear power reactors poses a severe disposal problem that could be a significant factor in considering new nuclear plant construction. The Nuclear Waste Policy Act of 1982 (NWPA, P.L. 97-425) (1982) commits the federal government to provide for permanent disposal of spent nuclear fuel in return for a fee on nuclear power generation. The US administration's decision to allow the beginning operation of the nation's nuclear waste repository at Yucca Mountain, Nevada, has created a new situation that needs to be considered before deciding to construct new nuclear power reactors is taken in the future. According to the document entitled "Nuclear Power: Outlook for New U.S. Reactors" (2007 RL33442), more than 50,000 metric tons of spent fuel are stored in pools of water or shielded casks at nuclear facility sites waiting for their final disposal. It is important to single out that US nuclear power plants produce an average of 2000 metric tons of spent fuel per year.

Undoubtedly, the storage of the spent nuclear fuel in the nuclear power reactor sites, within 30 years after the reactor's license expired, is one of the possibilities that the utilities have as an alternative now to proceed with the construction of new nuclear power plants. The NCR estimates an operation license period of 70 years for the new types of nuclear power reactors. Therefore, NRC does not consider the lack of a permanent repository for spent fuel to be an obstacle to nuclear plant licensing (Parker and Holt, 2007).

Finally, it is important to single out that two of the most important factors that need to be solved, as soon as possible, to encourage investors to finance the construction of new nuclear power reactors in the USA are the reduction of the long construction time of a nuclear power reactor, as well as the decrease of the initial capital cost associated with the construction of this type of power plants.

## New United States nuclear regulatory commission's regulations

What has to be done by the US authorities to reduce construction time and avoid a long waiting period for the beginning of a nuclear power reactor's operation in the future? The answer to this question is the following: in 2015, the US Government Accountability Office concluded that obtaining certification from the NRC for a new nuclear power reactor is "a multi-decade process, with costs up to US$1-billion to US$2-billion, to design and certify or license" (Bryce, 2020).

According to the document entitled "Nuclear Power: Outlook for new US Reactors" (2007), until 1989, licensing a new nuclear power plant involved a two-step process:

1. NRC issued a construction permit allowing an applicant to begin building a nuclear power reactor;
2. NCR issued an operating license permitting the nuclear power plant to generate electricity for sale.

This procedure resulted in cases where completed or nearly completed nuclear power reactors awaited years to be granted operating licenses — delays that drove up the affected plants' costs. In 1989, NRC issued regulations to streamline this process in three ways, according to Parker and Holt (2007):

- Early Site Permit Program (ESPR) allowing utilities to get their proposed nuclear power reactor sites approved by the NRC before a decision is made on whether or not to build the nuclear power plant;
- Standard Design Certification (SDC) for advanced nuclear power reactor designs allowing vendors to get their designs approved by NRC for use in the USA, so utilities can then deploy them primarily "off the shelf";
- Combined Construction and Operating License (COL) providing a "one-step" approval process, in which all licensing hearings for a proposed nuclear power plant are expected to be conducted before construction begins. The COL would then allow a completed plant to operate if inspections, tests, analyses, and acceptance criteria (ITAAC) were met. That is intended to reduce the chances of regulatory delays after a nuclear power plant is completed (Parker and Holt, 2007) (Fig. 7.27).

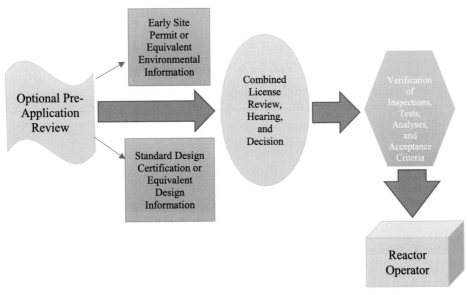

**FIG. 7.27** Relationship between Combined Licenses, Early Site Permits, and Standard Design Certification approved by NRC. *Source: US Nuclear Regulatory Commission information and Author design.*

After the license is issued, which could take three and a half years to be approved, the utility must decide whether to begin building the nuclear power plant or cancel the project. Current projections of nuclear power plant construction schedules assume that a nuclear power plant can be built between four and seven years. However, some nuclear power plants constructed during the last years took more time to be concluded in other countries. At the end of the construction, the NRC verifies that the new nuclear power reactors meet all the US competent authorities' requirements to allow the facility to start its operation as soon as possible. Overall, the process is anticipated that could take up to five years to be concluded. The US government has indeed enacted legislation to speed up the development and licensing process. Nevertheless, according to an official with Oregon-based NuScale Power, "the process is still very long, very tedious and very expensive" (Bryce, 2020).

## The nuclear industry in the United States

The US nuclear industry is one of the most advanced industries in the world. All necessary elements and components of a nuclear power plant can be produced in the country. Four companies supplying nuclear systems currently operate in the USA. These companies are the following:

- Westinghouse Corporation, which built the majority of the PWRs;
- Combustion Engineering (CE) and Babcock & Wilcox (B&W);
- Westinghouse and CE, which are now part of Westinghouse BNFL while Areva NP now owns elements of B&W's nuclear technology;
- General Electric, which designed all presently operating BWRs in the USA.

The NRC approves at least two nuclear power reactor designs for construction in the USA. These are the following:

- The Westinghouse System 80 + and AP 600;
- A General Electric and Hitachi ABWR designs of 1300–1500 MWe.

It is important to single out that in September 2009, GE Nuclear Hitachi Energy informed that it has presented to the NRC the revised design of its new ESBWR. The presentation marks an important landmark in advance in the process of NCR approval of the reactor of 1520 MWe that two American companies have selected for the possible construction of new nuclear power reactors in the future. Following GE-Hitachi, the new design improves significantly different aspects related to the reactor's safety, and it has economic advantages over the previous design.

There are currently eight nuclear power reactor designs that are either undergoing certification or pre-certification procedures with the NRC, some of them produced by US companies. It is anticipated that other designs will join this process soon. The following six nuclear power reactor designs are currently undergoing pre-certification:

- General Electric ESBWR;
- Areva NP SWR-1000;
- General Atomics GT-MHR;

- Atomic Energy of Canada ACR-700 advanced CANDU design;
- Eskom PBMR;
- Westinghouse BNFL IRIS.

It is important to note that the US administration has adopted some initiatives to encourage the US private power industry to consider new nuclear power plants' construction during the coming decades. Some of these initiatives have been described in previous paragraphs. Other initiatives have been adopted for the same purpose. One of these initiatives provided by the Energy Policy Act is the adoption of the so-called "Nuclear Production Tax Credit." Through this initiative, a 1.8 cents per kWh tax credit for up to 6000 MW of new nuclear power capacity for the first eight years of operation, up to US\$125 million annually per 1000 MW, is offered by the US authorities to the companies that request a license construction for a new nuclear power plant. An eligible nuclear power reactor under this initiative must be entered into service before January 1, 2021.

Another initiative to reduce cost and avoid unnecessary delays in the construction and the beginning of the operation of new nuclear power plants is the adoption of the so-called "Regulatory Risk Insurance." Continuing concern over potential regulatory delays, despite the streamlined licensing system now available in the country described in previous paragraphs, prompted the US Congress to include an insurance system that would cover some of the principal and interest on the debt and extra costs incurred in purchasing replacement power because of licensing delays. The first two new nuclear power reactors licensed by NRC that meet other criteria established by DoE could be reimbursed for all such costs, up to US\$500 million apiece, whereas each of the next four newly licensed reactors could receive 50% reimbursement of up to US\$250 million (Parker and Holt, 2007).

## The uranium resources and the mining industry in the United States

Uranium mining in the country today is undertaken only by few companies on a relatively small scale reaching, in 2019, a total of 67 tons, a decrease of 96.5% since the peak registered in 2014. Many companies undertake uranium exploration, often going over areas that were mined in the 1950s–1980s. According to the World Nuclear Association (2020b), uranium production in the USA is from one mill (White Mesa, Utah) fed by four or five underground mines and several in-situ leach operations (see Fig. 7.28).

According to Fig. 7.28, the USA's uranium production decreased 24.3-fold during the period 2010–19, falling from 1630 tons in 2010 to only 67 tons in 2019. The significant reduction in the uranium mining was the result of an Interior Secretary order "banning new hard rock uranium mining in about 4000 km$^2$ of land in Arizona for 20 years, which sterilized 145,000 tU of known resources, according to the Nuclear Energy Institute (NEI) and also much prospective land. The land is not within the Grand Canyon National Park or the buffer zone protecting the national park. The industry contends that uranium exploration and mining here would not compromise the Grand Canyon watershed. In March 2013, a US District Court judge declined to overturn the mining ban, and in October 2014, the US District Court affirmed the ban. That was then appealed, and in December 2017, the US Court of Appeals for the Ninth Circuit upheld the earlier decisions" (World Nuclear Association, 2020b).

7. The use of nuclear energy for electricity generation

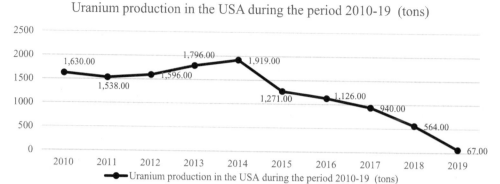

FIG. 7.28   Uranium production in the USA during the period 2010–19 (tons). *Source: World Nuclear Association, 2020. US Uranium Mining and Exploration and Author own work.*

In 2016, the EIA estimated that US uranium resources were 25,400 tU recoverable at US$78 per kg, 64,000 tU at US$130 per kg, and 139,500 tU at US$260 per kg. Estimated reserves for mines in production were 6167 tU at US$130 per kg.

The US companies carrying out uranium activities in the USA during the period 2015–19 are listed in Table 7.11.

Based on the data included in Table 7.11, the following can be stated: Six US companies out of nine planned no uranium production in 2019. None of the US companies included in Table 7.10 increased uranium production in 2019. The decreased in uranium production during the years 2018 and 2019 was 8.4-fold.

TABLE 7.11   US companies in uranium production (tons) during the period 2015–2019.

| Company | ISL operations, hard rock mill | 2016 | 2017 | 2018 | 2019 |
|---|---|---|---|---|---|
| Cameco | Smith Ranch – Highland, WY | 358 | 118 | 0 planned | 0 planned |
| Cameco | Crow Butte, NE | 89 | 29 | 0 planned | 0 planned |
| Uranium One | Willow Creek, WY | 23 | 38 | 0 planned | 0 planned |
| Ur-Energy | Lost Creek, WY | 216 drummed (207 captured) | 98 drummed (102 captured) | 110 drummed (116 captured) | 20 |
| Peninsula/Strata | Lance, WY (Ross unit) | 49 | 44 drummed | 55 | 21 |
| UEC | Hobson – La Palangana, TX | 0 | 0 | 0 | 0 |
| EFRC (Uranerz) | Nichols Ranch, WY | 129 | 100 | 54 | 27 |
| EFRC (Mestena U) | Alta Mesa, TX | 0 | 0 | 0 | 0 |
| EFRC | White Mesa mill + third party | 262 | 138.5 + 365.5 | 136 + 163 | 0 |
| **Total** | | **1126** | **931** | **564** | **67** |

*Source: World Nuclear Association, 2020. US Uranium Mining and Exploration.*

# United States trade within the nuclear power sector

According to the United States Trade Representative Office, the USA is the world's largest economy and one of the largest exporters and importers of goods and services. Trade is critical to America's prosperity - fueling economic growth, supporting good jobs at home, raising living standards, and helping Americans provide for their families with affordable goods and services.

USA trade policy has varied widely through various American historical and industrial periods. As a major developed nation, the USA has relied heavily on importing raw materials and exporting finished goods. Because of the American economy and industry's significance, much weight has been placed on trade policy by elected officials and business leaders (World Business Leaders Urge Trade Ministers to Seize the Opportunity to Resurrect the Doha Round-Trade Resources Center-Business Roundtable, 2006; Akhtar, 2018).

According to the US Department of State, the USA and Canada share the longest international border on the planet (5525-miles), and the bilateral relationship is one of the world's closest and most extensive. This excellent good relationship between the two countries is reflected "in the high volume of bilateral trade – nearly US$2 billion a day in goods and services – and people-to-people contact – about 400,000 people, cross between the two countries every day" (US Department of State, 2020).

Based on the US Census Bureau (2020), in 2019, Canada ranked second according to its total trade with the USA.[x] The total trade between these two countries reached, in 2019, US$612,428.8 or 14.8% of the total. The USA exported goods to Canada in 2019 for US$292,693.1, ranked first according to the level of its exports (17.8% of the total). At the same time, the USA ranked third for its level of imported goods for US$319,735.7 (12.8% of the total, after China with18.1%, and Mexico with14.3%). The trade deficit reached, in 2019, US$27,042.6. According to its trade deficit level, in 2019, Canada ranked ninth, after China, Mexico, Japan, Germany, Vietnam, Ireland, Italy, and Malaysia.

The USA and Canada's trade deficit evolution during the period 2015–19 is shown in Fig. 7.29.

Based on Fig. 7.29, the following can be stated: the trade deficit between the USA and Canada increased by 75% during the period under consideration, rising from US$15,449.9 in 2015 to US$ 27,042.6 in 2019. Except for 2016, the trade deficit between the two countries increased each year within the period 2015–19.

Finally, it is important to stress that Canada is the single largest foreign energy supplier to the USA and operates an integrated electricity grid under jointly developed reliability standards. In the nuclear sector, the USA has signed several bilateral agreements with Canada and is importing uranium from this country for the operation of its nuclear power plants. Despite having USA a large nuclear industry, the country has not exported nuclear power reactors to Canada because it produces its own nuclear power reactors.

---

[x] The USA exported to Canada vehicles, machinery, electrical machinery, food/agricultural products, mineral fuels and plastics. Automobile manufacturing, for instance, can involve highly integrated production and assembly of auto parts across the border. USA services exports to Canada include travel and intellectual property such as software and media. The USA imported from Canada, in the same year, mineral fuels, vehicles, agricultural products, machinery and plastics (US-Canada Trade Facts, 2015).

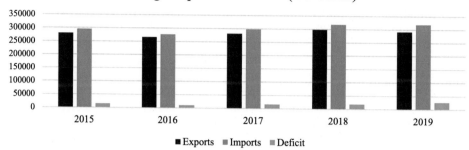

**FIG. 7.29** Evolution of the trade deficit between the USA and Canada during the period 2015–19. *Source: US Census Bureau, 2020. Department of Commerce; US administration and Author own work.*

## The public opinion on the use of nuclear energy for electricity generation in the United States

US public opinion supports the use of nuclear energy for electricity generation, and this support is continuing an upward trend during the last years. In 2005, a national survey was carried out by Bisconti Research, Inc., in which 1000 adults age18 and older participated. As a result of this survey, around 70% of the public supports using nuclear energy for electricity generation in the country in the future. The survey found that 85% of the participants favor that the current nuclear power plant license is renovated. According to the result of this survey, planning for the future has become a fundamental value in energy policy considerations. Seventy-seven percent agreed that electric companies should build new nuclear power reactors when needed in the future. A total of 71% approve companies participating in federal site approval reviews for possible construction of new units at nuclear power plant sites nearest to where they live. Also, 69% said that if a new nuclear power reactor were needed to supply electricity, they would find it acceptable to add new units at the nearest existing nuclear power plant (Nuclear Energy Institute, 2005).

Another NEI survey of opinions toward nuclear energy was conducted from September 21 to October 1, 2016, with a nationally representative sample of 1000 adults. The margin of error is plus or minus 3% points. The survey continues NEI's 33-year public opinion tracking program, which provides an analysis of long-term trends. One of the primary outcomes of this survey is that public opinion continues supporting the use of nuclear energy for electricity generation in the USA. Among the public at large, 65% favor the use of nuclear power for electricity generation, and 32% oppose, with 25% firmly in favor, 40% somewhat in favor, 20% somewhat opposed, 12% vehemently opposed, and 4% unable to answer. In NEI's Fall 2015 survey, 64% reported favorability toward nuclear energy, with 26% firmly in favor. As can be easily seen from both studies' outcomes, the number of persons who favor the use of nuclear energy for electricity generation in the USA increased by 1%. In comparison, the number of persons who strongly support using this type of energy source for this specific purpose increased by 1%. On the subject of license renewal and new nuclear power plants construction, the survey found:

- Eighty percent support license renewal of nuclear power plants that continue to meet federal safety standards;
- Seventy-one percent support preparing now so that new nuclear power plants could be built if needed in the next decade;
- Fifty-five percent of the general public believe the USA should build more nuclear power plants;
- Seventy-three percent of the people think that as countries worldwide build new nuclear power plants, the US nuclear industry must continue to play a leading role in world markets (Nuclear Energy Institute, 2016).

On the importance of using nuclear energy for electricity generation, NEI's survey found that the public sees nuclear power as necessary for the future but has mixed views about the technology's safety. These views are the following:

- Seventy-five percent of the general public believes that nuclear energy will be important in meeting the nation's electricity needs in the years ahead;
- Forty-five percent of the public believes that nuclear energy's importance will increase in the years ahead. Fewer believe nuclear energy's importance will decrease (17% of the people);
- On a seven-point scale rating the safety of nuclear power plants, 57% of the public gave a high rating versus 23% who gave a low rating;
- Sixty-eight percent of the public agrees that nuclear power plants operating in the USA are safe and secure (Nuclear Energy Institute, 2016).

The evolution of the percentage of people who favor and against the use of nuclear energy for electricity generation in the USA during the period 2005–16 is shown in Fig. 7.30.

As can be easily seen from Fig. 7.30, the evolution of the percentage of people in favor of nuclear energy for electricity generation in the last years moves between 64% and 74%, while the percentage of people against move between 23% and 34%. It can be stated that during

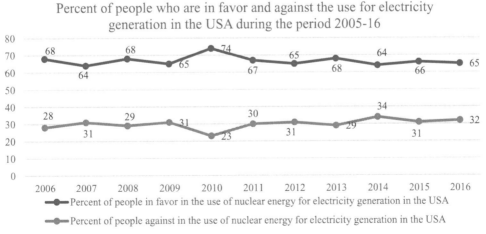

**FIG. 7.30** Evolution of the percentage of people in favor and against nuclear energy for electricity generation in the USA during the period 2005–16. *Source: Nuclear Energy Institute, 2016. National Public Opinion Survey on Nuclear Energy and Author own work.*

the whole period, the number of people in favor of nuclear energy for electricity generation is much higher than the number of people that are against it. It is expected that this situation will not change during the coming decades.

## The future of nuclear energy in the United States

The USA is a leader in the production, supply, and consumption of energy. US energy companies produce oil, natural gas, and coal. They also produce electricity from clean energy sources, including wind, solar, hydropower, geothermal, and nuclear power. US energy companies further transmit, distribute, and store energy through complex infrastructure networks supported by emerging products and services such as smart grid technologies. Growing consumer demand and world-class innovation – combined with a competitive workforce and supply chain capable of building, installing, and servicing all energy technologies – make the USA one of the world's most attractive energy markets. According to the International Energy Agency, total investment in the US energy sector was valued at US$350 billion in 2018 (the second-largest in the world). That same year, total foreign direct investment in the US nuclear industry reached US$172.8 billion (The Energy Industry in the United States, 2019).

However, nuclear power in the USA and several other countries using nuclear energy for electricity generation worldwide are facing, at this moment, a very difficult time. "Westinghouse's bankruptcy culminates the collapse of potential US strategic leadership in world nuclear energy. Regaining the strategic power will be technically straightforward but politically difficult" (Hargraves, 2017). The USA has faltered in many aspects of nuclear technology, allowing other nations to become world leaders in nuclear and energy diplomacy.

In April 2021, the USA had 94 nuclear power reactors in operation in 30 states with a net electrical capacity of 96.55 GW and generated, in 2019, nearly 20% of US generation (19.7%). In 2020, the country had two nuclear power reactors under construction with a net electrical capacity of 2234 MW. These two units are expected to be connected to the electrical grid in 2021.

The current US fleet of nuclear power reactors is split more or less 50–50% between regulated and unregulated assets. Of the total capacity of existing US nuclear power reactors, around 8 GW have announced an early retirement, and 47 GW are at risk for economic reasons, particularly those small-single unit power plants. Some large single-unit power plants are shutting down 30 years before their useful life expires. Assuming a 60-year run life for existing nuclear power reactor generation, many more gigawatts of nuclear generation will retire within the next couple of decades (though some nuclear power plants may get licenses extended to 80 years) (Sharp and Kuczynski, 2016).

From the total nuclear power capacity, only 22 GW is financially favorable, while another 22 GW's financial status is unknown because they are in the rate base. If the present challenges are not appropriately addressed by the US administration and the nuclear energy industry, then the future of nuclear energy may be far less promising than ever. The USA's prominent role that the country played in the past in promoting and using nuclear energy for electricity generation at national and world levels is very low. This role will likely be much lower in the future than it is today. It is important to stress that if present challenges facing the US nuclear industry are not addressed properly, then there is a risk that the USA could lose between 50% and 60% of the current number of nuclear power reactors; this means between 48 and 58 units.

Nuclear power plants across the USA are in great trouble, shutting down several nuclear power reactors well in advance of their license expiration, primarily for economic reasons. Small single-unit power plants are at particular risk, but all nuclear power plants "face challenges from low demand growth, low natural gas prices, subsidies for renewables, and increased operational costs. Nuclear power plant retirements have serious regional economic and employment impacts; they also tend to increase reliance on natural gas, which in turn increases greenhouse gas emissions" (Sharp and Kuczynski, 2016).

It is important to be aware that keeping operating nuclear power plants in the country preserves zero-carbon power in the near term and maintains the potential to add new nuclear technologies now under development. This action could unlock several gigawatts of additional capacity and paves the way for advanced solutions in the long term. Government interventions to help the existing nuclear power plants to continue operating could be materialized, among others, in the following manner, according to Sharp and Kuczynski (2016):

- Providing tax credits or accelerated depreciation for nuclear power plants;
- Establishing national carbon pricing;
- Speeding up the evaluation and implementation of accident-tolerant fuel;
- Addressing price formation issues;
- Independent System Operators and Regional Transmission Organizations could start pricing more attributes of nuclear electricity generation, implement a carbon adder and low-carbon dispatch, or let nuclear power plants ramp to avoid negative priced hours;
- Simplifying the NRC process for the approval of the construction of nuclear power plants and reducing its cost;
- Supporting the research and development of a new generation of nuclear power reactors;
- Increasing nuclear safety in the current nuclear power reactors in operation in the country;
- Solving the management of nuclear waste, particularly the technological solutions of the final disposal of high-level nuclear waste;
- Reducing the proliferation risk associated with certain parts of the nuclear fuel cycle;
- Ensuring the stability of the electrical grid and the supply of energy according to demand.

It is also important to single out that most of the relevant policy pieces are "at the state level, where solutions could include clean energy standards, power purchase agreements, re-regulation, subsidization, or carbon pricing."[y] There are several state and federal policy

---

[y] State solutions could take many forms. One option is the signing of power purchase agreements (PPAs). Another option is re-regulation in competitive states, or a hybrid to regulate generation only. States could place a value on diversity and security of fuel, which would benefit the use of nuclear power plant for electricity generation over natural gas-fired power plants that rely on interstate pipelines. States committed to competitive markets could address the range of out-of-market subsidies, by getting rid of them or adding ones for the use of nuclear energy for electricity generation. States could include the use of nuclear energy under their Renewable Portfolio Standards (though renewables do not want to compete with nuclear). In addition, states could enact carbon pricing to help make existing nuclear units economic. Although existing carbon prices have not helped the nuclear power reactors fleet because they are too low, a carbon price as low as US$16 per ton could save existing nuclear power plants, at least in some states (Sharp and Kuczynski, 2016).

approaches that could promote the construction of new nuclear power plants. However, to achieve that goal, such approaches "has to be made more clearly, more forcefully, and in a way that does not frame the solution as a binary choice between renewables and nuclear. While there does not appear to be sufficient urgency or political will in the USA to undertake the degree of mobilization of new conventional nuclear power plants needed to pursue advanced nuclear technology, advanced reactors that use new types of coolants, operate at different pressures and temperatures, or are smaller and more modular could represent the future of nuclear power" (Sharp and Kuczynski, 2016).

Though US officials regularly self-congratulate themselves about US gold-standard nuclear policies globally, the country hardly advances industrial nuclear technology and sells little of it, nor has little strategic influence.

Without a robust nuclear power industry and international trade, the USA has lost negotiating leverage. For example, the new 123 agreement with Vietnam does not prohibit it from enriching or reprocessing uranium or other fuels to be permitted to trade with US suppliers. The renewed agreement with South Korea weakened fuel manufacturing limitations and offered some spent fuel processing assistance at US national labs (Hargraves, 2017). Without concrete actions by the US administration and the nuclear industry to enhance the US nuclear sector, the country will lose its past strategic position in the world nuclear sector.

# References

Anon., 2012. A Closer Look at Canada's Nuclear Plants. CBC News. January 9, 2012.

Anon., 2002. A Technology Roadmap for Generation IV Nuclear Energy Systems. USS DOE Nuclear Energy Research Advisory Committee, and the Generation IV International Forum. GIF-002-00, December 2002.

Anon., 2020. Advanced Nuclear Power Reactors. World Nuclear Association (WNA).

Akhtar, S.I., 2018. USS Trade Policy Functions: Who Does What. Congressional Research Service, Washington, DC.

Anon., 2020. Annual Energy Outlook 2020 With Projections to 2050. US Energy Information Administration.

Anon., 2006. Basic Infrastructure for a Nuclear Power Project. IAEA, Vienna. IAEA-TECDOC-1513. June 2006.

Bel, H.T.I., 2000. Inspector General Report—Ind. Office of the Inspector General (OIG), US Nuclear Regulatory Commission (NRC).

Blade, T., 2004. Davis-Besse Stirs Again. https://www.toledoblade.com/opinion/editorials/2004/03/11/Davis-Besse-stirs-again/stories/200403110093.

Anon., 2019. BP Statistical Review of World Energy 2019, sixty-eighth ed.

Anon., 2020. BP Statistical Review of World Energy 2020, sixty-ninth ed.

Bryce, R., 2020. To Lead on Climate, Canada Should Invest in the Next Generation of Nuclear Reactors. Globe and Mail.

Anon., 2001. Canada's Nuclear Reactors: How Much Safety Is Enough? (pdf)., p. 11.

Anon., 2012. Canadian Environmental Assessment Act. Government of Canada.

Canadian Nuclear Association, 2017. Canada's Nuclear Energy Future. vol. 3 Public Newsletter. Issue 1.

Anon., 2020. Canadian Nuclear Safety Commission-Nuclear Power Plant. Government of Canada.

Anon., 1998. Choosing the Nuclear Power Option: Factors to be Considered. IAEA, Vienna. STI/PUB/1050. January 1998.

Anon., 2007. Considerations to Launch a Nuclear Power Programme. IAEA, Vienna. GOV/INF/2007/2.

Anon., 2003. Country Nuclear Power Profiles-Canada. IAEA.

Anon., 2018. Country Nuclear Power Profiles-Canada. IAEA.

Anon., 2019. Energy and the Economy. Natural Resources of Canada.

Anon., 2019. Energy Fact Book 2019-20 Canada. Natural Resources Canada.

Anon., 2005. Energy Policy Act (EPA). USA Government.

Friends of the Earth, 2020. Small Modular Nuclear Reactors: Climate Change.

Geuss, M., 2018. Mapping What It Would Take for a Renaissance for Nuclear Energy. ARS Technica.

Government of Canada Response to Recommendations in the Standing Committee on Natural Resources' Report, n.d. http://www.ourcommons.ca/content/Committee/421/RNNR/GovResponse/RP9142298/421_RNNR_Rpt05_GR/421_RNNR_Rpt05_GR-e.pdf. at 9.

Hanebach, M., McDonald, J., 2017. The Future of the Nuclear Sector in Canada: Industry and Government in Unison Once Again. Gowling WLG.

Hargraves, R., 2017. 'Energy Is Worth a War': USS Nuclear Supremacy Is Collapsing, But There Is a Way to Win It Back. Business Insider.

Hesketh, K., Worrall, A., Weaver, D., 2004. Future Challenges for Nuclear Energy in Europe. Europhysics News. November/December 2004; Article available at http://www.europhvsicsnews.org. https://doi.org/10.1051/epn:2004611.

Ingersoll, E., Gogan, K., Herter, J., Foss, A., 2020. Cost and Performance Requirements for Flexible Advanced Nuclear Plants in Future US Power Markets. Lucid Catalyst.

Jedicke, P., 2016. The NRX Incident. cns-snc.ca; Retrieved 2016-08-22.

MacKinnon, B.-J., 2012. Nuclear Commission Says Point Lepreau Leaks 'Unsettling'. CBC News.

Meirion, J., 2009. New UKK Nuclear Stations Unlikely to be Ready on Time. BBC.

Morales Pedraza, J., 2011. Is nuclear power a realistic alternative to the use of fossil fuels for the production of electricity? In: Advances in Energy Research. vol. 4. Nova Science Publisher.

Morales Pedraza, J., 2012. Nuclear Power: Current and Future Role in the World Electricity Generation. Nova Science Publishers, ISBN: 978-1-61728-504-2.

Morales Pedraza, J., 2013. World major nuclear accidents and their negative impact on the environment, human health and public opinion. Int. J. Energy Environ. Econ. 21 (2), 131–169.

Morales Pedraza, J., 2015. The current status and perspectives for the use of small modular reactors for electricity generation. In: Advances in Energy Research. vol. 21. Nova Science Publishers (Chapter 1).

Morales Pedraza, J., 2017. Small Modular Reactors for Electricity Generation: An Economic and Technologically Sound Alternative. Edition 2017, Springer International Publishing, https://doi.org/10.1007/978-3-319-52216-6. ISBN: Print: 978-3-319-52215-9, Online: 978-3-319-52216-6.

Anon., 2017. Nuclear Energy. Natural Resources Canada.

Nuclear Energy Data, 2019. Nuclear Technology Development and Economics. NEA-OECD.

Nuclear Energy Institute, 2005. National Public Opinion Survey on Nuclear Energy.

Nuclear Energy Institute, 2016. National Public Opinion Survey on Nuclear Energy.

Anon., 2020. Nuclear Power in Canada. World Nuclear Association.

Nuclear Power in the USA, 2008. Promoting the Peaceful Worldwide Use of Nuclear Power as a Sustainable Energy Resource. World Nuclear Association. June 2008.

Anon., 2007. Nuclear Power: Outlook for New US Reactors. EveryCRSReport.com (RL33442).

Anon., 2000. Nuclear Safety and Control Act. Canadian Nuclear Safety Commission; Government of Canada.

Anon., 1982. Nuclear Waste Policy Act of 1982 (NWPA, P.L. 97-425).

Anon., 2017. Ontario's Nuclear Generating Facilities: A History and Estimate of Unit Lifetimes and Refurbishment Costs. Appendix 2 Power for the Future. https://web.archive.org/web/20170404155726/https://www.pembinafoundation.org/reports/appendix2.pdf.

Ontario Power Generation, 2020. OPG Statement about Pickering Nuclear. Ontario Power Generation. https://www.opg.com/media_release/opg-statement-about-pickering-nuclear/.

Parker, L., Holt, M., 2007. Nuclear Power: Outlook for New USS Reactors. Energy Policy Resources, Science and Industry Division. CRS Report for Congress; March 9, 2007.

Anon., 1957. Price–Anderson Nuclear Industries Indemnity Act. USA Administration.

Anon., April 2021. PRIS—Power Reactor Information System. www.iaea.org/pris. © IAEA.

Schneider, M., Froggatt, A., 2007. The World Nuclear Industry Status Report 2007. Greens-EFA Group in the European Parliament, Brussels. November 2007.

Scott, M., 2018. Future of USS nuclear power fleet depends mostly on natural gas prices, carbon policies. In: Today in Energy. US EIA.

Anon., 2006. Selecting Strategies for the Decommissioning of Nuclear Facilities: A Status Report. NEA-OECD, Paris.

Sharp, P., Kuczynski, S., 2016. The Future of Nuclear Energy in the United States. The Aspen Institute, Energy Environment Program.

Slater-Thompson, N., 2014a. Today in Energy. Almost All US Nuclear Plants Require Life Extension Past 60 Years to Operate Beyond 2050. US EIA.

Slater-Thompson, N., 2014b. Today in Energy. Nuclear Regulatory Commission Resumes License Renewals for Nuclear Power Plants. US EIA.

Sovacool, B.K., 2009. The Accidental Century—Prominent Energy Accidents in the Last 100 Years. Archived 2012-08-21 at the Wayback Machine.

Sovacool, B.K., 2010. A critical evaluation of nuclear power and renewable electricity in Asia. J. Contemp. Asia 40 (3), 393–400. August 2010.

Taylor, D., 2011. Manitoba's Forgotten Nuclear Accident. www.winnipegfreepress.com. Retrieved 2016-08-22.

Anon., 2016. The Canadian Nuclear Industry and Its Economic Contributions. Natural Resources Canada.

The Energy Industry in the United States, 2019. Energy Industry Spotlight; Select USA. https://www.selectusa.gov/energy-industry-united-states.

Anon., 2018. The Future of Nuclear Energy in a Carbon-Constrained World: An Interdisciplinary MIT Study. MIT Energy Initiative.

Anon., 2017. The Nuclear Sector at a Crossroads: Fostering Innovation and Energy Security for Canada and the World. Report of the Standing Committee on Natural Resources.

Things to Consider Before Investing in Nuclear Energy for Electricity, n.d. SolaPV.

Anon., 2020. Uranium in Canada. World Nuclear Association.

Anon., 2020. Uranium Mines and Mills. Canadian Nuclear Safety Commission; Government of Canada.

US Census Bureau (2020); Department of Commerce; US Administration.

US Department of State, 2020. USA-Canada Relations. US Department of State Homepage.

Anon., 2020. US Nuclear Power Policy. World Nuclear Association.

US-Canada Trade Facts, 2015. US Trade Representative. https://ustr.gov/countries-regions/americas/canada.

Wien Automatic Planning System (WASP), n.d. IAEA.

Anon., 2020. Wikipedia CANDU Reactor.

Anon., 2020. Wikipedia Nuclear Energy Policy of the United States.

Anon., 2020. Wikipedia Nuclear Power in Canada.

Anon., 2020. Wikipedia Nuclear Power in Finland.

Winfield, M., Jamison, A., Wong, R., Czajkowski, P., 2006. Nuclear Power in Canada: An Examination of Risks, Impact, and Sustainability. The Pembita Institute, ISBN: 0-921719-87-6. December 2006.

Anon., 2020. Workforce Development for Teachers and Scientists. Office of Science, US DoE.

Anon., 2006. World Business Leaders Urge Trade Ministers to Seize the Opportunity to Resurrect the Doha Round-Trade Resources Center-Business Roundtable.

World Nuclear Association-Nuclear Power in the USA, 2020. https://www.world-nuclear.org/information-library/country-profiles/countries-t-z/usa-nuclear-power.aspx.

World Nuclear Association, 2020b. Uranium Mining and Exploration.

World Nuclear Association, 2021. Uranium in Canada.

Anon., 2020. World Nuclear News.

# 8

# Conclusion

Without a doubt, humankind is currently immersed in the fourth energy revolution. This new energy revolution began in the 1970s and had two main components: energy and the environment. It is based on the following three elements:

- The use of renewable energy sources for electricity generation and heating and their decentralized use;
- The rational and efficient use of energy;
- The conservation of the environment.

In 1973, the oil price forced industrialized countries to develop new energy options. In 1978, a further increase in the oil price was registered. This new increase in oil price encouraged the adoption of programs to develop new energy technologies. The aim was to increase the use of renewable energy sources for electricity generation and heating in many countries worldwide.

In the middle of the decade of the 1980s, with the oil price decreasing, renewable energy re-emerged as an effective energy option to reduce the high level of contamination of many cities in several countries worldwide. Furthermore, the development of new energy technologies since the 1990s made, in many cases, the use of renewable energy sources more competitive from the economic point of view compared with other energy sources.

Today the international community, already aware of the environmental problems generated by the use of fossil fuels for electricity generation and heating, faces the challenge of diversifying its energy matrix to reduce the contamination of cities and regions. The goal is to replace the use of fossil fuels for electricity generation and heating, particularly oil and coal, with all available renewable energy sources to reduce the high level of contamination of many cities in several countries all over the world. In 2020, the USA was the second most contaminating country at the world level. This situation is caused by the high participation of fossil fuels, particularly coal, within the country's energy mix due to implementing the Trump energy policy that prioritizes the use of fossil fuels for electricity generation and heating over renewables. However, today, and despite the Trump energy policy, renewable energy sources are one of the main energy sources used in the USA to generate electricity and heating. In

2019, the electricity generated by renewables represented 17.6% of total US electricity generation in that year.

One of the primary causes of the increased use of renewable energy sources worldwide has been thanks to the concerns of environmental groups and community associations on the negative impact on the environment and population due to the use of fossil fuels for electricity generation and heating. These groups and associations have raised the alarm on the high emissions of polluting particles as a consequence of the use of fossil fuels for electricity generation and heating in many countries. These particles have been directly linked to world climate change in recent decades.

The growing energy demand has undoubtedly forced governments and the private industry to continuously search for new energy sources for electricity generation and heating. All this, coupled with continuous or repeated increases in the prices of fossil fuels, or their instability during recent decades, has strengthened the role of renewable energy sources as an attractive generating option for many countries worldwide. However, it is important to single out that regardless of the increased use of renewable energy sources for electricity generation and heating worldwide in recent years, fossil fuels are the largest energy source used for electricity generation in the USA. Natural gas was the largest energy source used for this purpose within fossil fuels, with about 38% of US electricity generation in 2019. Coal was the second-largest energy source for US electricity generation in 2019, with about 23%.

According to the paper entitled "Electricity in the United States (2020)", the three main energy sources used in the country for electricity generation and heating are the following:

- Fossil fuels (coal, natural gas, and oil);
- Nuclear energy;
- Renewable energy sources.

Most US electricity is generated with steam turbines using fossil fuels, nuclear energy, bioenergy, particularly biomass, geothermal energy, and solar thermal energy. "Other major electricity generation technologies include gas turbines, hydro turbines, wind turbines, and solar photovoltaics" (Electricity in the United States, 2020). The USA generated 87% of the regional total.

In 2019, in Canada, according to the BP Statistical Review of World Energy 2020 (2020) report, the total electricity generation by all energy sources reached 660.4 TWh or 13% of the regional total. By energy sources, the electricity generation in Canada in 2019 was the following:

- Oil power plants generated only 4.1 TWh or 0.6% of the total;
- Natural gas power plants generated 69.3 TWh or 10.5% of the total;
- Coal power plants generated 54.6 TWh or 8.3% of the total;
- Nuclear power plants generated 100.5 TWh or 15.2% of the total[a];
- Renewables energy sources generated 432 TWh or 65.4% of the total.

[a] According to PRIS-IAEA April 2021, the electricity generated by nuclear power plants in Canada reached 95,469 GWh or 14.9% of the total.

In 2019, in the USA, according to the BP Statistical Review of World Energy 2020 report, the total electricity generation by all energy sources reached 4401 TWh or 87% of the regional total. By energy sources, the electricity generation in the USA in 2019 was the following:

- Oil power plants generated only 20 TWh or 0.5% of the total;
- Natural gas power plants generated 1700.9 TWh or 38.5% of the total;
- Coal power plants generated 1053.5 TWh or 24% of the total;
- Nuclear power plants generated 852 TWh or 19.4% of the total[b];
- Renewable energy sources generated 775 TWh or 17.6% of the total.

Without a doubt, renewable energy sources represent a clean and minimally invasive alternative energy source for electricity generation and heating to the environment and population globally and for the North American region. They are abundant energy sources that reduce dependence on fossil fuels, which to date turn out to be vital energy sources for electricity generation and heating for many countries worldwide. An indirect advantage of using renewable energy sources for electricity generation and heating in many countries today is the increased property value associated with renewable energy projects. It should also be noted that all non-conventional projects present a high degree of citizen acceptance by the Communities surrounding them, be they environmental, political, social, union, or just people from the neighborhood.

For this reason, renewable energy sources have become a new and important energy source for private companies in the forestry, paper, or cellulose sectors. In addition, biomass power plants use their industrial waste as energy generation sources, thus lowering costs optimizing operational resources.

As for weaknesses, renewable energy sources show three major comparative flaws concerning their conventional peers. The first one is that, until today, they generate energy at a higher financial cost. The second one is that they are largely circumscribed to weather conditions, particularly wind or solar energies. The third one is the connectivity or accessibility of the projects, measured with respect to urban centers or the crossing of transmission networks to which they can connect. The reason is that they are energy sources found many times in complex places such as deserts, coastal plains, the coastline, mountain ranges, or places located far away from urban centers. In some of these places, the loss of electricity transmission to consumers could be very high.

Different types of renewable energy sources can be used for electricity generation and heating at the world level and in the North American region. These are the following:

- Hydro;
- Wind energy;
- Solar energy;
- Bioenergy;
- Geothermal energy.

[b] According to PRIS-IAEA April 2021, the electricity generated by nuclear power plants in the USA reached 809,409 GWh or 19.7% of the total.

According to IRENA Renewable Energy Statistics 2020 report, the participation of all renewable energy sources connected to the electrical grid worldwide was the following:

- Hydropower is the world's leading renewable energy source for electricity generation, with around 64.8% share (4,267,085 GWh) of the total renewable output (6,586,124 GWh). In 2019, North America had a hydropower plant capacity of 183,822 MW, generating 698,754 GWh in 2018. In the case of hydropower plant pure pumped-storage, in 2019, the world capacity installed was 120,844 MW, generating 117,869 GWh in 2018. In 2019, the hydropower plant pure pumped-storage capacity installed in North America reached 19,326 MW, generating 21,614 GWh in 2018.

  In Canada, hydro is the primary energy source used for electricity generation, and this situation will not change during the next decades. In 2019, Canada had a hydropower plant installed capacity of 81,053 MW, generating 381,750 GWh in 2018 or 54.6% of the regional total (698,754 GWh).

  In the case of hydropower plant pure pumped-storage, in 2019, the capacity installed in the North American region reached 19,326 MW, generating 21,614 GWh in 2018. In 2019, the capacity installed in Canada of this type of hydropower plant reached 174 MW, generating 111 GWh in 2018.

  In 2019, the USA had a hydropower plant installed capacity of 102,769 MW, generating 317,004 GWh or 45.4% of the regional total in 2018. In 2019, in the USA, the capacity installed of hydropower plant pure pumped-storage reached 19,152 MW, generating 21,503 GWh in 2018;

- Wind power continues to be number two within all renewable energy sources at the world level. It has a share of around 19.2% (1,262,914 GWh) of the total electricity produced in the world by all renewable energy sources. In 2019, the total wind farm capacity reached 116,997 MW in the North American region, generating 307,682 GWh in 2018.

  In 2019, the USA had a wind farm capacity of 103,584 MW. It generated, in 2018, 275,834 GWh, representing 89.6% of the total regional electricity generation using this type of renewable energy source (307,683 GWh).

  In 2019, Canada had a wind farm capacity of 13,413 MW and generated, in 2018, 31,848 GWh, representing 10.4% of the regional total.

  In 2019, the USA had a wind farm capacity of 103,555 MW and generated, in 2018, 275,834 GWh, representing 89.6% of the regional total.

  In North America, the onshore wind farm capacity increased 23.7-fold during the period 2010–19, generating 307,581 GWh in 2018. In Canada, the onshore wind farm installed capacity increased approximately 3.3-fold in 2019, generating 1,194,718 GWh in 2018.

  In the specific case of offshore wind farms, the capacity installed during the period 2010–19 increased 9.2-fold, generating 68,196 GWh in 2018. Only the USA has a very small offshore wind farm capacity installed since 2016 (29 MW) within the region, generating 102 GWh in 2018;

- Solar power climbed to number three among all renewable energy sources used for electricity generation and heating at the world level, with a share of around 8.5% (562,033 GWh). In the North American region, solar park installed capacity reached 65,608 MW in 2019, generating 88,986 GWh in 2018.

In the USA, the solar park installed capacity reached 62,298 MW in 2019 and generated, in 2018, 85,184 GWh from solar parks (95.7% of the regional total, 88,986 GWh). In the same year, Canada generated 3802 GWh, representing 4.3% of the regional total, with a solar park installed capacity of 3310 MW in 2019.

In solar PV, the total installed capacity in the North American region reached 63,850 MW in 2019, generating 85,046 GWh in 2018.

In Canada, the solar PV installed capacity reached 3310 MW in 2019, generating 3802 GWh in 2018.

In the USA, the solar PV installed capacity reached 60,540 MW in 2019, generating 81,244 GWh in 2018.

In CSP, the total installed capacity in the North American region reached 1758 MW in 2019, generating 3940 GWh in 2018.

No CSP facilities were operating in Canada in 2019;

- Bioenergy falls to number four among all renewable energy sources used for electricity generation at the world level, with a share of around 7.9% (522,552 GWh). In the North American region, installed capacity reached 15,825 MW in 2019, generating 78,546 GWh in 2018.

In the USA, bioenergy installed capacity reached 12,450 MW in 2019 and generated, in 2018, 67,885 GWh using bioenergy sources (86.4% of the regional total, 78,546 GWh). Canada generated, in the same year, 10,661 GWh, representing 13.6% of the regional total;

- Geothermal energy is number five among all renewable energy sources used for electricity generation at the world level, with a share of around 1.3% (88,408 GWh). However, in the North American region, Canada has no geothermal power plant in operation. On the other hand, the USA generated, in 2018, 18,773 GWh from geothermal power plants (100% of the regional total).

Hydropower is the most important renewable energy source used in the North American region for electricity generation and heating compared to other renewable energy sources. In Canada, in 2019, hydropower plants generated 382 TWh, representing 58.5% of the regional total. Concerning 2018, the electricity generation by hydropower plants in Canada decreased by 1.1%. In the same year, hydropower plants generated 271.2 TWh in the USA, representing 41.5% of the regional total. Concerning 2018, the electricity generation using hydropower plants in the USA decreased by 6.4%.

According to the Electricity in the United States (2020) report, it is important to single out that hydropower plants produced about 7% of total US electricity generation and about 38% of electricity generation from renewable energy sources in 2019.

Wind energy takes advantage of the wind's mechanical force, which moves an internal turbine, leading the energy to a generator to transform it into electricity. Wind turbines need constant winds, contrary to what is commonly believed, not necessarily strong, so their location is limited to certain national territory sectors.

Without a doubt, wind energy is a safe, constant, and renewable energy source. It is an energy source that does not pollute or wreak havoc on the environment or communities. The wind farms technically have a useful life period of more than 20 years of continuous use. They allow the transfer and reuse of the facilities when they reach their lifecycle end.

One of the main disadvantages of wind energy is its dependency on atmospheric conditions. It is considered one of the highest-cost energy sources compared to other energy sources used for electricity generation and heating. Besides, wind farms present the following disadvantages:

- A considerable alteration of the local landscape;
- From the environmental point of view, it has a negative impact on the migratory flows of birds in the area;
- The blades' rotation generates a constant noise product of the friction, which causes discomfort in the surrounding social communities;
- The construction materials and technical generation equipment could cause interferences in the radio and television transmission signals within the surrounding communities;
- Generation energy efficiency is considerably lower compared to conventional energy sources.

In 2019, according to the BP Statistical Review of World Energy 2020 report, wind farms in the North American region produced 308.6 TWh, representing 57.2% of the regional total (539.1 TWh). Canada produced 33.2 TWh, representing 10.8% of the regional total, and the USA generated 275.4 TWh or 89.2% of the regional total.

Solar energy is the most popular renewable energy source in the North American region, but still behind hydropower and wind farms. Solar parks can produce heat and electricity in unison. In the case of heat, it is stored in thermal receivers and is commonly used to heat water in homes, or on a smaller scale, for industrial use. For its part, electrical energy is generated thanks to the use of photovoltaic panels, which are made up of so-called "cells" that allow the capture of solar energy, and then transform it into electricity.

Solar energy is presented as an inexhaustible source of energy because it is provided by the Sun. It is not in any way harmful to the environment as it does not emanate any polluting residue during the solar park operation. However, it is important to single out that solar parks require a large amount of land to operate profitably. The ideal places for installing and operating solar parks are located, in most cases, far from the urban centers. Therefore its transmission and operability are more expensive than other renewable energy sources. Therefore, despite being nourished by an inexhaustible source such as the Sun, solar energy is the least profitable renewable energy source.

In 2019, according to the BP Statistical Review of World Energy 2020 report, solar parks in the North American region generated 98.1 TWh, representing 18.2% of all renewable generation in that year. Solar parks in Canada generated 3.9 TWh in 2018, representing around 4% of the regional total. The USA, in the same year, generated 94.3 TWh or 96% of the regional total. It is important to single out that solar energy provided about 2% of total US electricity in 2019.

Bioenergy, particularly biomass, is an organic matter of animal or plant origin used as fuel for electricity generation and heating by many countries worldwide. In the case of biofuels,[c] they can be used in the transportation sector as well. Biomass is used to obtain an energy

---

[c] To date, food crops such as corn, sugar, and vegetable oil, have been the primary source of biofuels for transportation, but increased use of biofuels has created more problems than solutions, rising food prices and food price volatility, and accelerated expansion of agriculture in the tropics, according to the paper entitled "Biomass Resources in the United States" (2012).

agent through its incineration. For the most part, it is used through a process called "electrical cogeneration," which consumes the remains of other industrial manufacturing processes (see Figs. 8.1 and 8.2). Cogeneration is the co-production of thermal and electrical energy from the same type of fuel. By disposing of a considerable amount of waste, biomass is economically convenient. Moreover, it is a non-polluting energy source compared to its peers that use fossil fuels.

Biomass energy sources can be stored and used according to the requirements without depending on climatic conditions like other renewable energy sources. However, biomass is an energy source that feeds on manufacturing waste, and it turns out to be uncompetitive if it is to be used at a high level of generation. Large amounts of biomass are required to obtain the same energy from other energy sources because biomass has a low caloric power. Biomass was the source of about 1% of total US electricity generation in 2019 and about 5% of total

## Photosynthesis

In the process of photosynthesis, plants convert radiant energy from the sun into chemical energy in the form of glucose—or sugar.

(water)   (carbon dioxide)   (sunlight)   (glucose)   (oxygen)

$$6\,H_2O \;+\; 6\,CO_2 \;+\; \text{radiant energy} \;\rightarrow\; C_6H_{12}O_6 \;+\; 6\,O_2$$

FIG. 8.1  Photosynthesis. *Source: EIA. https://www.eia.gov/energyexplained/biomass/.*

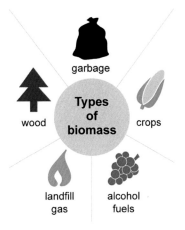

FIG. 8.2  Types of biomass. *Source: EIA. https://www.eia.gov/energyexplained/biomass/.*

primary energy use in the country. According to the Biomass Explained (2020) paper, "of that 5%, about 46% was from wood and wood-derived biomass, 45% was from biofuels (mainly ethanol), and 9% was from the biomass in municipal wastes."

The percentage shares of total US biomass energy use by the consuming sectors in 2019 were the following:

- The industrial sector: 49%;
- The transportation sector: 28%;
- The residential sector: 11%;
- The electric power sector: 9%;
- The commercial sector: 3%.

According to the above information, the following can be stated: the industrial and transportation sectors account for the largest percentage shares of total annual US biomass consumption. "The wood products and paper industries use biomass in combined heat and power plants for process heat and to generate electricity for their own use. Liquid biofuels (ethanol and biomass-based diesel) account for most of the transportation sector's biomass consumption. The residential and commercial sectors use firewood and wood pellets for heating. The commercial sector also consumes, and in some cases, sells renewable natural gas produced at municipal sewage treatment facilities and at waste landfills. The electric power sector uses wood and biomass-derived wastes to generate electricity for sale to the other sectors" (Biomass Explained, 2020).

Bioenergy comprises about 4% of Canada's total energy supply. In the specific case of biomass, it accounts for 23% of the national energy supply, the second largest after hydro (68%). "In 2018 there were 36 operational cogeneration units at pulp and paper mills and 41 Independent Power Providers (IPP) using biomass" (Renewable Energy Facts-Canada, 2020). The electrical capacity of pulp and paper cogeneration was 3427 MW, while heat capacity was 1348 MW. "IPP capacity for electricity and heat was 794 MW and 400 MW, respectively. In 2017, there were also 351 bioheat projects, of which 82% are less than 1 MW. Institutions, including schools and hospitals, are the strongest market for bioheat in Canada" (Renewable Energy Facts-Canada, 2020).

Historically, bioenergy consumption was very important for home energy use, as Canadians burned wood for heating and cooking. The contribution of forest biomass to Canada's energy supply has increased from between 3% and 4% in the 1970s to between 5% and 6% today. Changes in pulp and paper technology have resulted in most of this increase (Bioenergy From Biomass, 2020). In Canada, the pulp and paper industry within the Canadian industrial sector is the largest industrial consumer of bioenergy (Cruickshank et al., 2014).

A geothermal energy source is any energy that can be obtained by taking advantage of the Earth's crust's internal heat. It can be found in areas of very hot thermal waters that do not present greater depth, where the basal rock is drilled until reaching a depth where it can be used with better pressure. Geothermal power plants allow constant energy production since it is an inexhaustible energy source. Its generation process is not polluting as other energy sources such as oil and coal. Operating and maintenance expenses are low compared to other energy sources.

However, the construction of a geothermal power plant represents a high financial cost. Curiously, the generating process produces annoying odors in the environment since it works with steam and mineral water with high concentrations of sulfides. In addition, the sites suitable for constructing geothermal power plants are in desert and isolated places, making the electricity transfer cost highly expensive. The electricity transmission losses also are high.

It is important to know that they could pollute sources for nearby waters with arsenic and ammonia substances. According to the US Energy Information Administration source, the USA produced close to 16.7 billion kWh of geothermal power in 2018, which accounted for 0.4% of its total utility-scale power generation. In 2019, geothermal power plants produced about 0.5% of total US electricity generation, 0.1% higher than in 2018. There are currently geothermal power plants in seven US states (NS Energy, 2019). The two US states with the highest geothermal capacity are California with 2792 MW and Nevada with 805 MW.

Canada has substantial potential for geothermal energy development, but until 2020 no electricity has been generated from geothermal power plants (Wikipedia Geothermal Power in Canada, 2020).

According to the reports entitled "Renewables 2019. Analysis and Forecast to 2024 (2019)", "Energy, Electricity and Nuclear Power Estimates for the Period up to 2050" (2020), and the PRIS – Power Reactor Information System 2021 (2021), the future of renewable energy sources and nuclear energy in North America can be summarized as follows:

- Renewable power capacity is expected to expand by 50% between 2019 and 2024, led by solar PV. According to IEA sources, around 280 GW of electricity production capacity was added in 2020, a 45% increase from the amount added in 2019. That was the biggest gain since 1999 and equivalent to total power production capacity in the 10 Southeast Asian nations in the ASEAN trade block. The increase in 2020 will follow by an increase of about 270 GW of renewable capacity to be added in 2021 and almost 280 GW in 2022, as nations step up their shift to renewable energy production to meet their obligations to reduce the production of greenhouse gasses.
- Renewable energy accounted for 90% of new electricity generation installed last year. The foreseen increase of 1200 GW is equivalent to the US's total installed power capacity today. Solar PV alone accounts for almost 60% of the expected growth, with onshore wind representing 25%. Expansion of distributed solar PV in the North American region is twice as rapid between 2019 and 2024 as it was during 2013–18, mainly driven by the USA. Some 100 million solar rooftop systems for homes could be operating worldwide by 2024. These residential systems are expected to represent around 25% of total distributed solar PV capacity, with deployment expanding rapidly in many countries due to favorable policy designs and distributed solar PV's economic attractiveness. The top five markets for residential solar PV installations per capita in 2024 are Australia, Belgium, the USA (California), the Netherlands, and Austria;
- New wind turbine installation nearly doubled the level reported in 2020 to 114 GW. While the IEA expects the growth rate to slow down a bit in the coming years, it expects it will still be 50% higher than during the 2017–2019 period. Offshore wind contributes 4% of the increase in wind energy, with its capacity forecast to triple by 2024, stimulated by competitive auctions in the EU and expanding markets in China and the USA;
- Falling costs and more effective policies drive a significant upward revision in the forecast for renewable capacity deployment compared with last year's reports. Solar PV generation costs are estimated to decline by 15% to 35% for utility-scale and distributed

applications by 2024. Wind and solar PV developers are rushing to complete projects in the USA before federal tax incentives end. At the same time, corporate power purchase agreements and state-level policies contribute to growth.[d]

- Renewable electricity growth still needs to accelerate significantly to meet long-term sustainable energy goals. However, this growth is possible if governments address the three main challenges to faster deployment: policy and regulatory uncertainty, high investment risks in many developing economies, and system integration of wind and solar PV in some countries. Tackling these challenges underpins the forecast that total renewable capacity could increase more than 60% to 4000 GW by 2024, which is twice the size of today's global coal capacity.

- According to the IAEA-PRIS database, nuclear energy was the source of 19.7% of the total US electricity generation in 2019. In Canada, in 2019, nuclear energy generated 14.9% of the country's total electricity generation. According to the document entitled "Energy, Electricity and Nuclear Power Estimates for the Period up to 2050" (2020), since 1980, the share of fossil fuels in final energy consumption remained above 70%, with a slight reduction from 82% in 1980 to 73% in 2019. Of all fossil fuels, oil has the largest share, has remained at about 50% since 1980. With a share of 23%, natural gas was the second-largest energy source in 2019. Its share has remained relatively stable since 1980. From 1980 to 2010, the share of electricity gradually increased by eight percentage points. Its share in 2019 was about 20% of final energy consumption. Fossil fuels contributed more than 50% of the electricity produced in 2019. The share of coal has decreased by more than 50% since 1980, whereas the share of natural gas has more than doubled. The share of oil has decreased from 10% in 1980 to around 1%, where it remains.
Nuclear is the largest low-carbon energy source. Its share nearly doubled from 1980 to 1990 and has remained stable at almost 20% since 1990. The share of hydro has decreased by six percentage points over the past 39 years. The share of wind has increased since 2000, stabilizing at about 2%.

- Final energy consumption is expected to remain almost constant up to 2050. A slight increase of about 1% is expected by 2030, followed by a slight decrease over the next 20 years. Electricity consumption is expected to continue to grow. By 2030, it is projected to increase by nearly 10% from 2019 levels, reaching an increase of almost 25% by 2050. The share of electricity in the final energy consumption is expected to increase by about five percentage points by 2050 gradually. Total electrical generating capacity is projected to increase by almost 15% by 2030 and 23% by 2050.
Significant changes in nuclear electrical generating capacity are projected over the next three decades under one of the IAEA's two different scenarios. In the second scenario, that capacity is expected to remain relatively stable. In the first scenario, nuclear electrical generating capacity is projected to remain roughly constant, with a slight decrease between 3% and 4% by 2050. As a result, the share of nuclear in total electrical capacity is expected to decrease by one percentage point by 2030 and by almost two percentage points by 2050. However, in the second scenario, nuclear electrical generating capacity

---

[d] Installation of new solar PV photovoltaic panels at the world level should also continue to break records, hitting 160 GW by 2022.

will decrease by 20% by 2030 from the current capacity and by 2050 by around 33%. The share of nuclear in total electrical capacity is projected to decrease by about two percentage points by 2030 and almost six percentage points by 2050.

- The heat generated from renewable energy is expected to expand by 20% between 2019 and 2024. China, the EU, India, and the USA are responsible for 66.7% of the global increase in renewable heat consumption over the period 2019–24. However, renewables share global heat consumption is only marginally from 10% to 12% in 2024.
- China leads to biofuel production growth for the first time. As a result, total biofuel output is forecast to increase by 25% by 2024. The USA and Brazil are expected to provide 66.7% of total biofuel production in 2024.

Summing up, the following can be stated: it is expected that fossil fuels will reduce their role in the North American region's energy mix during the coming years. In contrast, renewable energy sources will increase their role. Furthermore, it is also foreseen a decreased role of nuclear energy in the North American region's energy mix during the same period, due to the closure of several nuclear power plants in the USA and Canada, and the small number of new nuclear power plants under construction (two in the USA only) with a total net capacity of 2234 MW(e).

Between 2030 and 2050, two scenarios can be expected to occur. In one of the scenarios, covering until 2030, it is foreseen that about 20% of nuclear power reactors currently operating in the region would be retired by 2030, with nuclear power reactor additions of only about 0.5 GW(e) of capacity. Between 2030 and 2050, it is expected that significantly more capacity will be retired than is added, resulting in a net reduction of about 50 GW(e), according to Energy, Electricity and Nuclear Power Estimates for the Period up to 2050 (2020).

# References

Bioenergy From Biomass., 2020. Government of Canada.

Biomass Explained., 2020. US Energy Information Administration (US EIA).

Biomass Resources in the United States., 2012. Union of Concerned Scientists.

BP Statistical Review of World Energy 2020., 2020, sixty-ninth ed.

Cruickshank, W.H., Robert, J.E., Silversides, C.R., 2014. Biomass Energy. The Canadian Encyclopedia.

Electricity in the United States., 2020. US Energy Information Administration. Electricity explained.

Energy, Electricity and Nuclear Power Estimates for the Period up to 2050., 2020. 1011-2642 International Atomic Energy Agency (IAEA), ISBN: 978-92-0-118120-6. Reference data series No 1, 2020 edition; IAEA-RDS-1/40.

NS Energy, 2019. Profiling the Top-Producing Geothermal Energy Plants in the US. NS Energy Staff Writer.

PRIS – Power Reactor Information System 2021 (www.iaea.org/pris), © IAEA, 2021.

Renewable Energy Facts-Canada., 2020. Government of Canada.

Renewables 2019., 2019. Analysis and Forecast to 2024. IEA.

Wikipedia Geothermal Power in Canada., 2020.

# Index

Note: Page numbers followed by *f* indicate figures, and *t* indicate tables.

Printed in the United States
by Baker & Taylor Publisher Services